Optimization in Chemical Engineering

Optimization is used to determine the most appropriate value of variables under given conditions. The primary focus of using optimization techniques is to measure the maximum or minimum value of a function depending on the circumstances. Any engineering discipline involving design, maintenance and manufacturing requires certain technical decisions to be taken at different stages. The primary outcome of taking these decisions is to maximize the profit with minimum utilization of resources.

This book presents a detailed explanation of problem formulation and problem solving with the help of algorithms such as secant method, Quasi-Newton method, linear programming and dynamic programming. It covers important chemical processes such as fluid flow systems, heat exchangers, chemical reactor and distillation systems with the help of solved examples.

It begins by explaining the fundamental concepts followed by an elucidation of various modern techniques including trust-region methods, Levenberg-Marquardt algorithms, stochastic optimization, simulated annealing and statistical optimization. It studies multi-objective optimization technique and its applications in chemical engineering. The knowledge of such a technique is necessary as most chemical processes are multiple input and multiple output systems.

The book also discusses theory and applications of various optimization software tools including LINGO, MATLAB, MINITAB and GAMS. It is designed as a coursebook for undergraduate and postgraduate students of chemical engineering and allied branches including biotechnology, food technology, petroleum engineering and environmental science.

Suman Dutta is Assistant Professor at the Department of Chemical Engineering, Indian School of Mines, Dhanbad. He was a visiting researcher at the Centre for Water Science in Cranfield University, UK. He teaches courses on chemical engineering thermodynamics, chemical reaction engineering, fluid mechanics, process modeling and optimization and process instrumentation and control. His areas of research include wastewater treatment, membrane technology, advanced oxidation process, photocatalysis, process simulation and optimization.

Optimization in Chemical Engineering

Suman Dutta

CAMBRIDGE
UNIVERSITY PRESS

4843/24, 2nd Floor, Ansari Road, Daryaganj, Delhi - 110002, India

Cambridge University Press is part of the University of Cambridge.

It furthers the University's mission by disseminating knowledge in the pursuit of education, learning and research at the highest international levels of excellence.

www.cambridge.org
Information on this title: www.cambridge.org/9781107091238

© Suman Dutta 2016

This publication is in copyright. Subject to statutory exception and to the provisions of relevant collective licensing agreements, no reproduction of any part may take place without the written permission of Cambridge University Press.

First published 2016

Printed in India by Thomson Press India Ltd., New Delhi 110001

A catalogue record for this publication is available from the British Library

ISBN 978-1-107-09123-8 Hardback

Cambridge University Press has no responsibility for the persistence or accuracy of URLs for external or third-party internet websites referred to in this publication, and does not guarantee that any content on such websites is, or will remain, accurate or appropriate.

To my father
Late Sukumar Dutta
and
God almighty

Contents

List of Figures		*xiii*
List of Tables		*xvii*
Preface		*xix*

1. A Brief Discussion on Optimization

1.1	Introduction to Process Optimization	1
1.2	Statement of an Optimization Problem	2
1.3	Classification of Optimization Problems	3
1.4	Salient Feature of Optimization	8
1.5	Applications of Optimization in Chemical Engineering	9
1.6	Computer Application for Optimization Problems	10
Summary		10
Review Questions		10
References		11

2. Formulation of Optimization Problems in Chemical and Biochemical Engineering

2.1	Introduction		12
2.2	Formulation of Optimization Problem		12
2.3	Fluid Flow System		13
	2.3.1	Optimization of liquid storage tank	13
	2.3.2	Optimization of pump configurations	14
2.4	Systems with Chemical Reaction		17
	2.4.1	Optimization of product concentration during chain reaction	18
	2.4.2	Optimization of gluconic acid production	20
2.5	Optimization of Heat Transport System		21
	2.5.1	Calculation of optimum insulation thickness	21
	2.5.2	Optimization of simple heat exchanger network	24
	2.5.3	Maximum temperature for two rotating cylinders	26
2.6	Calculation of Optimum Cost of an Alloy using LP Problem		28

	2.7	Optimization of Biological Wastewater Treatment Plant	30
	2.8	Calculation of Minimum Error in Least Squares Method	31
	2.9	Determination of Chemical Equilibrium	33

Summary — 35
Exercise — 35
References — 39

3. Single Variable Unconstrained Optimization Methods

	3.1	Introduction	40
	3.2	Optimization of Single Variable Function	41
		3.2.1 Criteria for optimization	41
		3.2.2 Classification of unconstrained minimization methods	47
	3.3	Direct Search Methods	48
		3.3.1 Finding a bracket for a minimum	48
		3.3.2 Unrestricted search method	49
		3.3.3 Exhaustive search	51
		3.3.4 Dichotomous search	53
		3.3.5 Interval halving method	56
		3.3.6 Fibonacci method	59
		3.3.7 Golden section method	62
	3.4	Direct Root Methods	64
		3.4.1 Newton method	65
		3.4.2 Quasi-Newton method	66
		3.4.3 Secant method	67
	3.5	Polynomial Approximation Methods	68
		3.5.1 Quadratic interpolation	69
		3.5.2 Cubic interpolation	70

Summary — 72
Exercise — 72
References — 73

4. Trust-Region Methods

	4.1	Introduction	74
	4.2	Basic Trust-Region Method	75
		4.2.1 Problem statement	75
		4.2.2 Trust-Region radius	76
		4.2.3 Trust-Region subproblem	78
		4.2.4 Trust-Region fidelity	78

	4.3	Trust-Region Methods for Unconstrained Optimization	79
	4.4	Trust-Region Methods for Constrained Optimization	80
	4.5	Combining with Other Techniques	82
	4.6	Termination Criteria	83
	4.7	Comparison of Trust-Region and Line-Search	83

Summary   84
Exercise   84
References   84

5. Optimization of Unconstrained Multivariable Functions

5.1	Introduction		86
5.2	Formulation of Unconstrained Optimization		87
5.3	Direct Search Method		87
	5.3.1	Random search methods	87
	5.3.2	Grid search method	90
	5.3.3	Univariate method	93
	5.3.4	Pattern search methods	94
5.4	Gradient Search Method		99
	5.4.1	Steepest descent (Cauchy) method	100
	5.4.2	Conjugate gradient (Fletcher–Reeves) method	102
	5.4.3	Newton's method	104
	5.4.4	Marquardt method	106
	5.4.5	Quasi-Newton method	109
	5.4.6	Broydon–Fletcher–Goldfrab–Shanno method	113
5.5	Levenberg–Marquardt Algorithm		114

Summary   116
Review Questions   116
References   117

6. Multivariable Optimization with Constraints

6.1	Formulation of Constrained Optimization		119
6.2	Linear Programming		122
	6.2.1	Formulation of linear programming problems	122
	6.2.2	Simplex method	127
	6.2.3	Nonsimplex methods	133
	6.2.4	Integer linear programming	139
6.3	Nonlinear Programming with Constraints		144
	6.3.1	Problems with equality constraints	144

		6.3.2	Problems with inequality constraints	149
		6.3.3	Convex optimization problems	151

Summary 154
Review Questions 154
References 156

7. Optimization of Staged and Discrete Processes

	7.1	Dynamic Programming		157
		7.1.1	Components of dynamic programming	158
		7.1.2	Theory of dynamic programming	159
		7.1.3	Description of a multistage decision process	160
	7.2	Integer and Mixed Integer Programming		166
		7.2.1	Formulation of MINLP	167
		7.2.2	Generalized Benders Decomposition	169

Summary 176
Exercise 176
References 178

8. Some Advanced Topics on Optimization

	8.1	Stochastic Optimization		180
		8.1.1	Uncertainties in process industries	180
		8.1.2	Basic concept of probability theory	182
		8.1.3	Stochastic linear programming	186
		8.1.4	Stochastic nonlinear programming	191
	8.2	Multi-Objective Optimization		193
		8.2.1	Basic theory of multi-objective optimization	197
		8.2.2	Multi-objective optimization applications in chemical engineering	202
	8.3	Optimization in Control Engineering		206
		8.3.1	Real time optimization	206
		8.3.2	Optimal control of a batch reactor	208
		8.3.3	Optimal regulatory control system	212
		8.3.4	Dynamic matrix control	214

Summary 218
Review Questions 218
References 219

9. Nontraditional Optimization

	9.1	Genetic Algorithm		222
		9.1.1 Working principle of GAs		223

		9.1.2	Termination	228
	9.2	Particle Swarm Optimization		229
		9.2.1	Working principle	230
		9.2.2	Algorithm	231
		9.2.3	Initialization	231
		9.2.4	Variants of PSO	232
		9.2.5	Stopping criteria	235
		9.2.6	Swarm communication topology	236
	9.3	Differential Evolution		241
		9.3.1	DE algorithm	241
		9.3.2	Initialization	242
		9.3.3	Mutation	243
		9.3.4	Crossover	244
		9.3.5	Selection	245
	9.4	Simulated Annealing		245
		9.4.1	Procedure	246
		9.4.2	Applications of SA in chemical engineering	253
Summary				253
Exercise				254
References				255

10. Optimization of Various Chemical and Biochemical Processes

10.1	Heat Exchanger Network Optimization		258
	10.1.1	Superstructure	259
	10.1.2	Problem statement	260
	10.1.3	Model formulation	260
10.2	Distillation System Optimization		263
10.3	Reactor Network Optimization		267
10.4	Parameter Estimation in Chemical Engineering		271
	10.4.1	Derivation of objective function	271
	10.4.2	Parameter estimation of dynamic system	273
10.5	Environmental Application		276 |//
Summary			281
Review Questions			281
References			282

11. Statistical Optimization

11.1	Design of Experiment		284
	11.1.1	Stages of DOE	285
	11.1.2	Principle of DOE	286
	11.1.3	ANOVA study	289
	11.1.4	Types of experimental design	291
11.2	Response Surface Methodology		296
	11.2.1	Analysis of a second order response surface	301
	11.2.2	Optimization of multiple response processes	303

Summary — 305
Review Questions — 305
References — 305

12. Software Tools for Optimization Processes

12.1	LINGO	307
12.2	MATLAB	316
12.3	MINITAB®	323
12.4	GAMS	333

Summary — 342
Review Questions — 342
References — 342

Multiple Choice Questions – 1 — 343

Multiple Choice Questions – 2 — 349

Multiple Choice Questions – 3 — 355

Index — 359

List of Figures

1.1	Convex function	4
1.2	Concave function	5
1.3	Local and global optimum points	6
1.4	Classification of optimization problem	7
1.5	Conversion of $f(x)$ to $-f(x)$	8
1.6	Plot of objective function vs. decision variable	8
2.1	Liquid storage tank	13
2.2	Configuration of an L level pump network	15
2.3	Concentration vs. time plot for a series reaction in PFR	18
2.4	Objectives used during optimization of gluconic acid production	21
2.5	Changes of heat flux with insulation thickness	22
2.6	Heat exchanger network with three heat exchangers	24
2.7	Rotating cylinder (with temperature and velocity profile)	26
2.8	Biological wastewater treatment plant	30
2.9	Least square method	31
3.1	Global and relative optimum points	41
3.2	Undefined derivative at x^*	43
3.3	Inflection or saddle point	44
3.4	Unimodal functions	47
3.5	Work done vs. intermediate pressure graph	51
3.6	Exhaustive search (see Example 3.3)	52
3.7	Interval halving method	57
3.8	Convergence process of Newton method	66
3.9	Convergence process of secant method	67
3.10	Quadratic interpolation	70
4.1	Trust region	77
5.1	Contour representation for random jumping method	88
5.2	Contour representation for grid search method	91
5.3	Contour representation for Cr(VI) removal	92
5.4	Contour for quadratic function	94
5.5	Pattern search method	94
5.6	Gradient search method	99

5.7	Direction of movement of any point	100
6.1	Graphical representation of feasible region	120
6.2	Constrained optimization problem	120
6.3	Unbounded feasible region	121
6.4	Graphical representation of problem 6.12a–6.12d	127
6.5	Ellipsoid method	135
6.6	Interior point method	136
6.7	Karmarkar's region inversion	137
6.8	Integer linear programming	140
7.1	Single-stage decision process	161
7.2	Multi-stage decision process	161
7.3	The annual net profit vs. time plot	163
7.4	Representation of 5 stage dynamic programming	164
7.5	Generalized Benders Decomposition method	175
8.1	Fluidization column	181
8.2	Multi-objective optimization	195
8.3	Pareto optimal set	195
8.4	Utopia point	196
8.5(a)	Convex objective space	199
8.5(b)	Non-convex objective space	199
8.6	Flowchart for evolutionary algorithm	201
8.7	Utopia-tracking approach	202
8.8	PID controller as MOO problem	205
8.9	Plant decision hierarchy	207
9.1	Parent chromosomes	226
9.2	Single point crossover	226
9.3	Multipoint crossover	226
9.4	Mutation operation	227
9.5	Flowchart of genetic algorithm	228
9.6	Movement of particles in PSO	230
9.7(a)	von Neumann topology	236
9.7(b)	Star topology	236
9.7(c)	Wheel topology	237
9.7(d)	Circle topology	237
9.7e	Pyramid topology	237
9.8	Flowchart for differential algorithm	242
9.9	Flowchart of simulated annealing	247
9.10	Probability of accepting vs temperature plot	253
10.1	HEN superstructure	259
10.2	Continuous distillation column	264
10.3	Reactor network superstructure	269
11.1	Full factorial design with three variables	292
11.2	Fractional factorial design with three variables	292

11.3	Central composite design	294
11.4	Box–Behnken design	295
11.5(a)	Response surface formed from Eq. 11.22a	298
11.5(b)	Contour representation of Fig. 11.5a	299
11.6(a)	Response surface formed from Eq. 11.22b	299
11.6(b)	Contour representation of Fig. 11.6a	300
11.7(a)	Response surface formed from Eq. 11.22c	300
11.7(b)	Contour representation of Fig. 11.7a	301
12.1	LINGO main window	308
12.2	Lingo Model-Lingo1 window with a maximization model	309
12.3	Toolbar of LINGO	309
12.4	LINGO error message box	309
12.5	Solver status window	310
12.6	LINGO solution report window	311
12.7	LINGO solver status window	315
12.8	Plot of objective 1 and objective 2	321
12.9	Pareto front for obj1 and obj2	322
12.10	MINITAB main window	324
12.11	MINITAB window for selecting DOE	325
12.12	MINITAB window with B–B design	326
12.13	MINITAB window with data for B–B design	327
12.14	MINITAB window for analyzing the response surface design	328
12.15	MINITAB window for selecting "response"	328
12.16	MINITAB window with results	329
12.17(a)	Response surface of Eq. 12.2 at pH = 0	330
12.17(b)	Response surface of Eq. 12.2 at TiO_2 = 0	331
12.17(c)	Response surface of Eq. 12.2 at Time = 0	331
12.18(a)	Contour of the response Eq. 12.2 at pH = 0	332
12.18(b)	Contour of the response Eq. 12.2 at TiO_2 = 0	332
12.18(c)	Contour of the response Eq. 12.2 at Time = 0	333
12.19	Main window of GAMS	334

List of Tables

2.1	Composition and cost of copper alloys	29
3.1	Values of stationary points	46
3.2	Values of successive iterations with accelerated step size	50
3.3	The available interval of uncertainty after different trials	52
3.4	Work done with different intermediate pressure (Exhaustive search method)	53
3.5	Final intervals of uncertainty for different pairs of experiments	54
3.6	Final interval of uncertainty by golden section method	63
5.1	Value of objective function at different grid points	92
6.1	Subproblems for branch and bound method	143
6.2	Results of the subproblems in Table 6.1	143
7.1	Values of the decision variables (as per Eq. 7.14)	164
7.2	Values of the decision variables (as per Eq. 7.15)	165
7.3	Values of the decision variables (as per Eq. 7.16)	165
7.4	Values of the decision variables (as per Eq. 7.17)	165
7.5	Values of the decision variables (as per Eq. 7.18)	166
8.1	Number of worker present with probability	185
9.1	Decimal to binary conversion	224
9.2	Single and multiple point crossover	227
9.3	Chromosomes after crossover	227
9.4	Calculated values of different iteration	252
11.1	Single factor experiment with 5 level of the factor and 5 replicates	287
11.2	Arrangement of experimental run after randomization	288
11.3	Reaction rate data at different pH	288
11.4	Three variable B–B design	295
11.5	Values of the coefficient (Anupam *et al.*)	302
12.1	Widget capacity data	313
12.2	Vendor widget demand	313
12.3	Shipping cost per widget ($)	313
12.4	Description of MATLAB function used for optimization	317
12.5	Levels of independent variables for B–B design	325
12.6	Matrix of B–B design	326
12.7	ANOVA for percentage dye removal	329

12.8	Different variable type and their GAMS keyword	335
12.9	Different relational operator in GAMS	335
12.10	Available solution procedure in GAMS	335
12.11	Availability and cost of petroleum stocks	338
12.12	Specification and selling price	339

Preface

Optimization in the field of chemical engineering is required to utilize the resources in an efficient way as well as to reduce the environmental impact of a process. Application of optimization processes helps us achieve the most favorable operating conditions. Maximum profit is achievable if a process plant runs at optimum conditions. Knowledge of optimization theory as well its practical application is essential for all engineers.

The idea of this book came to my mind long back, perhaps six years ago. Then I started working on it; selecting topics to be included, collecting research papers, and preparing the manuscript. Many people helped me during this process, especially while collecting research articles from different sources: Sudip Banerjee, Arindam Chatterjee, D.K. Sandilya to name a few. I received very useful suggestions from reviewers of this manuscript.

This book contains detailed theory and applications of optimization in chemical engineering and related fields. Prerequisites for this book include some understanding of chemical engineering, biotechnology and mathematics. This book has been divided into twelve chapters. It contains various classical methods for optimization; it also introduces some of the recently developed topics in optimization. Examples from the field of chemical engineering and biochemical engineering are discussed throughout the book.

Chapter 1 discusses the classification and fundamentals of optimization methods. It also includes the salient features of optimization. This chapter also lists different types of objective functions and conditions for optimization. Chapter 2 gives emphasis to different chemical engineering processes and problem formulation procedures for optimization application. This chapter includes objective function formulation of fluid flow system, heat transfer equipments, mass transfer equipments, and reactors. One dimensional unconstrained problem formulation and optimization have been discussed in Chapter 3. This chapter includes different methods like Newton's method, Quasi-Newton method, Secant method etc. Chapter 4 discusses the Trust-Region methods for both constrained and unconstrained optimization problems. An overview of optimization of multivariable unconstrained functions is given in Chapter 5. This chapter comprises various search methods (i.e. random search, grid search), gradient method, Newton's method etc. Chapter 6 discusses the optimization methods for multivariable functions with constraints. This chapter contains both linear programming and non-linear programming. Optimization of staged and discrete processes has been discussed in Chapter 7. This includes dynamic programming, integer and mixed integer programming. Chapter 8 contains some advanced topics on optimization. This chapter discusses stochastic optimization, multiobjective optimization and optimization problems related to control systems. Most chemical process involve highly nonlinear equations that are difficult to optimize using simple and traditional optimization techniques. Chapter 9 discusses some nontraditional optimization methods like Genetic Algorithm (GA), Particle Swarm optimization, Simulated annealing etc. Chapter 10 elucidates the practical application of optimization theory in various chemical and biochemical processes. Chapter 11 describes different statistical optimization methods. This chapter contains response surface methodology with examples from chemical engineering and biotechnology. Chapter 12 gives an overview of different optimization software tools. This chapter elucidates software for optimization such as LINGO, MATLAB, MINITAB and GAMS. A large number of multiple-choice questions are included at the end of this book. I hope this book will be helpful for students at undergraduate and graduate levels. Students will benefit if they go through the theories and solved examples side by side.

I am grateful to LINDO Systems Inc., MathWorks Inc., Minitab Inc., and GAMS for giving permission to include material and screenshots in this book. I am thankful to Gauravjeet Singh Reen for his support during the preparation of the manuscript. I must convey my gratitude to all members of Cambridge University Press for their kind cooperation. I am indebted to my family members for their kind cooperation. I am also thankful to all my colleagues, friends, and well-wishers.

Readers of this book are requested to send their comments and suggestions for further improvement.

Chapter 1

A Brief Discussion on Optimization

1.1 Introduction to Process Optimization

Optimization is a technique of obtaining the best available output for a system or process. The passion for optimality is inherent in human race. Knowingly or unknowingly, we are optimizing our efforts in daily life. This method is studied in every field of engineering, science, and economics. Optimization has been used from ancient times, mostly, by using analytical methods. There are many practical mathematical theories of optimization that have been developed since the sixties when computers become available. The main purpose of the theories is to develop reliable and fast methods to reach the optimum of a function by arranging its evaluations intelligently. Most of the modern engineering and planning applications incorporate optimization process at every step of their complex decision making process. Therefore, this optimization theory is significantly important for them. All of these fields have the same intention like maximization of profit or minimization of cost. As a process engineer, it is our primary concern to utilize resources carefully with minimum loss. An optimized process that use minimum input (raw material, energy, labor etc.) and gives maximum output (product quality and quantity, most environmental friendly) is always favorable. George E Davis extols, "The aim of all chemical procedures should be the utilization of everything and the avoidance of waste. It is often cheaper to prevent waste than to attempt to utilize a waste product." With proper design of optimized process, wastage of natural resources can be minimized. We can recall the famous quote by Dante *"All that is superfluous displeases God and Nature, All that displeases God and Nature is evil."* In nature, everything follows the optimized way to reach the destination. Heat, water etc. flow through the minimum resistance path.

Chemical process industries consist of several "unit operations" and "unit processes" e.g., heat exchanger, distillation column, batch reactor, packed bed reactor, etc. It is the responsibility of a process engineer to run the plant at an optimum condition to obtain the maximum profit with minimum environmental impact. The real driving force for process optimization is efficiency.

Chemical companies realize that if they can run the plant more efficiently, it will improve their bottom line. We can optimize the process by considering the individual unit one by one, or by considering many units at a time (e.g., water distribution system with pumps and pipe line, heat exchanger network, reactor network). Environmental pollution is also a crucial issue for process industries. Sometimes environmental issues are embedded with the objective function and solved as a multiobjective optimization problem. Process optimization involves the determination of process parameters (temperature, pressure, pH, time etc.) that provides us maximum output. In the chemical industry, proper selection of batch time gives the maximum selectivity for a batch reactor. Maximum amount of heat recovery is possible by the optimization of HEN.

1.2 Statement of an Optimization Problem

All optimization problems can be presented by some standard form. Each and every optimization problem contains objective function(s) $f(X)$ which we need to optimize. The general form of optimization problem is:

Min/Max $f(X)$

subject to $g(X) = 0$;

$h(X) \geq 0$

The solution $X = \begin{Bmatrix} x_1 \\ x_2 \\ \vdots \\ x_n \end{Bmatrix}$, which gives us the optimum value of $f(X)$.

Various terms are used for optimization problems, these are:

Decision variable During the optimization process, we have to formulate the objective function that depends on some variables. In real problem, there are many variables that control the output of any system. However, for optimization techniques, we consider few variables over which the decision maker has control. Decision variables are the variables within a model that one can control. Decision variables are usually input to the model that can be changed by the decision maker with the aim of revising the response of the system. For example, a decision variable might be operating temperature of a reactor, diameter of pipe, number of plates in a distillation column.

Identifying and prioritizing key decision variables Any variable that we need be controlled is a decision variable. However, all variables are not equally important during the optimization process. Based on the Pareto ranking of effects on objective function, the key decision variables are chosen. The sensitivity of the objective function to changes in the variables is the key factor for deciding important variables. Variables with high sensitivity may be considered as decision variables, whereas less sensitive variables may be ignored.

Limit of decision variables Every decision variables have some upper and lower limit. Say, mass/mole fraction of any component in a mixture must have the value $0 \leq X \leq 1$. Theoretically, some variable may reach infinity (e.g., time, length); however, in real life, we can allow them to certain practicable values. A first order irreversible reaction takes infinite time for 100 per cent conversion. However, we allow some limited time for practical application.

Objective function Objective function (also read as "cost function") is the mathematical expression that we need to optimize. This describes the correlation between the decision variables and process parameters. In most of the chemical engineering optimization, the objective functions are either profit from the process or cost of production. However, there are many types of objective functions like error during curve fitting/parameter estimation, minimization of environmental impact.

Constraint Constraints are additional relations between the decision variables and process parameters other than the objective function. For instance, during the optimization of any blending process, summation of all components should be equal to unity i.e., $\sum_{i=1}^{n} x_i = 1$. The limit of any variable can be incorporated as constraints i.e., $0 \leq X \leq 1$. As stated the constraints may be equality, inequality (reaction occurs at acidic condition, pH < 7; catalyst is active above certain temperature). Operation of chemical processes are also susceptible to market constraints; say, availability of raw materials and demand of products.

1.3 Classification of Optimization Problems

Classification of optimization methods into different categories has been made based on the physical structure of the problem, type of constraints, nature of design variables, nature of algorithms and area of applications. Optimization techniques are also classified based on the permissible value of the design variables, separability of the functions and number of objective functions. These classifications are briefly discussed in this book.

Topological optimization and parametric optimization

Topological optimization deals with the arrangement of the process equipments during the plant design. Topological optimization should be considered first because the topological changes usually have a large impact on the overall profitability of the plant. For instance, addition of one heat exchanger can change the scenario of the whole heat integration of the process plant. Parametric optimization is easier to interpret when the topology of the flow sheet is fixed. Combination of both types of optimization strategies may have to be employed simultaneously.

Topological arrangement like elimination of unwanted by-product, elimination/rearrangement of equipment, alternative separation methods, and alternative reactor configurations may be employed for the improvement of heat integration. Whereas, parametric optimization consists of the optimization of operating conditions like temperature, pressure, concentration, flow rate etc.

Unconstrained and constrained optimization

If we need to optimize the objective function without any additional constraints, then it is called unconstrained optimization.

e.g., find the minimum of the function $f(x) = x^2 - 2x - 3$.

Whenever, the objective function is accomplished with another correlation (constrained function), these optimization problems are called constrained optimization problems.

e.g., find the minimum of the function $f(x) = x^2 - 2x - 3$

subject to $x \geq 2$

For the unconstrained problem, the solution is $f_{min} = -4$ whereas for the constrained problem the solution is $f_{min} = -3$

Linear and nonlinear programming

This classification is done based on the structure of the equations involved in an optimization problem. If all the equations (objective function and constraints functions) involved in any optimization problem are linear, it is called linear programming.

e.g., minimize the function $f(X) = 2x_1 + 3x_2$

subject to $x_1 - x_2 = 2$

is a linear programming problem where both the objective function and constraint function are linear function of decision variables.

When any of these functions is nonlinear, then this class of optimization is termed as nonlinear programming.

e.g., minimize the function $f(X) = x_1^2 + 2x_2^2$

subject to $x_1 - x_2 = 2$

the aforesaid problem is a nonlinear programming as the objective function is nonlinear.

Convex function and concave function

The ease of optimization problem depends on the structure of the objective function and the constraints. A single-variable function can easily be solved if it is either a convex or a concave function.

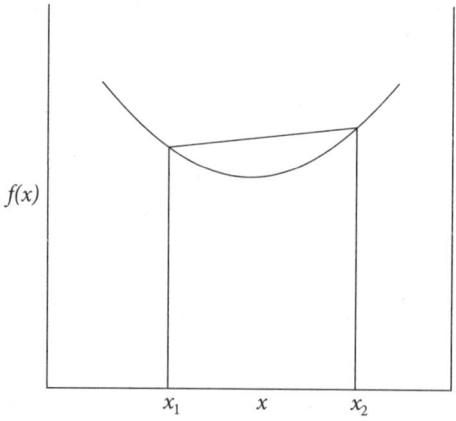

Fig. 1.1 Convex function

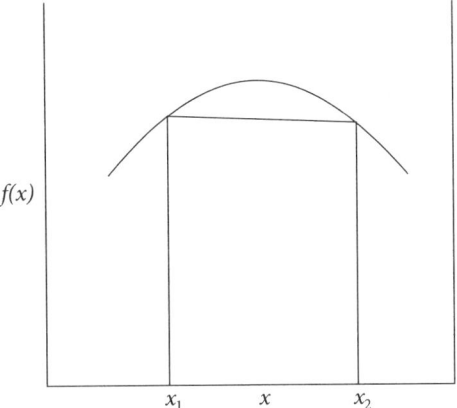

Fig. 1.2 Concave function

Draw a straight line between two points (x_1, x_2). If all point on the straight line $x_1\ x_2$ lie above curve $f(x)$, then it is called convex function (Fig. 1.1). When all point on the straight line $x_1\ x_2$ lie below curve $f(x)$, then it is called concave function (Fig. 1.2). The advantage of any convex or concave function is that it provides a single optimum point. Therefore, this optimum point is always a global optimum.

Continuous and discrete optimization

When the decision variables appear in any optimization problem are all real numbers (temperature, pressure, concentration, etc.) and the objective and constraint functions are continuous then these are called continuous optimization.

However, in some cases, we need to optimize an optimization problem where decision variables are discrete number these are termed as discrete optimization. Number of workers, number of batches per month, number of plate for distillation column etc., cannot be real numbers they are always integer, whereas the standard diameter of pipes are real numbers but, discrete in nature. The standard pipe available with OD (mm) of 10.3, 13.7, 17.1, 21.3, 26.7 etc. If we are interested to find the optimum pipe diameter, we have to perform a discrete optimization with these standard pipe diameters. Same heat exchanger can be used for both cooling and heating purpose: when we need to cool the process stream below the available temperature we have to use cooling water. On the other hand, if we need to heat the process stream we have to use hot water stream. For this situation, the presence of any stream (hot or cold) can be represented by binary variable 0–1.

Mixed integer optimization is another class of optimization method where both integer and real variables appear in the objective function (e.g., optimization of distillation column involves variables like temperature and number of tray/plate).

Single-objective and multi-objective optimization

When the optimization problem contains only one objective function it is known as single objective optimization. These types of problems are easy to solve. However, most of the chemical engineering problems involve more than one objective functions (e.g., optimization of yield and selectivity for a reactor). These problems need special consideration, as the objectives are conflicting.

Multi-objective optimization problem can be represented as

$$\min[F_1(X), F_2(X), \ldots, F_n(X)]$$

$$X \in R^m$$

$$h(X) = 0$$

$$g(X) \geq 0$$

The detail solution procedures are discussed in chapter 8.

Local and global search method

If the objective function possesses multiple stationary points, search algorithms have a tendency to find a stationary point nearer to the initial starting point. The algorithm is stuck in this local optimum point, they are not able to find the global optima. If the search process starts from point A, it will locate the point 1 as minimum (Fig. 1.3). Whereas, if it starts from point B, it will find point 2 as a minimum.

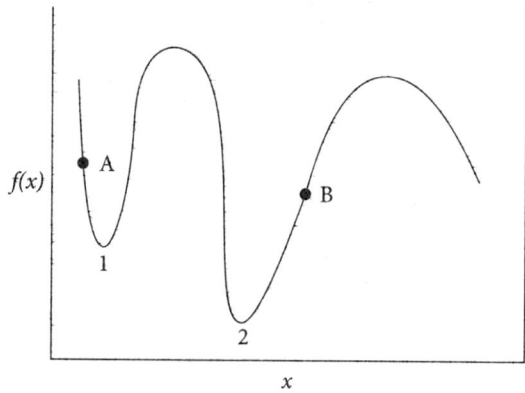

Fig. 1.3 Local and global optimum points

Global search algorithms are able to find the global optimum point. Global optimization algorithms are usually broadly classified into deterministic and stochastic methods.

Deterministic, stochastic and combinatorial optimization

Generally, optimization techniques fall into one of two categories, deterministic and stochastic. Deterministic methods use a simple algorithm, which start with some initial guess and then update iteratively to attain a solution. These techniques usually require the knowledge of the shape of the objective function. The examples of deterministic method are steepest descent method and

Powell's method (conjugate gradient method). Deterministic methods for optimization have primarily one strength and one weakness. The strength of deterministic optimization methods is that provided a good initial guess for an optimum solution they converge very quickly and with high precision. The weakness is that they have a tendency to get trapped in local minima if the initial guess is far from the optimum solution. On the other hand, stochastic optimization methods rely on randomness and retrials to better sample the parameter space in searching for an optimum solution. For these algorithms, search point moves randomly in the search space. The three most commonly used stochastic methods are Monte Carlo (MC), Simulated Annealing (SA), and the Genetic Algorithm (GA). Stochastic optimization problems are also used for optimization that attempt to model the process that has data with uncertainty by considering that the input is specified in terms of a probability distribution.

Combinatorial optimization is a topic that is discussed in applied mathematics and theoretical computer science. This technique consists of a finite set of objects from which we need to find an optimal object. Exhaustive search is not feasible for many such problems. This optimization technique works on the domain of those problems, in which the set of feasible solutions is either discrete or can be converted to discrete, and wherein the aim is to get the best solution. The traveling salesman problem (TSP) is the most common example of combinatorial optimization.

The following Fig. 1.4 shows the classification of optimization techniques.

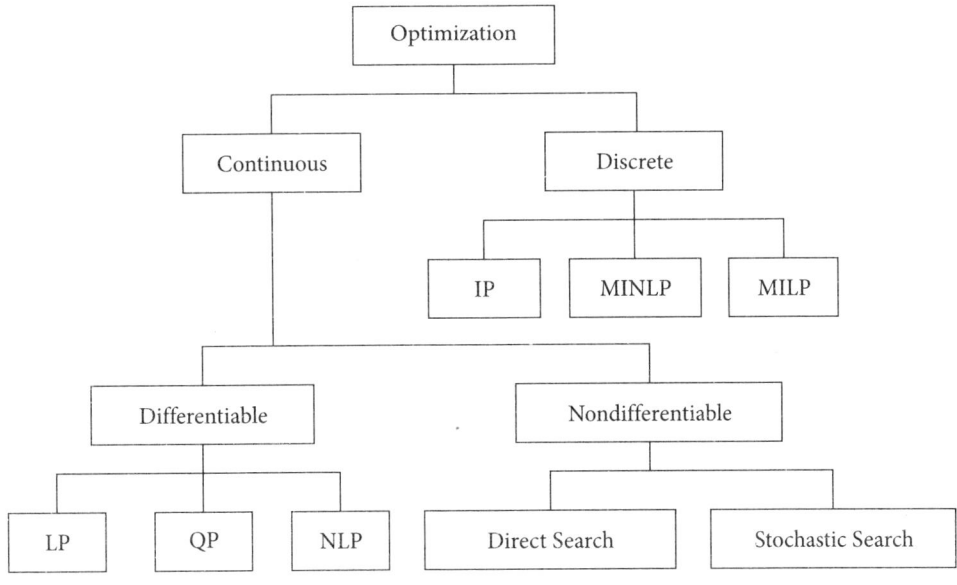

Fig. 1.4 Classification of optimization problem (IP: integer programming, MINLP: mixed integer non-linear programming, MILP: mixed integer linear programming, LP: linear programming, QP: quadratic programming, NLP: non-linear programming)

1.4 Salient Feature of Optimization

Optimization problem can be converted to a suitable form according to our convenience. The minimization of $f(X)$ can be transformed to maximization of $-f(X)$ as shown in Fig. 1.5.

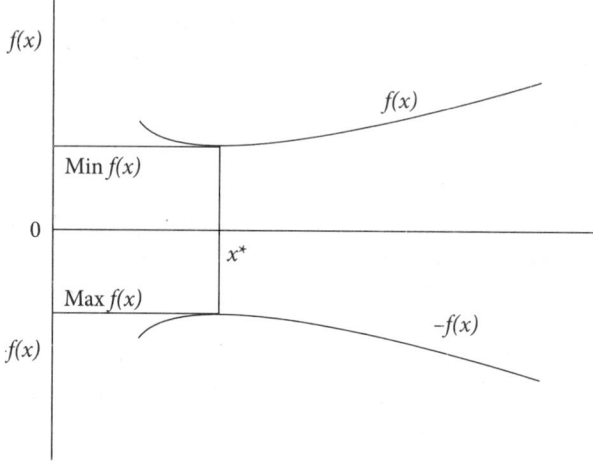

Fig. 1.5 Conversion of $f(x)$ to $-f(x)$

There is a common misconception that the first derivative of the objective function is zero at the optimum (minimum or maximum) point; this is not true for all cases. Consider an objective function as shown in Fig. 1.6. If we find the minimum and maximum values of that function within the range $[x_a, x_b]$, we will get the values f_a and f_b respectively. At these points x_a and x_b the value of the first derivative is not zero. Newton's method is not suitable for this situation as termination criteria does not work. This problem may be cursed with the entrapment at the local optimum point. Application of global optimization algorithm or multi-start algorithm may solve this problem. Solution of local search method may give us a wrong result. "Premature optimization is the root of all evil." Donald Ervin Knuth.

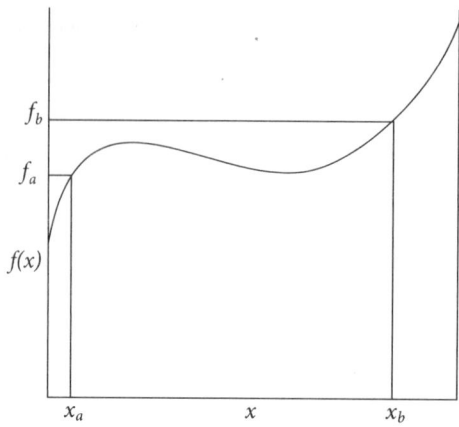

Fig. 1.6 Plot of objective function vs. decision variable

Several issues can make optimization problems practically complex and difficult to solve. Such complicating factors are the existence of multiple decision variables in the problem, multiple objective function in the problem. For solving any process optimization problem, we have to follow six important steps as described by Edger *et al.* These steps are:

1. Analyze the process properly to define the process variables and their characteristics. After that, prepare a list of all variables associated with the process.
2. Set the objective function in terms of the decision variables listed above along with coefficients involved with the process (heat conductivity, heat transfer coefficient, reaction rate constant etc.). Decide the criterion for optimization. This second step provides us the performance model; this model is also called the economic model when appropriate.
3. After developing the objective function, we need to check if any other process constraints are associated with this process. We have to include both equality and inequality constraints. For this purpose we generally use some well-known physical principles (conservation of mass and energy), empirical equations (Dittus–Boelter correlation for calculating heat transfer coefficient), implicit concepts (rules of thumb), and external restrictions. Identification of the independent and dependent variables has been done to determine the degrees of freedom.
4. When the formulated optimization problem is very large:
    (i) split the problem into small manageable parts or
    (ii) objective function and model equation can be simplified as per our convenience
5. Apply an appropriate optimization method to solve the problem. Selection of method depends on the structure of the problem. There is no such algorithm, which is suitable for all problems.
6. Check the sensitivity of the result to modify the coefficients in the problem

1.5 Applications of Optimization in Chemical Engineering

Optimization is an intrinsic part of design: the designer seeks the best, or optimum, solution of a problem [Coulson and Richardson]. The design of any chemical engineering systems can be formulated as optimization problems in which a measure of performance is to be optimized while satisfying all other constraints. In the field of chemical engineering, typical application of optimization problems arise in process identification, design and synthesis, model development, process control, and real-time optimization.

In this book, we have discussed many chemical engineering applications:

i. optimization of water storage tank
ii. optimization of water pumping network
iii. optimization of cost of an alloy
iv. optimization of heat exchanger network
v. optimization of chemical reactor and reactor network
vi. application of optimization in thermodynamics, determination of chemical equilibrium
vii. optimization of distillation system

viii. parameter estimation using least square/Levenberg–Marquardt method
ix. optimization of biological wastewater treatment plant
x. tuning of PID controller
xi. design of real-time controller
xii. optimization of air pollution control system
xiii. optimization of adsorption process
xiv. warehouse management
xv. optimization of blending process in petroleum refinery

1.6 Computer Application for Optimization Problems

In chemical engineering, optimization is a technique to find the most cost effective process under the given constraints. Most of the optimization problems in the chemical industry are very complicated in nature. There are a large number of variables with a large number of constraints involved in this process. Solving of these problems is difficult, sometimes impossible by hand. Numerical optimization helps us to solve these problems in an efficient way. There are many algorithms developed for solving complicated optimization problems. To execute these algorithms, we need the help of computer programming. Application of computer programming and software can solve these problems easily. However, the application of optimization is sometimes restricted by the lack of information as well as the lack of time to evaluate the problem. There are many commercial software available e.g., LINGO, MATLAB, MINITAB, GAMS etc. Chapter 12 discussed the detail of these software. Computer is also essential for real time optimization and controlling the process using advanced control systems (e.g., Model Predictive Control).

Summary

- This chapter explains different aspects of optimization methods. Classification of various algorithms has been made based on structure and application area of the optimization problem. There are six main steps for chemical process optimization. These steps of chemical process optimization have been discussed in this chapter. A list of chemical engineering application of optimization is given and the detail of those applications are discussed in the subsequent chapters.

Review Questions

1.1 Give a list of variables required to optimize a heat exchanger.
1.2 Why proper selection of decision variables is required for optimization process?
1.3 Devise a scheme for optimizing a reactor-separator process, which maximizes the yield.
1.4 How do you classify the optimization algorithms based on various aspects?
1.5 Define degree of freedom. What is the significance of degree of freedom during optimization of a chemical process?

1.6 Why topological optimization is required for the optimization of a chemical process plant?

1.7 Give some examples of combinatorial optimization in the field of chemical engineering.

1.8 How the local optimization can be avoided?

1.9 Limitations of any decision variable can be implemented in process optimization, give some examples in the area of chemical engineering.

1.10 What are advantages of computer applications for optimization?

References

Edgar, T. F., Himmelblau, D. M., Lasdon, L. S. 2001. *Optimization of Chemical Processes*, New York: McGraw-Hill.

Coulson, J. M., Sinnott, R. K., Richardson, J. F. 1999. *Coulson and Richardson's Chemical Engineering*, vol. 6, 2e: Butterworth–Heinemann.

Knuth, D. E., *Art of Computer Programming*, vol. 1, Fundamental Algorithms.

Chemical Engineers Who Changed the World, Claudia Flavell–While, www.tcetoday.com, accessed on March 2012.

Turton, R., Bailie R. C., Whiting, W. B., Shaeiwitz, J. A., Bhattacharyya, D. 2012. *Analysis, Synthesis, and Design of Chemical Processes,* Prentice Hall.

Chapter

2

Formulation of Optimization Problems in Chemical and Biochemical Engineering

2.1 Introduction

The major application of optimization in the chemical engineering field is minimizing energy consumption in process plant. Other applications include the optimum design of fluid flow systems, optimization of the separation process, and optimization of product concentration and reaction time in reacting systems. Estimation of the overall cost for any plant design is also a vital part. Optimization of process variables is the tool for the same. In biochemical engineering, optimization is required for finding the optimum operating conditions of a bioreactor, and parameter estimation in biochemical pathways. The process of optimization requires proper formulation of the objective function and necessary constraint functions. The objective function can be formulated in different ways based on the number of variables involve in the process. If the process mechanisms are known to us and the number of variables are less, then objective functions are formulated based on basic principles of science and technology (e.g., law of conservation, thermodynamic laws). Moreover, if the process mechanisms are complicated and not known to us, number of variables are large; statistical optimization methods like Response Surface Methodology (RSM) and Artificial Neural Network (ANN) are applicable for those processes.

2.2 Formulation of Optimization Problem

One of the important steps during the application of optimization technique to a practical problem is the formulation of objective functions. We have to develop the model equations based on the physical appearance of the system. When we are formulating the mathematical statement of the objective, we have to keep in mind that the complexity increases with the nonlinearity of the function. During optimization, more complex functions or more nonlinear functions are harder to solve. There are many modern optimization software (see chapter 12) that have been developed to solve highly nonlinear functions. Most of the optimization problems comprise one objective function. Even though some problems that involve multiple objective functions cannot be transformed into a single function with similar units (e.g., maximizing profit while simultaneously minimizing risk).

2.3 Fluid Flow System

Liquid storage and transportation trough pipeline is very common in chemical process industry. In this section, we will discuss about the design of optimization storage tank and optimum pump configuration.

2.3.1 Optimization of liquid storage tank

A cylindrical tank (Fig. 2.1) has a volume (V) that can be expressed by $V = (\pi/4)D^2L$, and we are interested to calculate the diameter (D) and height (H) that minimize the cost of the tank.

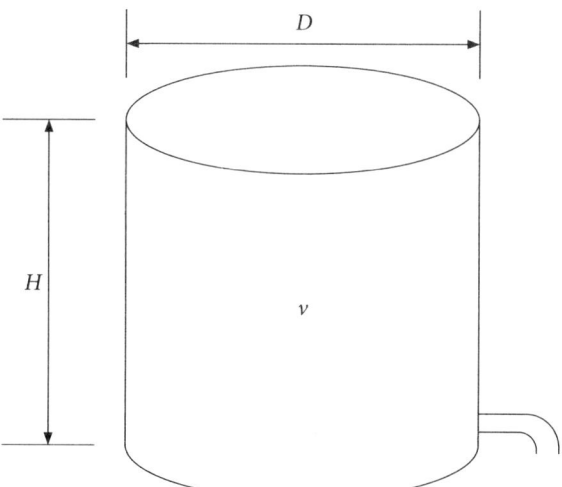

Fig. 2.1 Liquid storage tank

Cost of the tank is given by the by f; we will get the optimum design by solving the nonlinear problem:

$$\min_{H,D} f \equiv c_s \pi DH + c_t (\pi/2) D^2 \tag{2.1}$$

subject to $V = (\pi/4) D^2 H$, (2.2)

$D \geq 0, \ H \geq 0$ (2.3)

The cost of the tank depends on the amount of material needed which is proportional to its surface area and the cost per unit area of the tank's side is c_s, whereas for tank's top and bottom the cost per unit area is c_t.

We are able to simplify the problem by ignoring the bound constraints and eliminating the variable H from the above Eq. (2.1), giving us an unconstrained problem:

$$\min f \equiv 4c_s V/D + c_t (\pi/2) D^2 \tag{2.4}$$

Now, applying necessary condition for minimization; the objective function is differentiated with respect to D and setting the derivative to zero yields

$$\frac{df}{dD} = -4c_s V/D^2 + c_t \pi D = 0, \tag{2.5}$$

yielding $D = \left(\frac{4c_s V}{\pi c_t}\right)^{1/3}$, $H = \left(\frac{4V}{\pi}\right)^{1/3} \left(\frac{c_t}{c_s}\right)^{2/3}$ and the aspect ratio $H/D = c_t/c_s$.

We notice that the solution of this problem is obtained easily by using simple differential calculus. However, the generalization of this problem can make it more completed during analytical solution.

Note

The inequality constraints ($D \geq 0$, $H \geq 0$) are neglected as D and H both are positive, they remained satisfied with the solution. They are regarded as inactive constraints. If the situation were different from this case, we would need to apply more complicated optimality conditions.

We can easily eliminate one variable from the equality constraint that related H and D (Eq. (2.2)). This is often impossible for nonlinear equations and implicit elimination is required.

When we are considering these issues, particularly as they apply to very large and complicated problem, requires the numerical algorithms as discussed in chapter 3.

2.3.2 Optimization of pump configurations

There are many industries where energy consumption for fluid pumping is very high. For instance, in the pulp and paper industry, pumps consume approximately 10–20 per cent of the total electrical energy requirement. Minimization of the energy costs, as well as the total cost for the fluid pumping systems, is a major concern for these industries. In this section, we will develop the objective function for optimization of the fluid flow system. A typical problem of fluid pumping has been discussed by *T.* Westerlund *et al.* (Westerlund *et al.* 1994).

For a particular set of centrifugal pumps with given pressure rise (in terms of total head) and data for power requirement as a function of the capacity of these pumps, we have to choose the best pump or configuration of pumps coupled in parallel and/or series. This configuration should fulfill our requirements; the total required fluid flow and total pressure rise for the configuration. We are considering a very simple configuration with the total pump arranged to a L level pump configuration with $N_{p,i}$ parallel pumping lines of $N_{s,i}$ pumps in series in each line. Centrifugal pumps of the equal size and that have the same rotation speed (it may vary) are used on the same level (i). At each level, the rise in pressure head is equal to the total required pressure rise, whereas the total flow is the summation of flows through all these L levels. At each level, two real variables and two integer variables are considered for optimization. The real variables are the speed of rotation (ω_i) and the fraction (x_i) of the total flow ($\dot{V}_{tot}$). The integer variables are the number of parallel pumping lines ($N_{p,i}$) and the number of pumps in series ($N_{s,i}$). The Fig. 2.2 demonstrates an L level pump configuration.

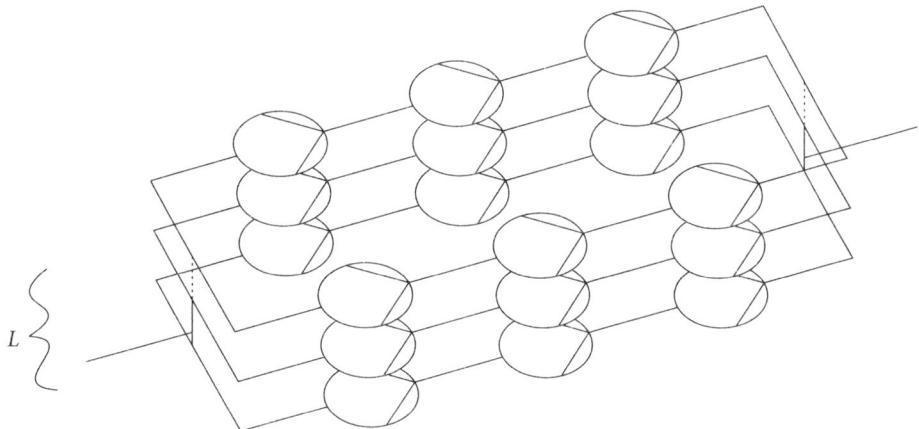

Fig. 2.2 Configuration of an L level pump network

We can represent the total cost of this pump arrangement by the following equation,

$$J = \sum_{i=1}^{L} \left(C_i + C_i' P_i \right) N_{p,i} N_{s,i} \tag{2.6}$$

where L signifies the number of levels, C_i is the yearly installment of the capital costs for a single pump, C_i' and P_i the energy and the power cost for every single pump on each level, respectively.

The power required for one pump (P_i) is usually a function of the rotational speed (ω_i) of the centrifugal pump, the flow ($\dot{V}_{tot}$) through the pump, and the density of the fluid to be pumped (ρ). Therefore, we can write

$$P_i = f_{1,i}\left(\dot{V}_i, \omega_i, \rho\right) \tag{2.7}$$

At any level i, the flow through a single pump is represented by the following equation

$$\dot{V}_i = \frac{x_i}{N_{p,i}} \dot{V}_{tot} \tag{2.8}$$

The pressure rise Δp_i is also a function of the flow ($\dot{V}_i$) through the pump and the rotational speed ω_i, as well as the fluid density (ρ); which can be written as

$$\Delta p_i = f_{2,i}\left(\dot{V}_i, \omega_i, \rho\right) \tag{2.9}$$

The rise in pressure over a single pump on each level can be demonstrated by the following equation,

$$\Delta p_i = \frac{1}{N_{s,i}} \Delta p_{tot} \tag{2.10}$$

Therefore, for a specified flow, number of pumps in series and the total rise in pressure, the rotational speed ω_i can be solved implicitly from Eqs (2.9)–(2.10). The functions, f_1 and f_2, are generally not known explicitly. However, the manufacturers are used to provide the data for power requirement and total head (pressure rise), at a constant rotational speed and a specific fluid (usually water) as a function of the capacity for a particular pump.

The data for pressure rise, given by the pump manufacturers, at a given constant rotation speed (ω_m), and for a fluid with the density (ρ_m), can be stated by the relation,

$$\Delta p_m = g_1\left(\dot{V}_m, \omega_m, \rho_m\right) \tag{2.11}$$

and the corresponding data for power requirement is given by the relation,

$$P_m = g_2\left(\dot{V}_m, \omega_m, \rho_m\right) \tag{2.12}$$

Now, by means of the proportionality relation as given in literature [Coulson and Richardsson (1985)] we get,

$$\dot{V}_i = \left(\frac{\omega_i}{\omega_m}\right)\dot{V}_m \tag{2.13}$$

$$\Delta p_i = \left(\frac{\omega_i}{\omega_m}\right)^2 \left(\frac{\rho}{\rho_m}\right) \Delta p_m \tag{2.14}$$

and

$$P_i = \left(\frac{\omega_i}{\omega_m}\right)^3 \left(\frac{\rho}{\rho_m}\right) P_m \tag{2.15}$$

Thus, the relations, $f_{1,i}$ and $f_{2,i}$, can be obtained by combining Eqs (2.13)–(2.15) with the relations, $g_{1,i}$ and $g_{21,i}$. The functions $g_{1,i}$ and $g_{21,i}$ refer to the "manufacturers" total head curve and the power curve for a particular type of pump used at level "i".

The result of the optimal configurations for pump network can be represented as MINLP problem. The statement of the optimization problem can be given as

$$\min_{N_{p,i}, N_{s,i}, x_i, \omega_i, i=1......L} \left\{ \sum_{i=1}^{L} \left(C_i + C'_i P_i \right) N_{p,i} N_{s,i} \right\} \quad (2.16)$$

subject to

$$\Delta p_i = f_{2,i} \left(\dot{V}_i, \omega_i \right) \quad (2.17)$$

$$\sum_{i=1}^{L} x_i = 1 \quad (2.18)$$

$$\omega_i - \omega_{i,\max} \leq 0 \quad (2.19)$$

where

$$P_i = f_{1,i} \left(\dot{V}_i, \omega_i \right) \quad (2.20)$$

$$\dot{V}_i = \frac{x_i}{N_{p,i}} \dot{V}_{tot} \quad (2.21)$$

$$\Delta p_i = \frac{1}{N_{s,i}} \Delta p_{tot} \quad (2.22)$$

ω_i and x_i are non-negative real variables and the variables, $N_{p,i}$ and $N_{s,i}$, are non-negative integers.

2.4 Systems with Chemical Reaction

Optimization of product concentration is a crucial job for process engineers. Maximum amount of product will produce with minimum loss of raw materials and energy. In this section, we will discuss two different types of optimization problems in two different reacting systems.

2.4.1 Optimization of product concentration during chain reaction

For determining the optimum product concentration of any series reaction, we need to develop the mathematical equation.

A series reaction $A \xrightarrow{k_1} B \xrightarrow{k_2} C$ has been carried out in a PFR; where $B \to C$ is undesired reaction with unwanted product C. If we allow small time for the reaction, production of B will be very less and if we allow large time then conversion of $B \to C$ will be high. Figure 2.3 gives a clear idea about the process. It shows that at time t^* the concentration of desired product (B) will be maximum.

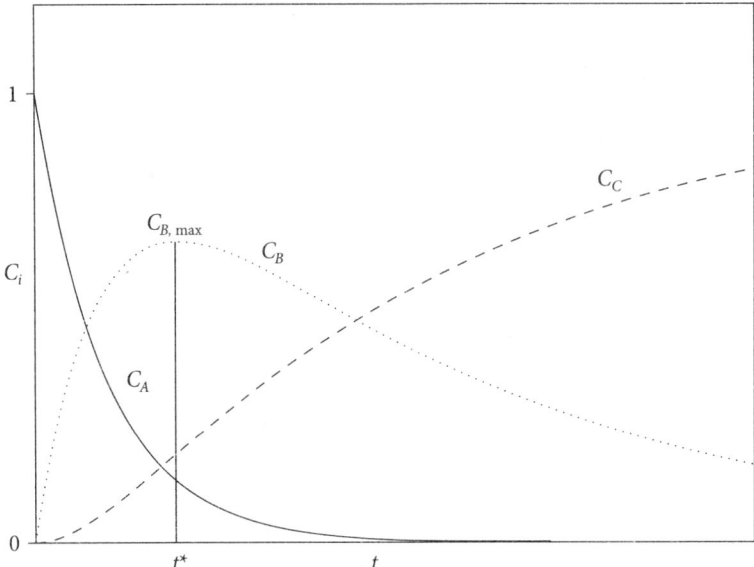

Fig. 2.3 Concentration vs. time plot for a series reaction in PFR

Now we have to develop the mathematical model equation (objective function) to find the optimum time t^*.

The rate of concentration change for A by the reaction $A \to B$ is

$$\frac{dC_A}{dt} = -r_A = -k_1 C_A \tag{2.23}$$

solving the equation we have

$$C_A(t) = C_{A_0} e^{-k_1 t} \tag{2.24}$$

the rate of concentration change for B by the reaction is

$$\frac{dC_B}{dt} = k_1 C_A - k_2 C_B \tag{2.25}$$

$$\frac{dC_B}{dt} = k_1 C_{A_0} e^{-k_1 t} - k_2 C_B \tag{2.26}$$

$$\frac{dC_B}{dt} + k_2 C_B = k_1 C_{A_0} e^{-k_1 t}, \tag{2.27}$$

this Eq. (2.27) is linear in C_B. Solution of this 1st order ordinary differential equation is

$$C_B(t) = C_{A_0} \frac{k_1}{k_2 - k_1} \left(e^{-k_1 t} - e^{-k_2 t} \right) \tag{2.28}$$

This Eq. (2.28) represents the concentration of component B with time. This is a single-variable (time t) optimization problem without any constraint.

The necessary condition for finding the relative maximum is

$$\left. \frac{dC_B}{dt} \right|_{t=t^*} = 0 \tag{2.29}$$

$$\frac{dC_B}{dt} = 0 = \frac{k_1 C_{A_0}}{k_2 - k_1} \left(-k_1 e^{-k_1 t} + k_2 e^{-k_2 t} \right) \tag{2.30}$$

which gives

$$t^* = \frac{\ln(k_2/k_1)}{k_2 - k_1} \tag{2.31}$$

and

$$C_B^* = C_{A_0} \left(\frac{k_1}{k_2} \right)^{k_2/(k_2 - k_1)} \tag{2.32}$$

and the sufficient condition is

$$\left. \frac{d^2 C_B}{dt^2} \right|_{t=t^*} < 0 \tag{2.33}$$

applying both necessary and sufficient condition we find the maximum concentration of component B and the optimum time t^*. This problem is considered as an isothermal PFR, if temperature changes with time objective function will change accordingly.

2.4.2 Optimization of gluconic acid production

Optimizing the production rate and product concentration in biochemical reaction is very important. In this section, we will develop a multi-objective optimization problem. This model simulates the production of gluconic acid by the fermentation of glucose in a batch stirred tank reactor using the microorganism *Pseudomona ovalis*. The mechanism of the overall biochemical reaction can be represented as follows [Johansen, T. A., and Foss, B. A. (1995)].

$$\text{Cells + Glucose + Oxygen} \rightarrow \text{More cells,} \tag{2.34a}$$

$$\text{Glucose + Oxygen} \rightarrow \text{Gluconolactone,} \tag{2.34b}$$

$$\text{Gluconolactone + Water} \rightarrow \text{Gluconic acid} \tag{2.34c}$$

The concentration of cells (X), gluconic acid (p), gluconolactone (l), glucose substrate (S), and dissolved oxygen (C) can be described by the following state-space model [Ghose, T. K., and Gosh, P. (1976)].

$$\frac{dX}{dt} = \mu_m \frac{SC}{k_s C + k_o C + SC} X \tag{2.35}$$

$$\frac{dp}{dt} = k_p l \tag{2.36}$$

$$\frac{dl}{dt} = v_l \frac{S}{k_l + S} X - 0.91 k_p l \tag{2.37}$$

$$\frac{dS}{dt} = -\frac{1}{Y_s} \mu_m \frac{SC}{k_s C + k_o C + SC} X - 1.011 v_l \frac{S}{k_l + S} X \tag{2.38}$$

$$\frac{dC}{dt} = K_L a (C^* - C) - \frac{1}{Y_o} \mu_m \frac{SC}{k_s C + k_o C + SC} X - 0.091 v_l \frac{S}{k_l + S} X \tag{2.39}$$

We can identify various objective criteria for optimizing the production of gluconic acid. Halsall-Whitney and Thibault [Hayley Halsall–Whitney and Jules Thibault (2006)] have described the multi-objective optimization scheme for this reacting system. They have concentrated on maximizing the overall rate of production (p_f/t_B) and p_f, the final concentration of gluconic acid, at the same time minimizing the concentration of final substrate (S_f) after the completion of the fermentation process. The simulations study may vary in terms of the inputs employed for defining the decision space. The choice of input variables included the duration of the fermentation process or batch time $t_B \in [5, 0.5h]$, the initial concentration of substrate $S_0 \in [20,0 \text{ g/L}]$, the concentration of initial biomass $X_0 \in [0.05, 1.0 \text{ UOD/mL}]$, and the oxygen mass transfer coefficient $K_{La} \in [50, 300 \text{ h}^{-1})]$. Figure 2.4 demonstrates the multi-input, multi-output optimization strategy used for optimizing the gluconic acid production.

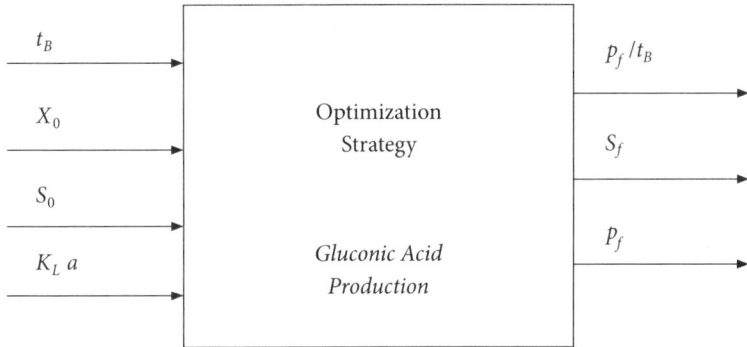

Fig. 2.4 Objectives used during optimization of gluconic acid production

2.5 Optimization of Heat Transport System

Optimization in heat transport system is required to achieve some purposes such as

i. minimize the heat loss
ii. optimum design of heat transfer equipments (e.g., heat exchanger, evaporator, condenser)
iii. optimization of Heat Exchanger Network (HEN)

2.5.1 Calculation of optimum insulation thickness

Addition of more insulation to a flat wall always decreases the heat transfer rate. The thicker the insulation, the lower the heat transfer rate. Since the area of heat transfer (A) is constant, therefore, addition of insulation always increases the thermal resistance of the wall without changing the convection resistance. However, in case of cylindrical or spherical shape, thermal resistance increases with increasing insulation thickness and convection heat transfer increases as surface area increases. Figure 2.5 shows variation of heat flux with insulation thickness. It shows that heat flux is maximum when the radius is r_{cr}. For current-carrying wires and cables, heat flux should be maximum to dissipate the heat produced. For finding the maximum heat flux and corresponding insulation thickness, we have to develop an objective function based on conduction and convection heat transfer theory.

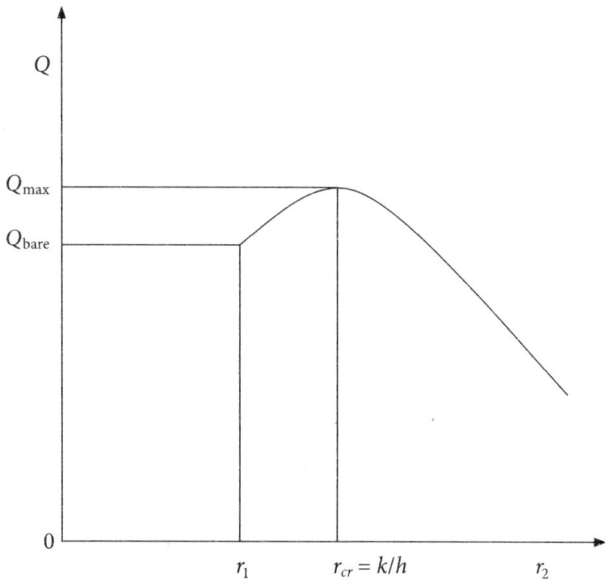

Fig. 2.5 Changes of heat flux with insulation thickness

heat transfer by conduction

$$q = \frac{2\pi L k (T_i - T_o)}{\ln \frac{r_i}{r_o}} \qquad (2.40)$$

and heat transfer by convection

$$q = 2\pi r_o L h (T_o - T_\infty) \qquad (2.41)$$

therefore, heat flux with the combined effect of conduction and convection is

$$q = \frac{T_i - T_\infty}{\frac{1}{2\pi L k} \ln \frac{r_o}{r_i} + \frac{1}{2\pi r_o L h}} \qquad (2.42)$$

the overall resistance for heat transfer is

$$R = \frac{1}{2\pi L k} \ln \frac{r_o}{r_i} + \frac{1}{2\pi r_o L h} \qquad (2.43)$$

To get maximum heat flux we have to minimize the overall resistance for heat transfer. This is a single-variable minimization problem where R will be minimized with respect to variable r_0. The necessary condition for solving this problem is

$$\frac{dR}{dr_o} = 0 \tag{2.44}$$

$$\frac{dR}{dr_o} = \frac{d}{dr_o}\left[\frac{1}{2\pi Lk}\ln\frac{r_o}{r_i} + \frac{1}{2\pi r_o Lh}\right] = 0 \tag{2.45}$$

$$\frac{1}{kr_o} - \frac{1}{hr_o^2} = 0 \tag{2.46}$$

$$r_o^* = r_{cr} = \frac{k}{h} \tag{2.47}$$

The sufficient condition for this problem is

$$\frac{d^2R}{dr_o^2} = \frac{d}{dr_o}\left[\frac{1}{kr_o} - \frac{1}{hr_o^2}\right] = -\frac{1}{kr_o^2} + \frac{2}{hr_o^3} = \frac{1}{r_o^2}\left(\frac{2}{hr_o} - \frac{1}{k}\right) \tag{2.48}$$

$$\left(\frac{d^2R}{dr_o^2}\right)_{r_o = r_{cr}} = \frac{h^2}{k^2}\left(\frac{2}{k} - \frac{1}{k}\right) = \frac{h^2}{k^3} > 0 \tag{2.49}$$

since, $d^2R/dr_o^2 > 0$, R is minimum and heat flux will be maximum

substituting the value of $r_{cr} = \frac{k}{h}$ in Eq. (2.43) we get the minimum value of R

$$R_{\min} = \frac{1}{2\pi Lk}\ln\frac{k}{hr_i} + \frac{1}{2\pi Lk} = \frac{1}{2\pi Lk}\left[\ln\frac{k}{hr_i} + 1\right] \tag{2.50}$$

and from Eq. (2.42) we have

$$q_{\max} = \frac{2\pi Lk(T_i - T_\infty)}{\left[\ln\frac{k}{hr_i} + 1\right]} \tag{2.51}$$

Equation 2.51 gives the maximum heat flux corresponds to critical insulation thickness.

2.5.2 Optimization of simple heat exchanger network

In this section, we are considering the optimization of a simple heat exchanger network presented in Fig. 2.6 that consists of three heat exchangers with two process streams. The process streams are defined as hot and cold streams with their inlet and outlet temperature. The hot stream that has fixed flow rate F and the heat capacity C_p requires to be cooled from T_{in} to T_{out} ($T_{in} > T_{out}$), whereas the cold stream with a fixed flow rate f and heat capacity c_p requires to be heated from t_{in} to t_{out} ($t_{out} > t_{in}$). This HE network is consists of three heat exchangers. Temperature of the steam is T_s and has a heat duty Q_h has been used by the heater, whereas the cooler with a heat duty Q_c uses cooling water at temperature T_w. However, a third heat exchanger is used to save a considerable amount of energy by transferring heat from the hot stream to the cold stream. This third heat exchanger has a heat duty Q_m with the exit temperatures of hot and cold streams, T_m and t_m, respectively.

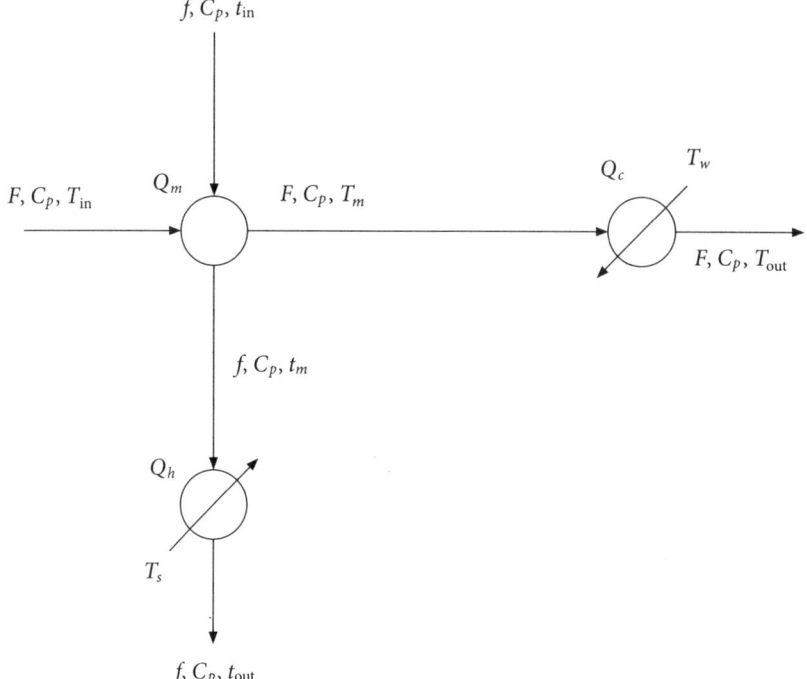

Fig. 2.6 Heat exchanger network with three heat exchangers

The model equations for this heat exchanger arrangement are described as follows:

- The energy balance equations for this HE network is represented by

$$Q_c = FC_p \left(T_{in} - T_{out} \right) \tag{2.52}$$

$$Q_h = fc_p \left(t_{out} - t_{in} \right) \tag{2.53}$$

$$Q_m = fc_p\left(t_m - t_{in}\right) = FC_p\left(T_{in} - T_m\right) \tag{2.54}$$

where the subscript c, h, and m indicate the cooler, heater and the heat exchanger.

- The capital cost of each heat exchanger depends on its area of heat exchange A_i, $i \in \{c,h,m\}$. Now we are considering a simple countercurrent, shell and tube heat exchanger that has an overall heat transfer coefficient, U_i, $i \in \{c,h,m\}$. The resultant equations for calculating the area are as follows:

$$Q_i = U_i A_i \Delta T_{lm}^i, \quad i \in \{c,h,m\}. \tag{2.55}$$

- The log-mean temperature difference (LMTD) ΔT_{lm}^i can be written as

$$\Delta T_{lm}^i = \frac{\Delta T_a^i - \Delta T_b^i}{\ln\left(\Delta T_a^i / \Delta T_b^i\right)}, \quad i \in \{c,h,m\}, \tag{2.56}$$

and

$$\Delta T_a^c = T_m - T_w, \quad \Delta T_b^c = T_{out} - T_w \tag{2.57}$$

$$\Delta T_a^h = T_h - t_m, \quad \Delta T_b^h = T_s - t_{out} \tag{2.58}$$

$$\Delta T_a^m = T_{in} - t_m, \quad \Delta T_b^m = T_m - t_{in} \tag{2.59}$$

In this optimization problem, our intention is to minimize the total cost (the capital cost of the heat exchangers as well as the energy cost) of this HE network. This gives us the following Nonlinear Programming (NLP):

$$\min \sum_{i \in \{c,h,m\}} \left(\hat{c}_i Q_i + \overline{c}_i A_i^\beta\right) \tag{2.60}$$

Subject to

$$Q_c = FC_p\left(T_{in} - T_{out}\right) \tag{2.61}$$

$$Q_h = fc_p\left(t_{out} - t_{in}\right) \tag{2.62}$$

$$Q_m = fc_p\left(t_m - t_{in}\right) = FC_p\left(T_{in} - T_m\right) \tag{2.63}$$

$$Q_i = U_i A_i \Delta T_{lm}^i, \quad i \in \{c,h,m\}. \tag{2.64}$$

$$\Delta T_{lm}^i = \frac{\Delta T_a^i - \Delta T_b^i}{\ln\left(\Delta T_a^i/\Delta T_b^i\right)}, \quad i \in \{c,h,m\}, \tag{2.65}$$

$$Q_i \geq 0, \ \Delta T_a^i \geq \varepsilon, \ \Delta T_b^i \geq \varepsilon, \ i \in \{c,h,m\} \tag{2.66}$$

Where the cost coefficients $\hat{c}_i$ and $\overline{c}_i$ represent the energy and amortized capital prices, the exponent $\beta \in [0,1]$ represents the economy of scale of the equipment, and a small constant $\varepsilon > 0$ is chosen such that the log-mean temperature difference does not become undefined. There is one degree of freedom for this example. For example, when we specify the heat duty Q_m, the temperatures of hot and cold stream and all other remaining parameters can be estimated.

Optimization method of a heat exchanger network model has been discussed in chapter 10 (see section 10.1). The total cost of the network is optimized using superstructure optimization.

2.5.3 Maximum temperature for two rotating cylinders

Two cylinders are rotating as shown in Fig. 2.7. There is a lubricant between these cylinders. Find the maximum temperature that can be achieved in the lubricant between these two cylinders. The outer cylinder that has a radius of 8 cm rotates at an angular velocity of 9000 revolution per minute, while the inner cylinder is fixed. Clearance between the cylinders (both at 30°C) is very small, 0.025 cm. The density, viscosity, and thermal conductivity of the lubricant are 1200kg/m³, 0.1kg/m.sec, and 0.13J/sec.m.°C respectively [Griskey, 2002].

For this system, as clearance is very small compared to their diameter, this rotating cylinders can be considered as an arrangement of two parallel plates where the upper plate (outer cylinder) moving with a velocity (the angular velocity times the radius $V_z = \Omega R$) of 12m/sec. The lower plate (inner cylinder) is fixed as shown in figure.

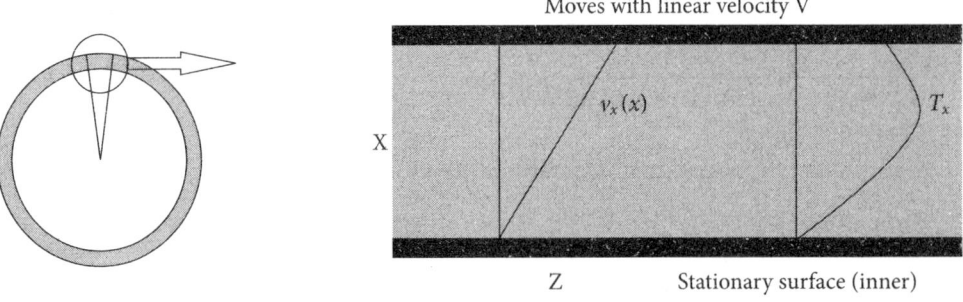

Fig. 2.7 Rotating cylinder (with temperature and velocity profile)

For the system given in Fig. 2.7, we can consider the energy balance equation and reduce it to a solvable form as described below.

The temperature profile for a Cartesian coordinate is as follows:

$$\rho \hat{C}_p \left(\cancel{\frac{\partial T}{\partial t}} + v_x \cancel{\frac{\partial T}{\partial x}} + v_y \cancel{\frac{\partial T}{\partial y}} + v_z \cancel{\frac{\partial T}{\partial z}} \right) = k \left[\frac{\partial^2 T}{\partial x^2} + \cancel{\frac{\partial^2 T}{\partial y^2}} + \cancel{\frac{\partial^2 T}{\partial z^2}} \right]$$

$$+ 2\mu \left\{ \left(\cancel{\frac{\partial v_x}{\partial x}} \right)^2 + \left(\cancel{\frac{\partial v_y}{\partial y}} \right)^2 + \left(\cancel{\frac{\partial v_z}{\partial z}} \right)^2 \right\} \qquad (2.67)$$

$$+ \mu \left\{ \left(\cancel{\frac{\partial v_x}{\partial y}} + \cancel{\frac{\partial v_y}{\partial x}} \right)^2 + \left(\cancel{\frac{\partial v_x}{\partial z}} + \frac{\partial v_z}{\partial x} \right)^2 + \left(\cancel{\frac{\partial v_y}{\partial z}} + \cancel{\frac{\partial v_z}{\partial y}} \right)^2 \right\}$$

eliminating the terms that are not required, we have

$$-k \frac{\partial^2 T}{\partial x^2} = \mu \left(\frac{\partial V_z}{\partial x} \right)^2 \qquad (2.68)$$

The velocity profile can be obtained from the z component of Equation of Motion. From this equation (considering gravitational and pressure effects both zero), we get

$$\frac{\partial^2 V_z}{\partial x^2} = 0 \qquad (2.69)$$

Integrating Eq. (2.69) with the boundary conditions

$$V_z = V \text{ (i.e., } \Omega R\text{)}, x = B \qquad (2.70)$$

$$V_z = 0, x = 0 \qquad (2.71)$$

we get the relation

$$\frac{V_z}{V} = \frac{x}{B} \qquad (2.72)$$

Substituting V_z from Eq. (2.72) to Eq. (2.68) we get

$$-k \frac{\partial^2 T}{\partial x^2} = \mu \frac{V^2}{B^2} \qquad (2.73)$$

Now, solving Eq. (2.73) with boundary conditions

$$T = T_0, x = 0 \qquad (2.74)$$

$$T = T_1, x = B \qquad (2.75)$$

we obtain

$$T = T_0 + (T_1 - T_0)\frac{x}{B} + \frac{\mu V^2}{2k}\frac{x}{B}\left(1 - \frac{x}{B}\right) \qquad (2.76)$$

or

$$\frac{T - T_0}{T_1 - T_0} = \frac{x}{B} + \frac{\mu V^2}{2k(T_1 - T_0)}\frac{x}{B}\left(1 - \frac{x}{B}\right) \qquad (2.77)$$

in the above equation, the dimensionless group is known as the Brinkman number (Br).

$$Br = \frac{\mu V^2}{k(T_1 - T_0)} = \frac{\text{Heat generated by viscous dissipation}}{\text{Conduction heat transfer}} \qquad (2.78)$$

this group signifies the impact of viscous dissipation effects.

for this present case, $T_1 = T_0$ and

$$T - T_0 = \frac{\mu V^2}{2k}\frac{x}{B}\left(1 - \frac{x}{B}\right) \qquad (2.79)$$

Equation (2.79) is the objective function for finding the maximum temperature. The maximum temperature will arise when $x = 0.5B$ that is, giving the largest value of $\frac{x}{B}\left(1 - \frac{x}{B}\right)$.

$$T = 30 + \frac{0.1 \times 12^2}{2 \times 0.13}\frac{1}{2}\left(1 - \frac{1}{2}\right) \qquad (2.80)$$

we get the maximum temperature $T = 43.85\ °C$.

2.6 Calculation of Optimum Cost of an Alloy using LP Problem

Two alloys A and B made of copper, zinc, lead, and tin are mixed to prepare C, a new alloy. The required composition of alloy C and the composition of alloys A, and B have shown in the following table (Table 2.1):

Table 2.1 Composition and cost of copper alloys

Alloy	Composition by weight			
	Copper	Zinc	Lead	Tin
A	78	12	6	4
B	62	20	16	2
C	≥ 72	≥ 15	≤ 10	≥ 3

If cost of alloy B is two times of alloy A, formulate the optimization problem for determining the amounts of A and B to be mixed to produce alloy C at a minimum cost.

Solution

Assume the amount of A, and B required for producing C are w_A and w_B respectively. And the corresponding costs per kg are c_A, and c_B. The production cost of C is c_C per kg.

The production cost of per kg C alloy is

$$c_C = c_A w_A + c_B w_B \tag{2.81}$$

We have to minimize the cost of C alloy
Therefore, we can write

$$\min_{w_A, w_B} c_C = c_A w_A + c_B w_B \tag{2.82}$$

with equality constraint

$$w_A + w_B = 1 \tag{2.83}$$

and inequality constraints

$$0.78 w_A + 0.62 w_B \geq 0.72 \tag{2.84}$$

$$0.12 w_A + 0.20 w_B \geq 0.15 \tag{2.85}$$

$$0.06 w_A + 0.16 w_B \geq 0.10 \tag{2.86}$$

$$0.04 w_A + 0.02 w_B \geq 0.03 \tag{2.87}$$

$$w_A, w_B, w_C \geq 0 \tag{2.88}$$

This problem can be solver using Linear Programming method.

2.7 Optimization of Biological Wastewater Treatment Plant

A rectangular tank has been made for biological treatment of wastewater (batch process). The dimensions of the tank are given in Fig. 2.8 (length x_1 meters, width x_2 meters, and height x_3 meters). The sides and bottom of the tank cost, respectively, Rs.1200/-, and Rs.2500/- per m² area. The operating cost for the tank is Rs.500/- for each batch of water treatment. A maintenance cost of Rs.100/- for every 10 batch is required. Assuming that the tank will have no salvage value, find the minimum cost for treatment of 1000 m³ of wastewater. Assume the salvage value of the tank is zero after 1000 m³ of wastewater treatment.

Solution

The total cost of water treatment =
    cost of the tank + operating cost of wastewater treatment + maintenance cost
    = (cost of sides + cost of bottom) + number of batch × (cost for each batch) + 100 × (number of batch)/10

$$f(X) = 1200(2x_1 x_3 + 2x_2 x_3) + 2500 x_1 x_2 + 500\left(\frac{1000}{x_1 x_2 x_3}\right) + 100\left(\frac{1000}{10 x_1 x_2 x_3}\right) \quad (2.89)$$

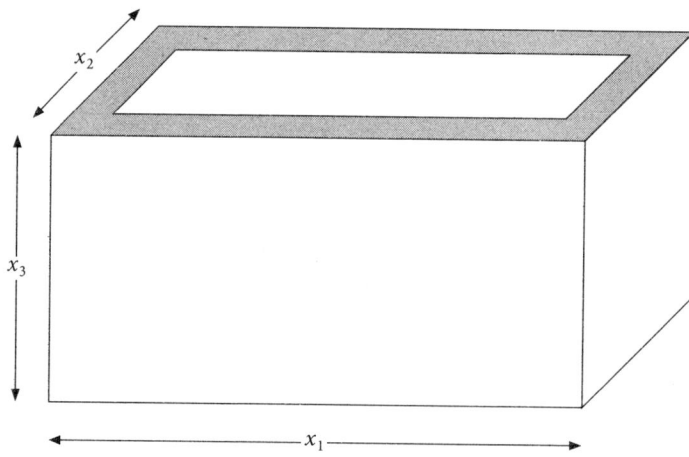

Fig. 2.8 Biological wastewater treatment plant

The above problem is an unconstrained multivariable problem that can be solve by **Geometric Programming**.

The statement of the problem is as follows:

$$\min f(X) = 1200(2x_1 x_3 + 2x_2 x_3) + 2500 x_1 x_2 + 500\left(\frac{1000}{x_1 x_2 x_3}\right) + 100\left(\frac{1000}{10 x_1 x_2 x_3}\right) \quad (2.90)$$

The detail algorithm for solving this problem has been discussed in chapter 5 and chapter 9.

2.8 Calculation of Minimum Error in Least Squares Method

In many applications in chemical and biochemical engineering, we need to find the best fit curve from our experimental results. The curve of best fit is that for which e's (error values) are as small as possible i.e., E, the sum of the squares of the errors is a minimum. This is known as the principle of least squares.

Suppose it is required to fit the curve

$$y = a + bx + cx^2 \qquad (2.91)$$

to a given set of observations $(x_1, y_1), (x_2, y_2), \ldots, (x_5, y_5)$. For any x_i, the observed value is y_i and the expected value is $\eta_i = a + bx_i + cx_i^2$ so that the error $e_i = y_i - \eta_i$ (see Fig. 2.9).

Therefore, sum of the squares of these errors is

$$E = e_1^2 + e_2^2 + \ldots + e_5^2 \qquad (2.92)$$

$$E = \left[y_1 - \left(a + bx_1 + cx_1^2\right)\right]^2 + \left[y_2 - \left(a + bx_2 + cx_2^2\right)\right]^2 + \ldots + \left[y_5 - \left(a + bx_5 + cx_5^2\right)\right]^2 \qquad (2.93)$$

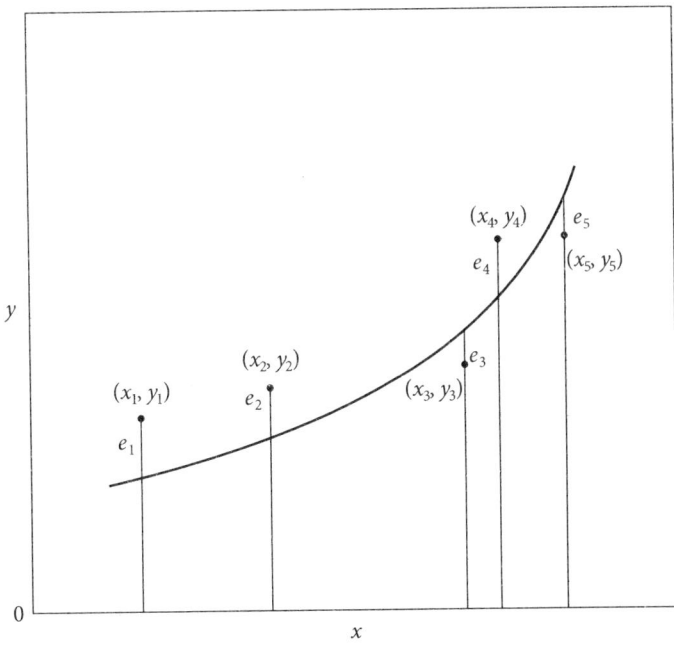

Fig. 2.9 Least square method

Equation (2.93) is the multi-variable (a, b, c) objective function from which E can be minimized.

For E to be minimum, we have

$$\frac{\partial E}{\partial a} = 0 = -2\left[y_1 - (a + bx_1 + cx_1^2)\right] - 2\left[y_2 - (a + bx_2 + cx_2^2)\right] - \ldots \\ -2\left[y_5 - (a + bx_5 + cx_5^2)\right] \qquad (2.94)$$

$$\frac{\partial E}{\partial b} = 0 = -2x_1\left[y_1 - (a + bx_1 + cx_1^2)\right] - 2x_2\left[y_2 - (a + bx_2 + cx_2^2)\right] - \ldots \\ -2x_5\left[y_5 - (a + bx_5 + cx_5^2)\right] \qquad (2.95)$$

$$\frac{\partial E}{\partial c} = 0 = -2x_1^2\left[y_1 - (a + bx_1 + cx_1^2)\right] - 2x_2^2\left[y_2 - (a + bx_2 + cx_2^2)\right] - \ldots \\ -2x_5^2\left[y_5 - (a + bx_5 + cx_5^2)\right] \qquad (2.96)$$

Equation (2.94) simplifies to

$$y_1 + y_2 + \ldots + y_5 = 5a + b(x_1 + x_2 + \ldots + x_5) + c(x_1^2 + x_2^2 + \ldots + x_5^2) \qquad (2.97)$$

$$\sum_{i=1}^{5} y_i = 5a + b\sum_{i=1}^{5} x_i + c\sum_{i=1}^{5} x_i^2 \qquad (2.98)$$

Equation (2.95) becomes

$$x_1 y_1 + x_2 y_2 + \ldots + x_5 y_5 = a(x_1 + x_2 + \ldots + x_5) + b(x_1^2 + x_2^2 + \ldots + x_5^2) \\ + c(x_1^3 + x_2^3 + \ldots + x_5^3) \qquad (2.99)$$

$$\sum_{i=1}^{5} x_i y_i = a\sum_{i=1}^{5} x_i + b\sum_{i=1}^{5} x_i^2 + c\sum_{i=1}^{5} x_i^3 \qquad (2.100)$$

Similarly, (2.96) simplifies to

$$\sum_{i=1}^{5} x_i^2 y_i = a\sum_{i=1}^{5} x_i^2 + b\sum_{i=1}^{5} x_i^3 + c\sum_{i=1}^{5} x_i^4 \qquad (2.101)$$

The Eqs (2.98), (2.100), and (2.101) are known as Normal equations and can be solved as simultaneous equations in a, b, c. The values of these constants when substituted in Eq. (2.91) give the desired curve of best fit.

For *n* number of data point similar equations can be written as

$$\sum_{i=1}^{n} y_i = na + b\sum_{i=1}^{n} x_i + c\sum_{i=1}^{n} x_i^2 \tag{2.102}$$

$$\sum_{i=1}^{n} x_i y_i = a\sum_{i=1}^{n} x_i + b\sum_{i=1}^{n} x_i^2 + c\sum_{i=1}^{n} x_i^3 \tag{2.103}$$

$$\sum_{i=1}^{n} x_i^2 y_i = a\sum_{i=1}^{n} x_i^2 + b\sum_{i=1}^{n} x_i^3 + c\sum_{i=1}^{n} x_i^4 \tag{2.104}$$

a, *b*, *c* are found from these equations.

2.9 Determination of Chemical Equilibrium

The following example is used to explain the application of nonlinear programming in chemical engineering. A mixture of various chemical compounds has been considered for this study. The problem on chemical equilibrium is proposed by Bracken and McCormick [Bracken and McCormick (1968)]. This problem is to define the mixture composition of different chemicals when the mixture is at the state of chemical equilibrium. According to the second law of thermodynamics, the free energy of a mixture of chemicals reaches its minimum value at a constant temperature and pressure when the mixture is in chemical equilibrium condition. Therefore, by minimizing the free energy of the mixture, we can determine the chemical composition of any mixture satisfying the state of chemical equilibrium. For describing this system, we will consider the following notations as given below

m : number of chemical elements in the mixture

n : number of compounds in the mixture

x_j : number of moles for compound j, $j = 1, ..., n$

s : total number of moles in mixture, $s = \sum_{i=1}^{n} x_j$

a_{ij} : number of atoms of element i in a molecule of compound j

b_i : atomic weight of element i in the mixture $i = 1, ..., n$

The constraint equations for the mixture are given below. All compounds should have a nonnegative number of moles.

$$x_j \geq 0, j = 1, ..., n \tag{2.105}$$

There is a material balance equation for each element. These equations are represented by linear equality constraint.

$$\sum_{j=1}^{n} a_{ij} x_j = b_i, \ i = 1, \ldots, m \qquad (2.106)$$

Here, the total free energy of the mixture is the objective function

$$f(x) = \sum_{j=1}^{n} x_j \left[c_j + \ln\left(\frac{x_j}{s}\right) \right] \qquad (2.107)$$

where

$$c_j = \left(\frac{F^0}{RT}\right)_j + \ln(P) \qquad (2.108)$$

and $\left(F^0/RT\right)_j$ is the model standard free energy function for the jth compound. P represents the total pressure in atmospheres. Our aim is to find out the parameters x_j that minimize the objective function $f(x)$ subject to the constraints non-negativity of mole number (as given by Eq. (2.105)) and linear balance (as given by Eq. (2.106)). Therefore, the optimization problem can be written as

$$\text{Min } f(x) = \sum_{j=1}^{n} x_j \left[c_j + \ln\left(\frac{x_j}{s}\right) \right] \qquad (2.107)$$

$$\text{subject to: } c_j = \left(\frac{F^0}{RT}\right)_j + \ln(P) \qquad (2.108)$$

$$\sum_{j=1}^{n} a_{ij} x_j = b_i, \ i = 1, \ldots, m \qquad (2.106)$$

$$x_j \geq 0, \ j = 1, \ldots, n \qquad (2.105)$$

Further study

Readers can find some advanced topic for further studies.

i. Optimization of energy consumption in refinery (Gueddar and Dua, 2012)
ii. Global optimization of pump configurations using Binary Separable Programming (Pettersson and Westerlund, 1997)
iii. Optimization of a large scale industrial reactor by genetic algorithms (Rezende et al. 2008)
iv. Economic Process Optimization Strategies (Buskies, 1997)

Summary

- This chapter gives us an idea about the formulation of the optimization problem. Various chemical engineering systems have been presented in this chapter. Optimization problems for fluid flow system, heat transport systems, and reactor systems are formulated to maximize the efficiency of the system. Optimization of composition during alloy preparation and blending is very crucial job for chemical engineers. We have discussed the optimization problem on alloy preparation. This chapter also includes optimization of biological wastewater treatment plant, parameter estimation using least square method, determination of chemical equilibrium. The solutions of the aforementioned are discussed in the subsequent chapters.

Exercise

Problem 2.1

We are interested to produce P in the reaction $A \to P$ using a continuous reactor at v = 240 liters/hr with C_{A_0} = 3 moles/liter. However, it is noticed that there is a second reaction $P \to R$ that can also occur. This undesired reaction produced undesired product R. It is found that both reactions are irreversible and first order with k_1 = 0.45 min^{-1} and k_2 = 0.1 min^{-1}. Derive the objective function for finding maximum yield of P.

Problem 2.2

For installation and operation of a pipeline for an incompressible fluid, the total cost (in dollars per year) can be represented as follows:

$$C = C_1 D^{1.5} L + C_2 m \Delta p / \rho$$

where

C_1 = the installed cost of the pipe per foot of length computed on an annual basis ($C_1 D^{1.5}$ is expressed in dollars per year per foot length, C_2 is based on $0.05/kWh, 365 days/year and 60 percent pump efficiency).

$\quad D$ = diameter (to be optimized)

$\quad L$ = pipeline length = 100 miles

$\quad m$ = mass flow rate = 200,000 lb/h

$\quad \Delta p = \left(2\rho v^2 L / D g_c\right) f$ = pressure drop, psi

$\quad \rho$ = density = 60 lb/ft3

$\quad v$ = velocity = $\left(4m\right) / \left(\rho \pi D^2\right)$

$\quad f$ = friction factor = $\left(0.046 \mu^{0.2}\right) / \left(D^{0.2} v^{0.2} \rho^{0.2}\right)$

$\quad \mu$ = viscosity = 1 cP

a. Find general expressions for D^{opt}, v^{opt}, and C^{opt}
b. For $C_1 = 0.3$ (D expressed in inches for installed cost), calculate D^{opt} and v^{opt} for the following pairs of values of μ and ρ; $\mu = 0.2, 10$ cP, $\rho = 50, 80$ lb/ft^3

Problem 2.3

A fertilizer producing company purchases nitrates, phosphates, potash, and an inert chalk base and produces four different fertilizers A, B, C, and D. The cost of these nitrates, phosphates, potash, and an inert chalk base are $1600, $550, $1100, and $110 per ton, respectively. The cost of production, selling price, and composition of the four fertilizers are given in the following table.

Table 2.2 Cost of production, selling price, and composition of fertilizers

Fertilizer	Production cost ($/ton)	Selling price ($/ton)	Percentage composition by weight			
			Nitrates	Phosphates	Potash	Inert chalk base
A	98	340	4.5	10	5.5	80
B	155	560	6	15	9	70
C	260	720	14	6	15	65
D	195	440	10	18	10	62

The supply of nitrates, phosphates, and potash is limited, no more than 1100 tons of nitrate, 2200 tons of phosphates, and 1600 tons of potash will be available for a week. The company is required to supply to its customers a minimum of 5200 tons of fertilizer A and 4100 tons of fertilizer D per week; however, it is otherwise free to produce the fertilizers in any quantities it satisfies. Formulate the problem (objective function and constrained functions) to find the quantity of each fertilizer to be produced by the company that maximize its profit.

Problem 2.4

Heavy fuel oil, initially semisolid at 15°C is to be heated and pumped through a 15 cm diameter (inside) pipe at the rate of 20000 kg/h. The pipe line is 1500 long and efficiently lagged. The cost of power for pumping is Rs.1.0 per kwh used with 50 per cent efficiency, while the cost of steam heat is Rs.40.0 per million kilocalorie. On the basis of the following data, calculate the economic (optimum) pumping temperature.

Data: Specific heat of oil = 0.5 kcal/kg °C ; Density of oil = 950 kg/m^3

$$H_{fi} = \frac{2 f v^2 L}{g_c D}$$ where $f = 4 \times Re^{-0.2}$, Re = Reynolds Number

Oil viscosity (kg/m.sec) $\mu = \dfrac{1.0}{dT}$

Where dT is the rise in oil temperature over 15°C

Problem 2.5

The topological optimization is discussed in chapter 1. Here, we will consider a topological optimization problem for a chemical process plant. The layout of the chemical process plant has been shown in Fig.

2.10. This plant consists of a water tank (*T*), a pump (*P*), a fan (*F*), and a compressor (*C*). The positions of the different units are also indicated in this figure in terms of their (*x*, *y*) coordinates. It has been decided to add a new heat exchanger (*H*) within this plant. Addition of new unit may cause congestion within the plant. It is decided to place *H* within a rectangular area given by $\{-15 \leq x \leq 15, -10 \leq y \leq 10\}$ to avoid congestion. Formulate the optimization problem to find the position of *H* to minimize the sum of its distances *x* and *y* from the existing units, *T, P, F,* and *C*.

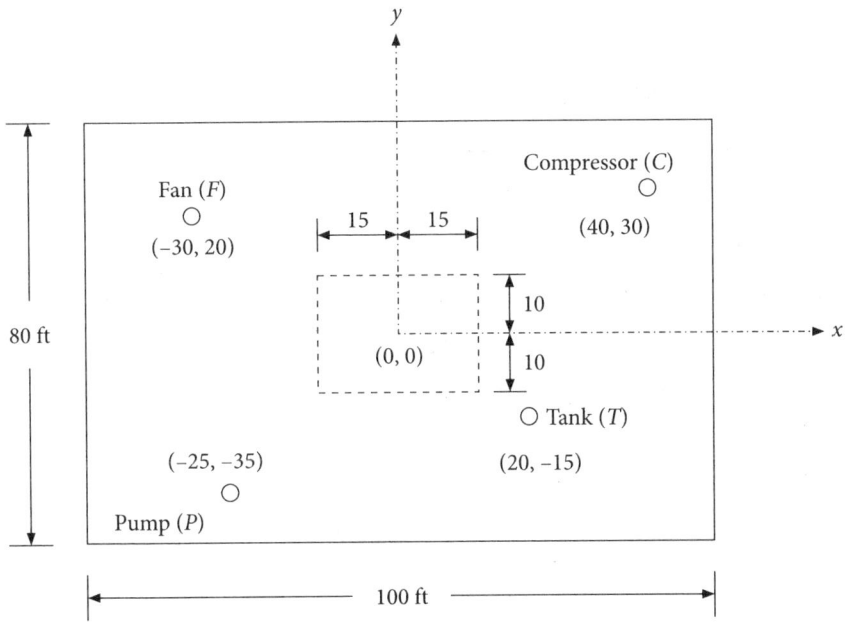

Fig. 2.10 Layout of a chemical processing plant (coordinates in ft)

Problem 2.6

Two grade of coal (*A, B*) are mixed to get a coal (*C*) for blast furnace. The composition of coals and cost of coals are given in the table below

Table 2.3 Composition and cost of coal

Coal	Composition (%) and cost (per ton) of coal			
	Carbon	Sulphur	Ash	Cost per ton
A	92	1	7	Rs. 2000/-
B	81	2	17	Rs. 1500/-
C	≥ 88	≤ 1.5	≤ 10	

Develop the optimization problem to determine the amounts of *A* and *B* to be mixed to produce *C* at a minimum cost.

Problem 2.7

An oil refinery has three process plants, and four grades of motor oil have been produced from these plants. The refinery is liable to meet the demand of customers. The refinery incurs a penalty for failing to meet the demand for any particular grade of motor oil. The capacities of the various plants, the costs of production, the demands of motor oil of the different grades, and the penalties have shown in the following table:

Table 2.4 The details of various plants

Process plant	Plant capacity (kgal/day)	Cost of production ($/day)			
		1	2	3	4
1	100	800	900	1000	1200
2	150	850	950	1150	1400
3	200	900	1000	1250	1600
Demand (kgal/day)		45	140	100	70
Penalty (for shortage of each kilogallon)		$9	$12	$15	$21

Formulate the optimization problem as an LPP for minimizing the overall cost.

Problem 2.8

An adiabatic two-stage compressor is used to compress a gas, which is cooled to the inlet gas temperature between the stages, the theoretical work can be expressed by the following equation:

$$W = \frac{kp_1 V_1}{k-1}\left[\left(\frac{p_2}{p_1}\right)^{(k-1)/k} - 2 + \left(\frac{p_3}{p_2}\right)^{(k-1)/k}\right]$$

Where, $k = C_p/C_v$; p_1 = pressure inlet; p_2 = intermediate stage pressure; p_3 = outlet pressure; V_1 = inlet volume

We are interested to optimize the intermediate pressure p_2 such that the work is a minimum.

Problem 2.9

A refinery produce three major products: gasoline, jet fuel and lubricants by distilling crude petroleum from two sources, Venezuela and Saudi Arabia. These two crudes have different chemical composition and therefore, provide different product mixes. From one barrel of Saudi crude, 0.25 barrel of gasoline, 0.45 barrel of jet fuel, and 0.2 barrel of lubricants are produced. Whereas, one barrel of Venezuelan crude produces 0.4 barrel of gasoline, 0.2 barrel of jet fuel, and 0.3 barrel of lubricants. The refinery losses 10 per cent of each barrel during the crude refining.

The crudes also differ in cost and availability: Up to 9,000 barrels per day of Saudi crude are available at the cost $20 per barrel; Up to 6,000 barrels per day of Saudi crude are also available at the lower cost $15 per barrel. The refinery has contracts with independent distributors to supply 2,000 barrels of gasoline per day, 1,500 barrels of jet fuel per day, and 500 barrels of lubricants per day. Formulate an optimization model in standard form to fulfill the requirements in the most efficient manner.

Problem 2.10

A chemical company has acquired a site for their new plant. They required to enclose that field with a fence. They have 700 meter of fencing material with a building on one side of the field where fencing is not needed. Determine the maximum area of the field that can be enclosed by the fence.

References

Bracken, J. and McCormick, G. P. 1968. *Selected Applications of Nonlinear Programming*, New York : John Wiley and Sons.

Buskies, U. 'Chemical Engineering Technology' *Economic Process Optimization Strategies*, 20(1997) 63–70.

Coulson, J. M. and Richardson, J. F. 1985. *Chemical Engineering*, vol. 1, Oxford: Pergamon Press.

Edgar, T. F., Himmelblau D. M., Lasdon L. S. *Optimization of Chemical Processes* (2e), New York: McGraw–Hil.

Ghose, T. K. and Gosh, P. (1976). 'Kinetic Analysis of Gluconic Acid Production by Pseudomonas Ovalis', *Journal of Applied Chemical Biotechnology*, 26, 768–77.

Grewal, B. S. *Higher Engineering Mathematics* (37e), Delhi-110006: Khanna Publishers, India.

Griskey, R. G. 2002. *Transport Phenomena and Unit Operations*: A Combined Approach, New York: John Wiley and Sons.

Gueddar, T., Dua V. *Novel model reduction techniques for refinery-wide energy optimization*, Applied Energy, 89(2012): 117–26.

Halsall–Whitney, H., Thibault, J. *Multi-Objective Optimization for Chemical Processes and Controller Design: Approximating and Classifying the Pareto Domain*, Computers and Chemical Engineering, 30(2006): 1155–68.

Johansen, T. A. and Foss, B. A. (1995). *Semi-Empirical Modeling of Non-linear Dynamic Systems through Identification of Operating Regimes and Locals Models.* In K. Hunt, G. Irwin, and K. Warwick (Eds.), Neural Network Engineering in Control Systems, Springer–Verlag, pp. 105–26.

Levenspiel, O. *Chemical Reaction Engineering* (3e), New Jersey: John Wiley and Sons, Inc.

Peters, M. S., Timmerhaus, K., West, R. E. *Plant Design and Economics for Chemical Engineers* (5e), New York: McGraw–Hil.

Pettersson, F. and Westerlund, T. 1997. *Global Optimization of Pump Configurations Using Binary Separable Programming*, Computers Chemical Engineering, vol. 21, No. 5, pp. 521–29.

Rao, S. S. *Engineering Optimization Theory and Practice*, Fourth Edition, New Jersey: John Wiley and Sons, Inc.

Rezende, M. C. A. F., Costa C. B. B., Costa A. C., Maciel M. R. W., Filho R. M. *Optimization of a Large Scale Industrial Reactor by Genetic Algorithms*, Chemical Engineering Science, 63(2008): 330–41.

Roy, G. K. 2000. *Solved Examples in Chemical Engineering* (7e), Delhi: Khanna Publishers

Schmidt, L. D. 2005. *The Engineering of Chemical Reactions* (2e), New York: Oxford University Press.

Westerlund, T., Pettersson, F., Grossmann, I. E. 1994. *Optimization of Pump Configurations as a MINLP Problem*, Computers Them. Engng., vol. 18, No. 9, pp. 845–58.

Chapter 3

Single Variable Unconstrained Optimization Methods

3.1 Introduction

A function $f(x)$ is described as the relation between the input and output variables of a system. In a very basic sense, it is a correlation that provides a unique value of $y = f(x)$ for every choice of x. In this situation, x is termed as the independent variable whereas, y is called the dependent variable. The dependent variables vary with the change of independent variables. The thermal conductivity of any material changes with temperature; here, temperature is independent variable whereas thermal conductivity is the dependent variable. Mathematically, we can represent a set $S \subset R$, where R Rrepresents the set of all real numbers. A correspondence or a transformation can be defined by certain correlation that gives a single numerical value to all points $S \in R$. This type of relationship is called a scalar function f defined on the set S.

When, the set $S = R$, we have an unconstrained function of one variable. On the other hand, we can define a function on a constrained region whenever S is a proper subset of R. For example,

$$f(x) = 2x^3 + 5x^2 - 3x + 2 \text{ for all } x \in R \tag{3.1}$$

is an unconstrained function, while

$$f(x) = 2x^3 + 5x^2 - 3x + 2 \text{ for all } x \in S = \{x \mid -5 \le x \le 5\} \tag{3.2}$$

is a constrained function.

In this chapter, we will discuss different methods for single variable unconstrained problem.

3.2 Optimization of Single Variable Function

An appropriate method for the optimization of a function with single variable is necessary for two main reasons:

i. A number of unconstrained problems intrinsically involve only one variable (Example: 3.3)
ii. One-dimensional search is repeatedly used during the optimization of unconstrained and constrained optimization problems (e.g., univariate method).

3.2.1 Criteria for optimization

A single variable function $f(x)$ is supposed to have a local or relative minimum at $x = x^*$ if $f(x^*) \leq f(x^* + h)$ for all positive and negative values of h which is sufficiently small. Correspondingly, a point x^* is said t have a local or relative maximum when $f(x^*) \geq f(x^* + h)$ for all values of h is very close to zero. A function $f(x)$ is called a absolute or global minimum at the point x^* when $f(x^*) \leq f(x)$ for all x; not only for all x close to x^* but, also in the domain over which $f(x)$ is defined. Likewise, if $f(x^*) \geq f(x)$ for all values of x in the domain then the point x^* will be a global maximum of $f(x)$. Figure 3.1 shows various local and global optimum points.

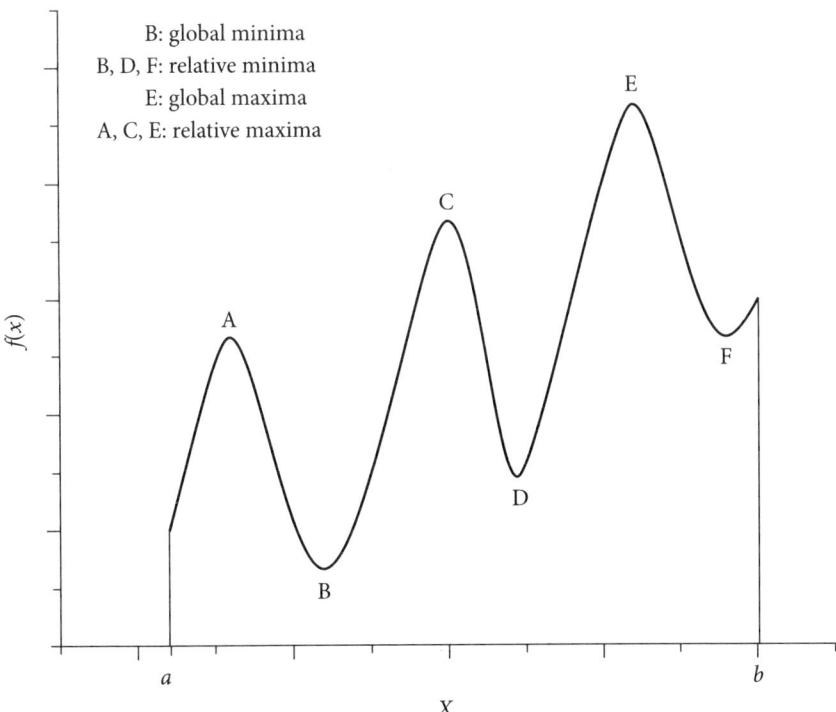

Fig. 3.1 Global and relative optimum points

For an optimization problem with single-variable, we need to find the value of $x = x^*$ in the interval $[a, b]$ such that x^* minimizes the function $f(x)$. This interval $[a, b]$ is considered for initial calculation of the static point. The global optimum point may be outside this interval. For a single variable function, the necessary and sufficient conditions for the relative minimum are given in the following two theorems.

Theorem 3.1

Necessary Condition

If a function $f(x)$ is defined in the interval $a \leq x \leq b$ and possesses a relative minimum at the point $x = x^*$, where $a \leq x^* \leq b$, and if the derivative $df(x)/dx = f'(x)$ exists as a finite number at $x = x^*$, then $f'(x) = 0$.

Proof

By the definition we know that

$$f'(x^*) = \lim_{h \to 0} \frac{f(x^* + h) - f(x^*)}{h} \quad (3.3)$$

exists as a definite number, which we require to shown to be zero. As the point x^* is a relative minimum, we can write

$$f(x^*) \leq f(x^* + h) \quad (3.4)$$

When, the all values of h are sufficiently close to zero, we have

$$\frac{f(x^* + h) - f(x^*)}{h} \geq 0 \text{ if } h > 0 \quad (3.5)$$

$$\frac{f(x^* + h) - f(x^*)}{h} \leq 0 \text{ if } h < 0 \quad (3.6)$$

Thus, Eq. (3.3) provides the limit as h tends to zero through positive values as

$$f'(x^*) \geq 0 \quad (3.7)$$

at the same time as it provides the limit as h approaches to zero through negative values as

$$f'(x^*) \leq 0 \quad (3.8)$$

both the Eqs (3.7) and (3.8) will be satisfied only when

$$f'(x^*) = 0 \quad (3.9)$$

Therefore, the theorem is proved.

Limitations

i. This theorem does not give any idea what happens when a maximum or minimum occurs at a point x^* where the derivative does not exist. For example, it is clear in Fig. 3.2 that the value of the derivative varies depending on whether h moves toward zero through negative or positive values, respectively.

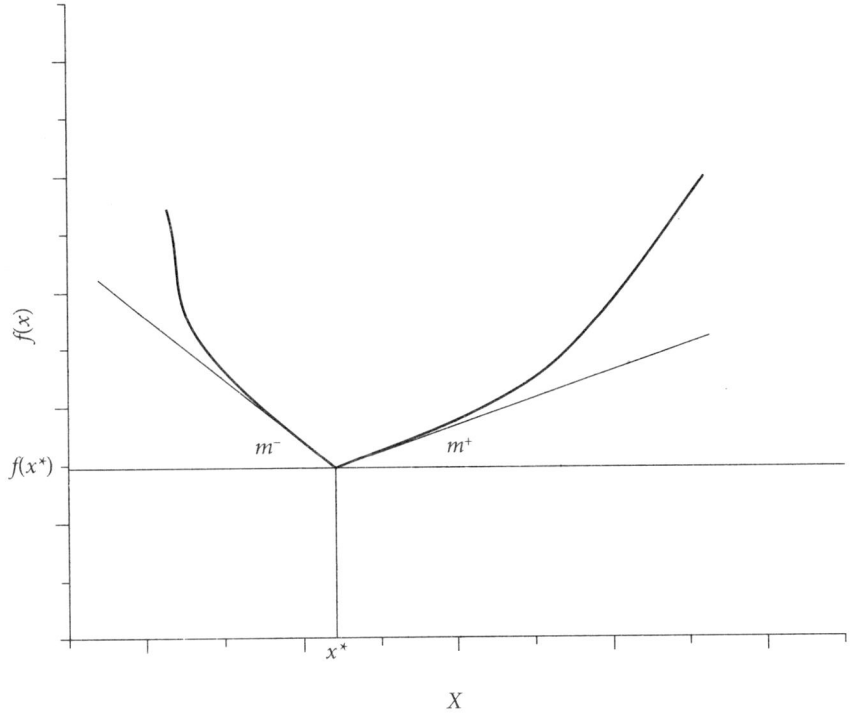

Fig. 3.2 Undefined derivative at x^*

The derivative $f'(x^*)$ fail to exist unless the numbers m^- and m^+ are equal. If $f'(x^*)$ does not exist, the Theorem 3.1 is not applicable.

A function $f(x)$, well defined at $x = x^*$. The point $x = x^*$ is called *critical point* if the derivative of $f(x)$ does not exist at that point. Consider, the function $f(x) = x^{2/3}$; derivative of this function is $f'(x) = \dfrac{2}{3} x^{-1/3}$. The derivative of $f(x)$ does not exist when $x = 0$ since, the denominator then takes the value 0. The point, $x = 0$ is therefore, a critical point of $f(x)$.

ii. A maximum or minimum can arises at an endpoint of the interval within which the function is defined; however, this theorem does not state what happens for this situation. In this case, the term

$$\lim_{h \to 0} \frac{f(x^* + h) - f(x^*)}{h} \tag{3.10}$$

exists only for negative values of h or for positive values of h, and for this reason the derivative is not defined at the endpoints.

iii. By this theorem, it is not inevitable that the function will have a maximum or minimum at every point wherever the derivative is zero. For instance, as shown in Fig. 3.3, the derivative of the function $f'(x^*) = 0$ at $x = 0$. However, this point is neither a maximum nor a minimum. Generally, a point x^* at which $f'(x^*) = 0$ is known as *stationary point*. A stationary point may be a maximum, minimum, or an inflection point.

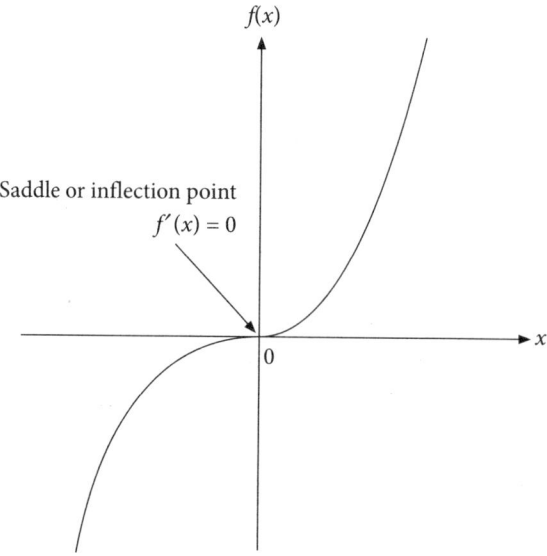

Fig. 3.3 Inflection or saddle point

Theorem 3.2

Sufficient Condition

We will apply the sufficient condition to get more idea about the stationary point.

Let $f'(x^*) = f''(x^*) = f'''(x^*) = \ldots\ldots\ldots = f^{(n-1)}(x^*) = 0$, however, $f^n(x^*) \neq 0$. Then, the point x^* gives

i. a minimum value of $f(x)$ when $f^n(x^*) > 0$ and n is even
ii. a maximum value of $f(x)$ when $f^n(x^*) < 0$ and n is even
iii. neither a minimum nor a maximum when n is odd

Proof

Taylor's theorem is applied with remainder after n terms, we can write

$$f(x^* + h) = f(x^*) + hf'(x^*) + \frac{h^2}{2!}f''(x^*) + \ldots\ldots + \frac{h^{n-1}}{(n-1)!}f^{(n-1)}(x^*) + \frac{h^n}{n!}f^{(n)}(x^* + \theta h)$$

for $0 < \theta < 1$ (3.11)

Since, $f'(x^*) = f''(x^*) = f'''(x^*) = \ldots\ldots\ldots = f^{(n-1)}(x^*) = 0$, Eq. (3.11) can be written as

$$f(x^* + h) - f(x^*) = \frac{h^n}{n!}f^{(n)}(x^* + \theta h) \quad (3.12)$$

As, $f^n(x^*) \neq 0$, there must have an interval around x^* at every point x of which the nth derivative $f^n(x)$ has the same sign, as of $f^n(x^*)$. Therefore, $f^{(n)}(x^* + \theta h)$ has the sign of $f^{(n)}(x^*)$ at every point $x^* + h$ of this interval. When n is even, $h^n/n!$ is positive irrespective of the sign of h whether it is positive or negative, and therefore, $f(x^* + h) - f(x^*)$ will have the similar sign as that of $f^{(n)}(x^*)$. Thus, x^* will be a relative maximum if $f^{(n)}(x^*)$ is negative and a relative minimum if $f^{(n)}(x^*)$ is positive. When n is odd, $h^n/n!$ changes its sign as the sign of h changes; for this reason the point x^* is neither a minimum nor a maximum. In this case, the point x^* is termed as *point of inflection* (see Fig. 3.3).

Example 3.1

Find all stationary points for the function (Eq. (3.13)), and then determine the maximum and minimum values of the function.

$$f(x) = 5x^6 - 36x^5 + \frac{165}{2}x^4 - 60x^3 + 36 \quad (3.13)$$

Solution

Since, $f'(x) = 30x^5 - 180x^4 + 330x^3 - 180x^2 = 30x^2(x-1)(x-2)(x-3)$

for finding stationary points $f'(x) = 0$

$30x^2(x-1)(x-2)(x-3) = 0$

$\Rightarrow x = 0, 1, 2, 3$

Stationary points are $x = 0$, $x = 1$, $x = 2$, and $x = 3$

Now,

$$f''(x) = 150x^4 - 720x^3 + 990x^2 - 360x$$

Table 3.1 Values of stationary points

Values of x^*	Values of $f(x^*)$	Values of $f''(x^*)$	Comments
0	36	0	we must investigate the next derivative
1	27.5	60	a relative minimum
2	44	−120	a relative maximum
3	5.5	540	a relative minimum

The third derivative of the function is

$$f'''(x) = 600x^3 - 2160x^2 + 1980x - 360$$

at $x = 0$, $f'''(x) = -360$

since $f'''(x) \neq 0$ at $x = 0$, and $n = 3$ is odd number; $x = 0$ is neither a minimum nor a maximum, and it is an inflection point.

Example 3.2

Find all local optima of the function

$$f(x) = x^{\frac{1}{3}}(x-1) \tag{3.14}$$

Solution

The first derivative of the function is

$$f'(x) = \frac{1}{3}x^{-\frac{2}{3}} \cdot (x-1) + x^{\frac{1}{3}} \cdot 1$$

$$= \frac{x+1}{3x^{\frac{2}{3}}} + x^{\frac{1}{3}}$$

$$= \frac{1+4x}{3x^{\frac{2}{3}}}$$

Applying necessary condition, we have $f'(x^*) = 0$

Therefore, $x = 0$, and $x = -1/4$

When $x = 0$, the denominator is zero; at this point $f'(x)$ is not defined. This is a critical point.

Second derivative of the function is

$$f''(x) = -\frac{2}{9}x^{-\frac{5}{3}} + \frac{4}{9}x^{-\frac{2}{3}} \qquad (3.15)$$

We cannot find the value of second derivative at point $x = -1/4$, therefore, it is difficult to say if the point is local maxima or local minima. This problem can be tackled by numerical optimization method.

3.2.2 Classification of unconstrained minimization methods

There are various techniques available for solving an unconstrained minimization problem. These techniques can be categorized into two broad classes such as direct search methods and descent methods. The direct search methods do not use the partial derivatives of the function; only the objective function values are require to find the minimum. Therefore, these methods are often called the non-gradient methods. As they utilize zeroth-order derivatives of the function, the direct search methods are also called zeroth-order methods. When the number of variables is relatively small, these methods are most suitable. In general, these methods are less effective than the descent methods. In addition to the function values, the descent methods require the first and in some instances the second derivatives of the objective function. Descent methods are usually more effective than direct search techniques as they use more information (through the use of derivatives) about the function for minimizing. The descent methods are also called gradient methods. Among these gradient methods, those utilizing only first derivatives of the function are known as first-order methods; those utilizing both first and second derivatives of the function are called second-order methods.

Unimodal function

A function is said to be an unimodal function when it consists of either one valley (minimum) or peak (maximum) only within the specified interval $[a, b]$. Therefore, a single-variable function is supposed to be unimodal when for any two specified points that lie on the same side of the optimum, the one that is closer to the optimum point gives the better value of the functional (smaller value for a minimization problem and larger value for a maximization problem). Mathematically this can be written as follows:

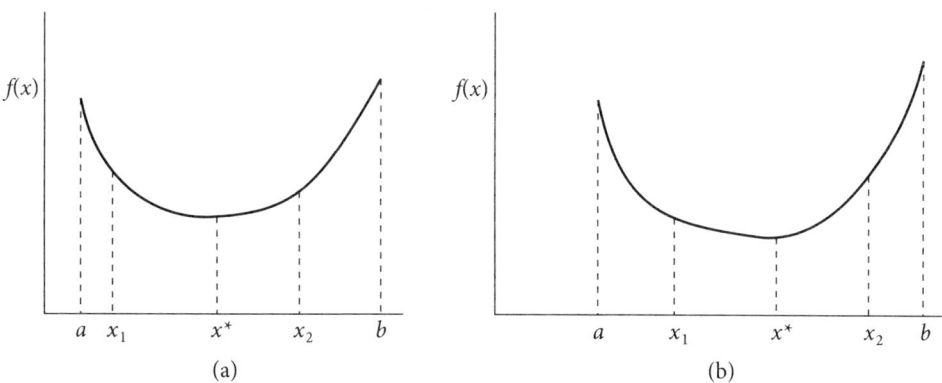

Fig. 3.4 Unimodal functions

Suppose $f(x)$ is strictly unimodal on the interval $a \leq x \leq b$ with a minimum at x^*. Let two points x_1 and x_2 be in the given interval such that $a < x_1 < x_2 < b$.

By comparing the values of the function at x_1 and x_2, the following conclusion can be drawn:

i. When $f(x_1) > f(x_2)$, then the minimum of $f(x)$ does not lie in the interval (a, x_1) or, $x^* \in (x_1, b)$ (see Fig. 3.4a).

ii. When $f(x_1) < f(x_2)$, then the minimum of $f(x)$ does not lie in the interval (x_2, b). In other words, $x^* \in (a, x_2)$ (see Fig. 3.4b).

Unimodality is a very significant functional property and almost all the search techniques currently in use for single-variable optimization require at least the assumption that the function is unimodal within the domain of interest. The usefulness of this characteristic lies in the fact that when $f(x)$ is unimodal we are required to compare $f(x)$ only at two different points for predicting in which of the subintervals defined by those two points the optimum does not lie.

3.3 Direct Search Methods

The search techniques, which find a single-variable optimum by successive elimination of subintervals to shrink the residual interval of search, are known as region elimination methods.

3.3.1 Finding a bracket for a minimum

Bracketing is significantly required for the optimization of a single-variable function when we have no idea about the range of the solution. Now we will discuss a methodical way of determining a range $a < x < b$ that contains a minimum of $f(x)$. This technique utilizes the slope $f'(x)$ to give us an indication whether the minimum lies to the right or left of an initial point x_0. When $f'(x_0)$ is positive then lower function values will be found for $x < x_0$, while $f'(x_0) < 0$ indicates lower values of $f(x)$ arise when $x > x_0$. Basically, the step size in this algorithm becomes larger and larger in a "downhill" direction until the function value starts to increase, which signifies that the minimum has been bracketed.

How to find a and b for bracketing a local minimum of f(x)

Select an initial point x_0 and a step length α (> 0)
Set $\delta = \text{sign}(f'(x_0)) \times [-\alpha]$
Repeat for $i = 0, 1, 2, \ldots$
$x_{i+1} = x_i + \delta$, $\delta = 2\delta$
until $f(x_{i+1}) > f(x_i)$
if $i = 0$ then set $a = x_0$ and $b = x_1$
if $i > 0$ then set $a = x_{i-1}$ and $a = x_{i+1}$

3.3.2 Unrestricted search method

The optimum solution lies within the restricted ranges of design variables for most of the practical applications. In some instances, this range is unknown to us, and therefore, we need to perform the search procedure without any restrictions on the values of the variables.

Search with Fixed Step Size

This is the most straightforward method to solve a problem where we start moving from an initial guess point in a suitable direction (negative or positive) with a fixed step size. The size of the step used should be small in comparison to the desired final accuracy. Even though this process is very simple to execute, it is not effective in many situations. This process is explained by the following steps:

1. Start with x_1, an initial guess point.
2. Calculate $f_1 = f(x_1)$.
3. Consider a step size s, find $x_2 = x_1 + s$.
4. Calculate $f = f(x_2)$.
5. When the problem is of minimization one and if $f_2 < f_1$, from the assumption of unimodality we get the indication that the desired minimum cannot lie at $x < x_1$. Therefore, the search procedure required to continue further through the points x_3, x_4 .... using the concept of unimodality while checking each pair of experiments. This search process is continued till the function value starts to increase at a point $x_i = x_1 + (i - 1)$.
6. The search process is terminated at point x_i, and either x_{i-1} or x_i is considered as the optimum point.
7. On the other hand, when $f_2 > f_1$, the search should be performed in the opposite direction at x_{-2}, x_{-3} ...,... points, where $x_{-j} = x - (j - 1)s$.
8. The desired minimum lies in between x_1 and x_2 when we get $f_2 = f_1$, and either x_1 or x_2 is considered as the minimum point. Sometimes we consider the middle point of $[x_1, x_2]$ as the minimum.
9. And when it occurs that f_{-2} and f_2 both are greater than f_1, it specifies that the desired minimum will lie in the double interval $x_{-2} < x < x_2$.

Search with Accelerated Step Size:

During the real life application, search with the fixed step size is very simple to execute. Despite this advantage, the major drawback arises owing to the unrestricted nature of the region within which the minimum is located. Sometimes, the number of iterations becomes very large. For example, when the optimum point of a specified function supposed to be at x_{opt} = 10, 000 and, without proper information about the position of the minimum, if we arbitrarily select x_1 and s as 0.0 and 0.1, respectively, we are require to calculate the function value 1,000,001 times to locate the minimum point. This method is involved with a huge amount of computational work. If we gradually increase the step size till the minimum point is bracketed, an obvious improvement can be done. A very common practice is that of doubling the size of the step during search process as long as it shows an upgrading of the objective function. This method can be improved in different ways. One such option is that, after bracketing the optimum in (x_{i-1}, x_i), decrease the step length.

By starting the search process either from x_{i-1} or x_i, the fundamental method can be performed with a reduced step size. This search process is repeated until the interval that is bracketed turns into adequately small. The search process with accelerated step size can be demonstrated by the following example.

Example 3.3

Air is to be compressed from 1 to 10 atm. pressure in a two-stage compressor. To increase the compression efficiency, the compressed air from the first stage of compression is cooled (it is passed through a heat exchanger) before entering the second stage of compression. For isentropic compression of air the total work input to a compressor (W) can be represented by

$$W = c_p T_1 \left[\left(\frac{p_2}{p_1} \right)^{(k-1)/k} + \left(\frac{p_3}{p_2} \right)^{(k-1)/k} - 2 \right] \tag{3.16}$$

where the specific heat of air at constant pressure is $c_p = 1.006$ kJ/kg. K, $k = 1.4$ is the ratio of specific heat at constant pressure to specific heat at constant volume of air, and the entering gas temperature $T_1 = 300$K. Find the intermediate pressure p_2 at which cooling is required to minimize the work input to the compressor. Also, calculate the minimum work required to operate the compressor. Use search method with accelerated step size. Starting point = 1.0 (as the inlet pressure is 1 atm.) and initial step size = 0.05

Solution

Substituting the value of different parameters we get

$$W = 301.8 \left[(p_2)^{0.286} + \left(\frac{10}{p_2} \right)^{0.286} - 2 \right] \tag{3.17}$$

We will start searching from the starting point = 1.0 (as the inlet pressure is 1 atm.) and with an initial step size = 0.05. The values of successive iterations are given in Table 3.2.

Table 3.2 Values of successive iterations with accelerated step size

i	Value of s	$x_i = x_1 + s$	$f_i = f(x_i)$	Is $f_i > f_{i-1}$?
1		1.0	281.27	--
2	0.05	1.05	277.43	No
3	0.1	1.1	273.93	No
4	0.2	1.2	267.80	No
5	0.4	1.4	258.26	No
6	0.8	1.8	246.29	No
7	1.6	2.6	236.69	No
8	3.2	4.2	238.14	Yes

Table 3.2 shows that the minimum point should be lie between 2.6 and 4.2. Figure 3.5 shows that minimum value is 3.15.

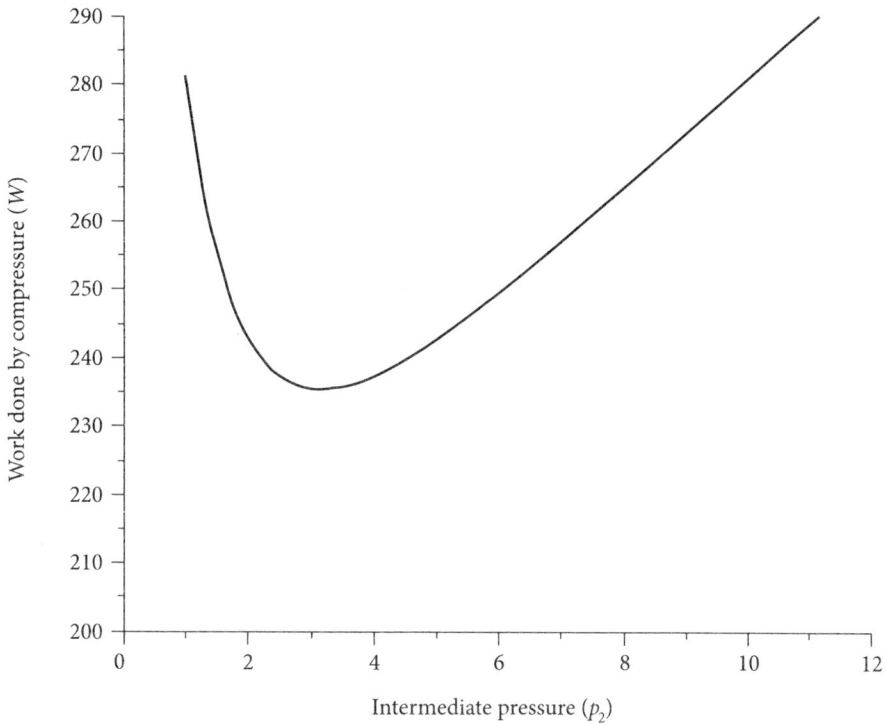

Fig. 3.5 Work done vs. intermediate pressure graph

3.3.3 Exhaustive search

In some search processes, we have plenty of knowledge about the region in which optimum is located. The exhaustive search technique can be applied for solving those problems wherever the interval in which the optimum point lies is finite. Suppose the starting and final points of the interval of uncertainty is represent by x_s and x_f respectively (see Fig. 3.6). The exhaustive search process includes the evaluation of the objective function at some equally spaced points within the interval (x_s, x_f). This number of points is predetermined by the user. Then reduce the interval of uncertainty by utilizing the concept of unimodality. Consider that the function (workdone by compressor) is defined on the interval (x_s, x_f). Eight equally spaced interior points $x_1 = 2$ to $x_8 = 9$ atm. pressure can be placed within this interval.

Considering that the values of the function are found as shown in Fig. 3.6, according to the assumption of unimodality, the minimum point must lie between points $x_2 = 3$ and $x_4 = 4$. Therefore, the interval (x_2, x_4) i.e., (3.0, 4.0) can be taken as the final interval of uncertainty. Generally, when the value of any function is calculated at n equally spaced points within the

52 Optimization in Chemical Engineering

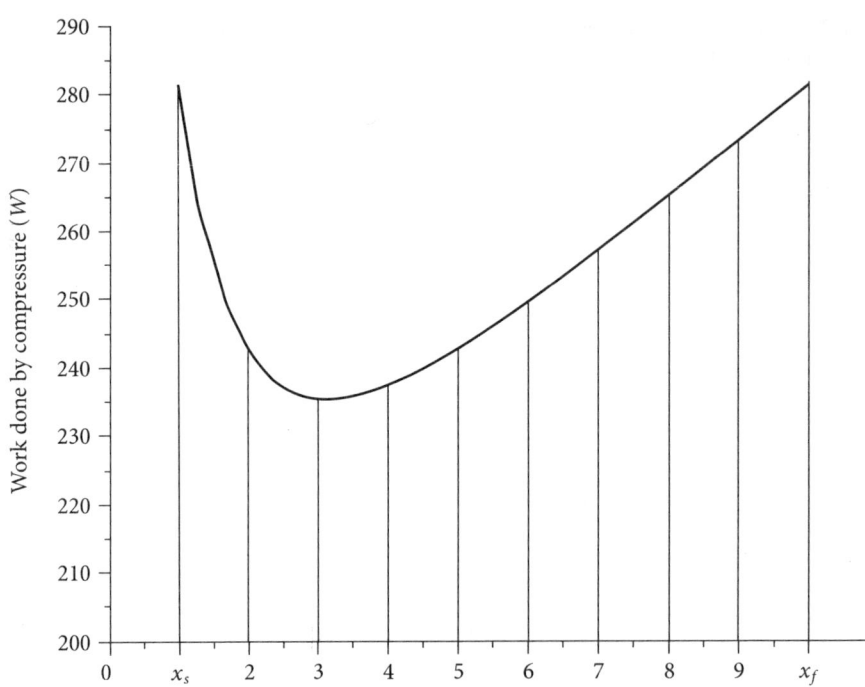

Fig. 3.6 Exhaustive search (see Example 3.3)

original interval of uncertainty of length $L_0 = x_f - x_s$, and if the point x_j shows the optimum value of the function (among these n values of the function), the final interval of uncertainty is given by

$$L_n = x_{j+1} - x_{j-1} = \frac{2}{n+1} L_0 \qquad (3.18)$$

The final interval of uncertainty will change with the number of points. For various numbers of trial points in the exhaustive search method, the available final interval of uncertainty is given in Table 3.3:

Table 3.3 The available interval of uncertainty after different trials

Number of trials	2	3	4	5	6	...	n
L_n/L_0	2/4	2/4	2/5	2/6	2/7	...	2/(n + 1)

This method is known as simultaneous search method since the function value is calculated at all n points simultaneously. Whereas, in sequential search methods the information obtained from the initial trials is utilized to place the subsequent experiments. The sequential search methods are relatively more efficient compared to the simultaneous search.

Example 3.4

Solve the Example 3.3 by exhaustive search method. Maximum error should be less than 10 per cent.

Solution

The middle point of the final interval of uncertainty is considered as the approximate optimum point, and then the maximum deviation will be $1/(n + 1)$ times the initial interval of uncertainty. Hence, to estimate the optimum within 10 per cent of the exact value, we have

$$\frac{1}{n+1} \leq \frac{1}{10} \text{ or } n \geq 9$$

We use Eq. (3.17) as the function that is to be minimize by considering $n = 9$, the following function values have been found out:

Table 3.4 Work done with different intermediate pressure (Exhaustive search method)

i	1	2	3	4	5	6	7	8	9
p_2	2	3	4	5	6	7	8	9	10
W	242.59	235.47	237.27	242.59	249.50	257.13	265.11	273.19	281.27

Table 3.4 shows that the optimum point should lay in between 3 and 4.

3.3.4 Dichotomous search

In exhaustive search method, we need to perform all the experiments simultaneously before taking any decision regarding the position of the optimum point. Unlike the exhaustive search method, dichotomous search method, interval halving method, Fibonacci method and the golden section method discussed in the following sections, are all sequential search methods. The experiments are placed sequentially for these methods. In these methods, the position of the any experiment is influenced by the outcome of previous experiment. During this dichotomous search process, two experiments are performed very close to the center of the interval of uncertainty. As the points very close to the centre, almost half of the interval of uncertainty is eliminated depending on the relative values of the objective function at those two points. The locations of these two experiments are estimated by the following equations.

$$x_1 = \frac{L_0}{2} - \frac{\delta}{2} \tag{3.19}$$

$$x_2 = \frac{L_0}{2} + \frac{\delta}{2} \tag{3.20}$$

where δ is a small positive number. We have to chose the value of δ in such a way that the result of two experiments will be significantly different. After the first experiment, the new interval of

54 Optimization in Chemical Engineering

uncertainty can be represented by $(L_0/2 + \delta/2)$. The process of dichotomous search works based on the performing a pair of experiments near the center of the existing interval of uncertainty. Therefore, the next pair of experiments is performed near the center of the residual interval of uncertainty. By this method, the interval of uncertainty becomes almost half after each pair of experiment. The following table shows the intervals of uncertainty at the end of different pairs of experiments:

Table 3.5 Final intervals of uncertainty for different pairs of experiments

Number of experiments	2	4	6
Final interval of uncertainty	$\dfrac{1}{2}(L_0 + \delta)$	$\dfrac{1}{2}\left(\dfrac{L_0 + \delta}{2}\right) + \dfrac{\delta}{2}$	$\dfrac{1}{2}\left(\dfrac{L_0 + \delta}{4} + \dfrac{\delta}{2}\right) + \dfrac{\delta}{2}$

usually, the final interval of uncertainty after performing n experiments (where n is an even number) is given by

$$L_n = \frac{L_0}{2^{n/2}} + \delta\left(1 - \frac{1}{2^{n/2}}\right) \tag{3.21}$$

The Example 3.5 is given to demonstrate this search method.

Example 3.5

Solve the problem given in Example 3.3 by dichotomous search method. Maximum error should be less than 10 per cent.

Solution

We can write the ratio of final to initial intervals of uncertainty by using Eq. (3.21).

$$\frac{L_n}{L_0} = \frac{1}{2^{n/2}} + \frac{\delta}{L_0}\left(1 - \frac{1}{2^{n/2}}\right) \tag{3.22}$$

where δ is a very small quantity, say 0.001, and n is the number of experiments. If we consider the middle point of the final interval as the desired optimum point, the requirement can be stated as

$$\frac{1}{2}\frac{L_n}{L_0} \leq \frac{1}{10}$$

$$\frac{1}{2^{n/2}} + \frac{\delta}{L_0}\left(1 - \frac{1}{2^{n/2}}\right) \leq \frac{1}{5}$$

Since, $\delta = 0.001$ and $L_0 = 9$, we have

$$\frac{1}{2^{n/2}} + \frac{0.001}{9}\left(1 - \frac{1}{2^{n/2}}\right) \leq \frac{1}{5}$$

$$\frac{1}{2^{n/2}} + \frac{1}{1111}\left(1 - \frac{1}{2^{n/2}}\right) \leq \frac{1}{5}$$

$$\frac{1110}{1111}\frac{1}{2^{n/2}} \leq \frac{1106}{5555}$$

$$2^{n/2} \geq \frac{5550}{1106} \approx 5.0$$

As we are performing two experiments each time, the n should to be even. From the above inequality, we get the minimum acceptable value of $n = 6$.

The search process is performed as follows. The first two experiments are conducted at

$$x_1 = \frac{L_0}{2} - \frac{\delta}{2} = 4.5 - 0.0005 = 4.4995$$

$$x_2 = \frac{L_0}{2} + \frac{\delta}{2} = 4.5 + 0.0005 = 4.5005$$

the function values at these points are given by

$$f_1 = f(x_1) = 239.6468$$

$$f_2 = f(x_2) = 239.6522$$

Since, $f_2 > f_1$, the new interval of uncertainty will be (1.0, 4.4995). The second pair of experiments is carried out at

$$x_3 = \frac{3.4995}{2} - \frac{0.001}{2} = 1.7498 - 0.0005 = 1.7493$$

$$x_4 = \frac{3.4995}{2} + \frac{0.001}{2} = 1.7498 + 0.0005 = 1.7503$$

the function values are calculated as follows

$$f_3 = f(x_3) = 247.4326$$

$$f_4 = f(x_4) = 247.4092$$

Since, $f_3 > f_4$, the new interval of uncertainty will be (1.7493, 4.4995).

The final set of experiments are performed at

$$x_5 = 1.7493 + \frac{2.7502}{2} - \frac{0.001}{2} = 1.7493 + 1.3751 - 0.0005 = 3.1239$$

$$x_6 = 1.7493 + \frac{2.7502}{2} + \frac{0.001}{2} = 1.7493 + 1.3751 + 0.0005 = 3.1249$$

The corresponding values of the function are

$$f_5 = f(x_5) = 235.3805$$

$$f_6 = f(x_6) = 235.3803$$

Since, $f_5 > f_6$, then the new interval of uncertainty is (3.1249, 4.4995). The middle point of this interval is considered as optimum point, and therefore,

$$x_{opt} \simeq 3.8122 \text{ and } f_{opt} \simeq 236.5745$$

3.3.5 Interval halving method

Interval halving method is a very useful and simple search method. In every stage of this method, exactly one-half of the current interval of uncertainty is eliminated. In the first stage of this process, three experiments are needed and two experiments in each succeeding stage. Because, in all the stages except the first stage of this process, value of the function will be available at the middle point of the interval of uncertainty (f_0). This method can be illustrated by the following steps:

Step 1 Split $L_0 = [a, b]$, the initial interval of uncertainty into four equal parts. The point x_0 is labeled as the middle point and the quarter-interval points x_1 and x_2 as shown in Fig. 3.7.

Step 2 The value of the function $f(x)$ has been calculated at these three interior points to get $f_1 = f(x_1), f_0 = f(x_0),$ and $f_2 = f(x_2)$.

Step 3 (a) When $f_2 > f_0 > f_1$, discard the interval (x_0, b), label x_1 and x_0 as the new x_0 and b, respectively, and move to step 4.

(b) If $f_2 < f_0 < f_1$, remove the interval (a, x_0), label x_0 and x_2 as the new a and x_0, respectively, and move to step 4.

(c) When $f_1 > f_0$ and $f_2 > f_0$, remove both the intervals (a, x_1) and (x_2, b), then label x_1 as new a and x_2 as new b, and move to step 4.

Step 4 Check whether $L = b - a$, the new interval of uncertainty, satisfies the convergence criterion $L \leq \varepsilon$, where ε is considered as a small quantity. Stop the process when it satisfies the convergence criterion. Or else, set the new $L_0 = L$ and go to step 1.

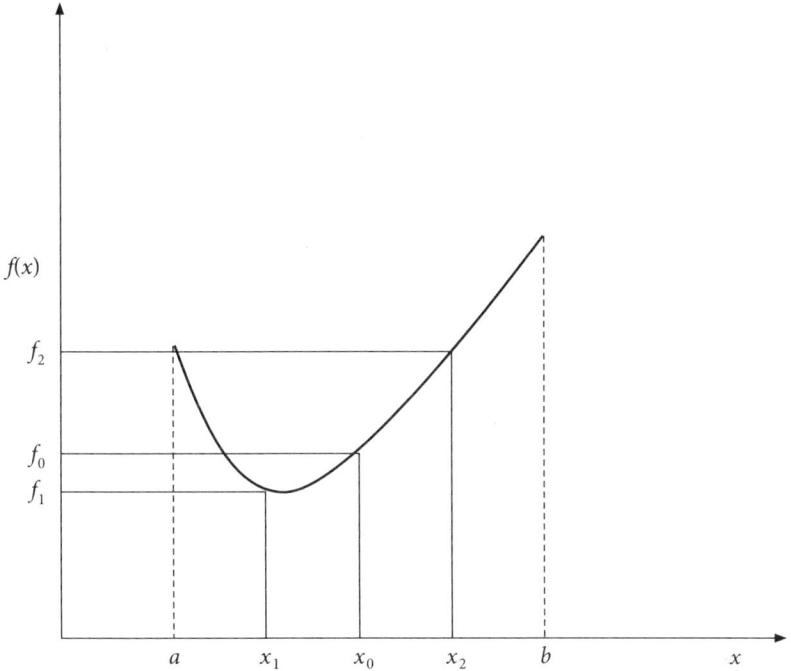

Fig. 3.7 Interval halving method

After the completion of n experiments ($n \geq 3$ and odd), the interval of uncertainty left over can be represented by

$$L_n = \left(\frac{1}{2}\right)^{(n-1)/2} L_0 \quad (3.23)$$

Example 3.6

Solve the Example 3.3 by interval halving method. Maximum error should be less than 10 per cent.

Solution

If we consider that the optimum point will be located at the middle point of the final interval of uncertainty, the required accuracy level can be reached when

$$\frac{1}{2}L_n \leq \frac{L_0}{10} \text{ or } \left(\frac{1}{2}\right)^{(n-1)/2} L_0 \leq \frac{L_0}{5}$$

for $L_0 = 9$, we have

$$\left(\frac{1}{2^{(n-1)/2}}\right) \leq \frac{1}{5} \text{ or } 2^{(n-1)/2} \geq 5$$

Since, n should to be odd number, the above inequality provides us the minimum acceptable value of n as 7. Considering the value of $n = 7$, the search process is carried out as follows. The first three experiments are placed at one-fourth, middle, and three-fourth points of the interval $L_0 = [a = 1, b = 10]$ as

$x_1 = 3.25, \; f_1 = 235.4011$

$x_0 = 5.50, \; f_0 = 245.9077$

$x_2 = 7.75, \; f_2 = 263.0979$

Since, $f_2 > f_0 > f_1$, we eliminate the interval $(x_0, b) = (5.50, 10.00)$, new $a = 1.00$ and $b = 5.50$
By dividing the new interval of uncertainty, $L_3 = [1.00, 5.50]$ into four equal parts, we have

$x_1 = 2.125, \; f_1 = 240.8034$

$x_0 = 3.25, \; f_0 = 235.4011$

$x_2 = 4.375, \; f_2 = 238.9937$

Since, $f_1 > f_0$ and $f_2 > f_0$, we remove both the intervals (a, x_1) and (x_2, b). Now the new interval is $L_5 = [a = 2.125, b = 4.375]$. This interval is again divided into four equal parts to obtain

$x_1 = 2.6875, \; f_1 = 236.2837$

$x_0 = 3.25, \; f_0 = 235.4011$

$x_2 = 3.8125, \; f_2 = 236.5755$

In this experiment we notice that $f_1 > f_0$ and $f_2 > f_0$ therefore, we remove both the intervals (a, x_1) and (x_2, b) to get the new interval of uncertainty as $L_7 = (2.6875, 3.8125)$. Considering the middle point of this interval as optimum, we have

$x_{opt} \simeq 3.25$ and $f_{opt} \simeq 235.4011$

3.3.6 Fibonacci method

The Fibonacci method is useful to find the minimum of a single variable function. This method can be applied even if the function is not continuous. Despite many advantages, Fibonacci method has some limitations like other elimination methods:

i. This is not a global search method. We should know the initial interval of uncertainty within which the optimum is located.
ii. The objective function being optimized should be unimodal within the initial interval of uncertainty.
iii. We are not able to locate the exact value of optimum by using this method. Only an interval will be estimated; this interval is known as the final interval of uncertainty. The length of the final interval of uncertainty can be controlled. This interval can be found as small as we require by increasing the computational effort.
iv. In this method, we need to specify the number of function evaluations to be made in the search process or we have to specify the required level of resolution.

This search process uses of the sequence of Fibonacci numbers $\{F_n\}$, for conducting the experiments. These numbers can be defined as

$$F_0 = F_1 = 1 \tag{3.24}$$

$$F_n = F_{n-1} + F_{n-2}, n = 2, 3, 4 \ldots \tag{3.25}$$

which gives us the sequence 1, 1, 2, 3, 5, 8, 13, 21, 34, 55, 89, …

Algorithm

Suppose the initial interval of uncertainty L_0 is defined by $a \leq x \leq b$ and n is the total number of experiments to be carried out. We can define

$$L_2^* = \frac{F_{n-2}}{F_n} L_0 \tag{3.26}$$

and locate the points x_1 and x_2 for the first two experiments. These points are placed at a distance of L_2^* from each end of L_0. Which yield

$$x_1 = a + L_2^* = a + \frac{F_{n-2}}{F_n} L_0 \tag{3.27}$$

$$x_2 = b - L_2^* = b - \frac{F_{n-2}}{F_n} L_0 = a + \frac{F_{n-1}}{F_n} L_0 \tag{3.28}$$

By using the unimodality assumption, remove one part of the interval. Then the remaining part is a smaller interval of uncertainty L_2 that can be written as

$$L_2 = L_0 - L_2^* = L_0\left(1 - \frac{F_{n-2}}{F_n}\right) = \frac{F_{n-1}}{F_n}L_0 \qquad (3.29)$$

and there is one experiment left in this search process. This experiment is placed at a distance of

$$L_2^* = L_0 \frac{F_{n-2}}{F_n} = L_2 \frac{F_{n-2}}{F_{n-1}} \qquad (3.30)$$

from one end and

$$L_2 - L_2^* = L_0 \frac{F_{n-3}}{F_n} = L_2 \frac{F_{n-3}}{F_{n-1}} \qquad (3.31)$$

from the other end. The 3rd experiment is now placed in the interval L_2 such a way that the current two experiments are positioned at a distance of

$$L_3^* = L_0 \frac{F_{n-3}}{F_n} = L_2 \frac{F_{n-3}}{F_{n-1}} \qquad (3.32)$$

from each end of the interval L_2. Again, we can reduce the interval of uncertainty to L_3 by using the unimodality property. The value of L_3 is represented by

$$L_3 = L_2 - L_3^* = L_2 - L_2 \frac{F_{n-3}}{F_{n-1}} = L_2 \frac{F_{n-2}}{F_{n-1}} = L_0 \frac{F_{n-2}}{F_n} \qquad (3.33)$$

We can continue the process of eliminating a certain interval and conducting a new experiment in the residual interval. The position of the jth experiment and the interval of uncertainty at the completion of j experiments are given by Eq. (3.34) and Eq. (3.35) respectively.

$$L_j^* = L_{j-1} \frac{F_{n-j}}{F_{n-(j-2)}} \qquad (3.34)$$

$$L_j = L_0 \frac{F_{n-(j-1)}}{F_n} \qquad (3.35)$$

For a n predetermined experiments, the ratio of the interval of uncertainty remaining after performing j experiments to the initial interval of uncertainty becomes

$$\frac{L_j}{L_0} = \frac{F_{n-(j-1)}}{F_n} \qquad (3.36)$$

and when $j = n$, we have

$$\frac{L_n}{L_0} = \frac{F_1}{F_n} = \frac{1}{F_n} \qquad (3.37)$$

The ratio L_n/L_0 will allow us to estimate the value of n, the number of experiments needed to reach the optimum point with any desired level of accuracy.

Example 3.7

Solve the problem given in Example 3.3 using Fibonacci method taking $n = 6$.

Solution

The given value of $n = 6$ and $L_0 = 9.00$, which yield

$$L_2^* = L_0 \frac{F_{n-2}}{F_n} = (9.00)\frac{5}{13} = 3.4615385$$

hence, the locations of the first two experiments are given by

$$x_1 = 1.00 + 3.4615385 = 4.4615385, \quad f_1 = 239.4438353$$

$$x_2 = 9.00 - 3.4615385 = 5.5384615, \quad f_2 = 246.1751813$$

Since, $f_1 < f_2$, we can remove the interval $[x_2, 9.00]$ by using the assumption of unimodality.

Placing the third experiment at

$$x_3 = 1.00 + (5.5384615 - 4.4615385) = 2.076923$$

the corresponding function value is

$$f_3 = 241.4471087$$

Since, $f_1 < f_3$, we eliminate the interval $[x_3, 1.00]$. The new interval is $[x_3, x_2]$.

The next experiment is placed at $x_4 = 3.153846$ with $f_4 = 235.3756528$

Here, $f_4 < f_1$, therefore, we remove the interval $[x_1, x_2]$. The position of the next experiment can be found as $x_5 = 3.3846155$, and the corresponding objective function value of $f_5 = 235.5338301$.

Since, $f_4 < f_5$, we eliminate the interval $[x_5, x_1]$. The new interval is $[x_3, x_5] = [2.076923, 3.3846155]$

The final experiment is positioned at $x_6 = 2.3076925$ with the corresponding objective function value of $f_6 = 238.7833231$.

Since $f_4 < f_6$, we remove the interval $[x_3, x_6]$. The new interval is $[x_6, x_5] = [2.3076925, 3.3846155]$

Considering the middle point as the optimum point

$x_{opt} \simeq 2.846154$ and $f_{opt} \simeq 235.7560691$

3.3.7 Golden section method

The golden section method is also a search technique similar to the Fibonacci method. The main dissimilarity is that the total number of experiments to be performed in the Fibonacci method is required to mention before starting the calculation, while this is not necessary in the golden section method. In the Fibonacci method, the total number of experiments (N) determines the location of the first two experiments. We start the golden section method with a presumption that we are ready to perform a quite large number of experiments. However, we are able to decide the total number of experiments during the computation. After completion of the various numbers of experiments, the intervals of uncertainty can be calculated as follows:

$$L_2 = \lim_{N \to \infty} \frac{F_{N-1}}{F_N} L_0 \tag{3.38}$$

$$L_3 = \lim_{N \to \infty} \frac{F_{N-2}}{F_N} L_0 = \lim_{N \to \infty} \frac{F_{N-2}}{F_{N-1}} \frac{F_{N-1}}{F_N} L_0 \tag{3.39}$$

$$\simeq \lim_{N \to \infty} \left(\frac{F_{N-1}}{F_N} \right)^2 L_0 \tag{3.40}$$

We can generalize this result to obtain

$$L_k = \lim_{N \to \infty} \left(\frac{F_{N-1}}{F_N} \right)^{k-1} L_0 \tag{3.41}$$

Using the relation

$$F_N = F_{N-1} + F_{N-2} \tag{3.42}$$

after dividing both sides of the above equation by F_{N-1} we obtain

$$\frac{F_N}{F_{N-1}} = 1 + \frac{F_{N-2}}{F_{N-1}} \tag{3.43}$$

By defining a ratio γ as

$$\gamma = \lim_{N \to \infty} \left(\frac{F_N}{F_{N-1}} \right) \tag{3.44}$$

Equation (3.43) can be expressed as

$$\gamma \simeq \frac{1}{\gamma} + 1 \tag{3.45}$$

that is,

$$\gamma^2 - \gamma - 1 = 0 \tag{3.46}$$

This gives the root $\gamma = 1.618$, and hence, Eq. (3.41) yields

$$L_k = \left(\frac{1}{\gamma} \right)^{k-1} L_0 = (0.618)^{k-1} L_0 \tag{3.47}$$

For large values of N, the ratios F_{N-2}/F_{N-1} and F_{N-1}/F_N in Eq. (3.40) have been taken to be same. The following table (Table 3.6) confirms the validity of this assumption:

Table 3.6 Final interval of uncertainty by golden section method

Value of N	2	3	4	5	6	7	8	9	10	∞
Ratio $\frac{F_{N-1}}{F_N}$	0.5	0.667	0.6	0.625	0.6156	0.619	0.6177	0.6181	0.6184	0.618

The term golden section is very old and it is found in the Euclidean geometry. According to Euclid's geometry, if we split a line segment into two unequal parts so that the ratio of the whole to the larger part is equal to the ratio of the larger to the smaller, the division is called the golden section and the ratio is called the golden mean.

Algorithm

The algorithm of golden section method is similar to the Fibonacci method except that the position of the first two experiments is given by

$$L_2^* = \frac{F_{N-2}}{F_N} L_0 = \frac{F_{N-2}}{F_{N-1}} \frac{F_{N-1}}{F_N} L_0 = \frac{L_0}{\gamma^2} = 0.382 L_0 \tag{3.48}$$

For stopping the process, the required accuracy level can be specified.

Example 3.8

Solve the Example 3.3 by golden section method using $n = 6$.

Solution

The first two experiments are placed according to the following calculation

$$L_2^* = 0.382 L_0 = (0.382)(9.00) = 3.438$$

thus, $x_1 = 4.438$ and $x_2 = 5.562$ with $f_1 = 239.3196483$ and $f_2 = 246.3396105$

Since, $f_1 < f_2$, we remove the interval $[x_2, 9.00]$ based on the unimodality assumption and get the new interval of uncertainty as $L_2 = [1.00, x_2] = [1.00, 5.562]$.

The third experiment is positioned at

$$x_3 = 1.00 + (x_2 - x_1) = 2.124 \text{ and } f_3 = 240.8162691$$

Since, $f_3 > f_1$, we can delete the interval $[x_1, x_2]$ and get a new interval of uncertainty as $[1.00, x_1] = [1.00, 4.438]$

The location of the next experiment is given by

$$x_4 = 3.314 \text{ and } f_4 = 235.4507654$$

Since, $f_3 > f_4$, we can delete the interval $[1.00, x_3]$ and a new interval of uncertainty as $[x_3, x_1] = [2.124, 4.438]$ is obtained.

The next experiment is positioned at

$$x_5 = 3.248 \text{ and function value } f_5 = 235.3999553$$

Since, $f_4 > f_5$, we can delete the interval $[x_4, x_1]$ and obtain a new interval of uncertainty as $[x_3, x_4] = [2.124, 3.314]$.

The final experiment is placed at

$$x_6 = 2.190 \text{ and corresponding function value } f = 240.0110286$$

Since, $f_6 > f_5$, we can remove the interval $[x_3, x_6]$ and obtain a new interval of uncertainty as $[x_6, x_4] = [2.190, 3.314]$.

If we consider the middle point of the final interval as optimum point, we have

$$x_{opt} \approx 2.752 \text{ and } f_{opt} \approx 236.0381082$$

3.4 Direct Root Methods

The necessary condition for $f(x)$ to possesses a minimum at the point x^* is that $f'(x^*) = 0$. In the direct root methods, we will try to find the solution (or roots) of the equation $f'(x) = 0$. In this section, we have discussed three root-finding methods namely, the Newton, the Quasi-Newton, and the secant methods.

3.4.1 Newton method

Originally the Newton method was proposed by Newton to solve nonlinear equations and afterward it is modified by Raphson, therefore, this method is also familiar as Newton–Raphson method in the field of numerical analysis.

Consider the quadratic approximation of the function $f(x)$ at $x = x_i$ using the Taylor's series expansion

$$f(x) = f(x_i) + f'(x_i)(x - x_i) + \frac{1}{2} f''(x_i)(x - x_i)^2 \tag{3.49}$$

For the minimum of $f(x)$, we set the value of the derivative in Eq. (3.49) equal to zero, then we get

$$f'(x) = f'(x_i) + f''(x_i)(x - x_i) = 0 \tag{3.50}$$

If the point x_i represents an approximation to the minimum of $f(x)$, Eq. (3.50) is rearranged to achieve an improved approximation as

$$x_{i+1} = x_i - \frac{f'(x_i)}{f''(x_i)} \tag{3.51}$$

Therefore, the Newton method as represented by Eq. (3.51) is similar as the utilization of quadratic approximation for the function $f(x)$ and applying the necessary conditions. This iterative process shown in Eq. (3.51) can be assumed to have converged whenever the value of the derivative, $f'(x_{i+1})$, is close to zero:

$$|f'(x_{i+1})| \leq \varepsilon \tag{3.52}$$

where ε is a small quantity. Figure 3.8 graphically explains the convergence process of the Newton method.

Notes

1. Both the first- and second-order derivatives of the function $f(x)$ have been used in this method.
2. When $f''(x_i) \neq 0$ [in Eq. (3.51)], the Newton iterative method possesses a very fast convergence property that is known as quadratic convergence.
3. When starting point is very close to the optimum point, Newton's method is suitable for that problem. The Newton iterative process might diverge when the starting point for the iterative process is far from the true solution x^*.

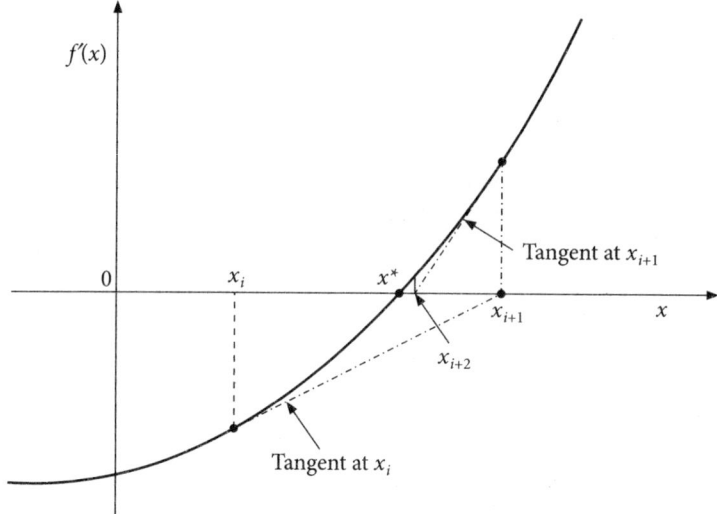

Fig. 3.8 Convergence process of Newton method

3.4.2 Quasi-Newton method

When the function $f(x)$ being minimized is not available in closed form or differentiation is also difficult, the derivatives $f'(x)$ and $f''(x)$ in Eq. (3.51) can be approximated using the finite difference formulas as

$$f'(x_i) = \frac{f(x_i + \Delta x) - f(x_i - \Delta x)}{2\Delta x} \qquad (3.53)$$

$$f''(x_i) = \frac{f(x_i + \Delta x) - 2f(x_i) + f(x_i - \Delta x)}{\Delta x^2} \qquad (3.54)$$

where Δx is a small step size. Substitution of Eqs (3.53) and (3.54) into Eq. (3.51) leads to

$$x_{i+1} = x_i - \frac{\Delta x \left[f(x_i + \Delta x) - f(x_i - \Delta x) \right]}{2 \left[f(x_i + \Delta x) - 2f(x_i) + f(x_i - \Delta x) \right]} \qquad (3.55)$$

The iterative process described by Eq. (3.55) is known as the Quasi-Newton method.

To check the convergence of this iterative process, the criterion given by Eq. (3.56) can be utilized:

$$\left| f'(x_{i+1}) \right| = \left| \frac{f(x_{i+1} + \Delta x) - f(x_{i+1} - \Delta x)}{2\Delta x} \right| \leq \varepsilon \qquad (3.56)$$

where ε is a very small quantity and a central difference formula has been utilized for calculating the derivative of f.

Notes

1. Equations (3.55) and (3.56) have been developed by using the central difference formulas. We can also utilize the forward or backward difference formulas for the same.
2. In each iteration, Eq. (3.55) requires the estimation of the function at the points $x_i + \Delta x$ and $x_i - \Delta x$ in addition to x_i.

3.4.3 Secant method

An equation similar to Eq. (3.50) is employed to the secant method

$$f'(x) = f'(x_i) + s(x - x_i) = 0 \tag{3.57}$$

where s is the slope of the line that connects the two points $(a, f'(a))$ and $(b, f'(b))$, where a and b represent two different approximations to the actual solution, x^*. From Fig. 3.9, we can define the slope s as

$$s = \frac{f'(b) - f'(a)}{b - a} \tag{3.58}$$

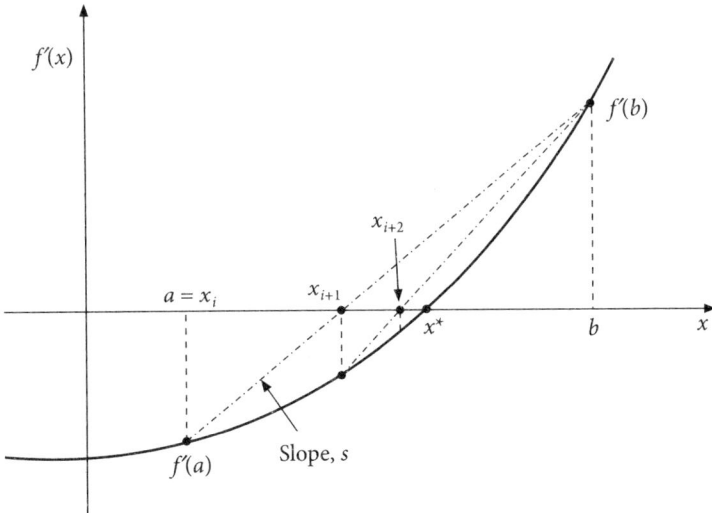

Fig. 3.9 Convergence process of secant method

Equation (3.57) approximates the function $f'(x)$ as a linear equation (secant) between the points a and b, and consequently the solution of Eq. (3.57) gives the new approximation to the root of $f'(x)$ as

$$x_{i+1} = x_i - \frac{f'(x_i)}{s} = a - \frac{f'(a)(b-a)}{f'(b) - f'(a)} \qquad (3.59)$$

The iterative process expressed by the Eq. (3.59) is called the secant method (Fig. 3.9). The secant method can also be considered as a Quasi-Newton method as the secant approaches the second derivative of $f(x)$ at a as b approaches a. This method can also be considered as a form of elimination process as part of the interval (a, a_{i+1}) is eliminated in every iteration as shown in Fig. 3.9. The iterative process can be executed by using the following steps [Rao (2009)].

Step 1 Set $x_i = a = 0$ and calculate $f'(a)$. The value of $f'(a)$ will be negative. Assume an initial trial step length t_0. Set $i = 1$.

Step 2 Calculate $f'(t_0)$.

Step 3 If $f'(t_0) < 0$, set $a = x_i = t_0$, $f'(a) = f'(t_0)$ new $t_0 = 2t_0$, and go to step 2.

Step 4 If $f'(t_0) \geq 0$, set $b = t_0$, $f'(b) = f'(t_0)$, and move to step 5.

Step 5 The new approximate solution of the problem can be found as

$$x_{i+1} = a - \frac{f'(a)(b-a)}{f'(b) - f'(a)} \qquad (3.60)$$

Step 6 Check the convergence criteria:

$$|f'(x_i + 1)| \leq \varepsilon$$

where ε is a small quantity. If the above convergence criteria is satisfied, consider $x^* \approx x_{i+1}$ as the solution and terminate the process. If not, move to step 7.

Step 7 If $f'(x_{i+1}) \geq 0$, set new $b = x_{i+1}$, $f'(b) = f'(x_{i+1})$, $i = i + 1$, and go to step 5.

Step 8 If $f'(x_{i+1}) < 0$, set new $a = x_{i+1}$, $f'(a) = f'(x_{i+1})$, $i = i + 1$, and go to step 5.

Comments

1. The secant method is the same as to considering a linear equation for $f'(x)$. This shows that the original function $f(x)$ is approximated by a quadratic equation.

2. In some cases we may come across the condition that the function $f'(x)$ varies very slowly with x. In this condition, we can identify it by observing that the point b remains unchanged for several consecutive iterations. When this situation is suspected, we can improve the convergence process by taking the next value of x_{i+1} as $(a + b)/2$ in place of finding its value from Eq. (3.60).

3.5 Polynomial Approximation Methods

There is another class of unidimensional minimization methods which locates a point x near x^* by interpolation and extrapolation using polynomial approximations as models of $f(x)$. This x is the value of the independent variable that corresponds to the minimum of $f(x)$. Both quadratic and cubic approximation have been proposed in this section. We are using only the function values and using both function and derivative values for quadratic and cubic approximation respectively. These approximation methods have higher efficiency than other methods for those functions where $f'(x)$ is continuous. These methods are extensively used to perform line searches for finding the multivariable optimizers.

3.5.1 Quadratic interpolation

The quadratic interpolation method starts with three points x_1, x_2 and x_3 in increasing order that might be placed in equal distance, however, the extreme points (x_1 and x_3) must bracket the minimum. It is known to us that a quadratic function $f(x) = a + bx + cx_2$ can be passed exactly through the three points, and that the differentiation of the function can be estimated to find the minimum of the approximating function, the derivative of the function should be equal to 0. Then by solving $f'(x) = 0$, we get

$$\tilde{x} = -\frac{b}{2c} \tag{3.61}$$

consider that $f(x)$ is evaluated at x_1, x_2, and x_3, to find $f(x_1) = f_1$, $f(x_2) \equiv f_2$, and $f(x_3) \equiv f_3$. Now, we have three linear Eqs (3.62), (3.63), and (3.64) with three unknown variables. The coefficients a, b and c can be calculated from the solution of these three linear equations

$$f(x_1) = a + bx_1 + cx_1^2 \tag{3.62}$$

$$f(x_2) = a + bx_2 + cx_2^2 \tag{3.63}$$

$$f(x_3) = a + bx_3 + cx_3^2 \tag{3.64}$$

by means of determinants or matrix algebra. Incorporating the values of b and c expressed in terms of x_1, x_2, x_3, f_1, f_2 and f_3 into the Eq. (3.61) yields

$$\tilde{x}^* = \frac{1}{2}\left[\frac{(x_2^2 - x_3^2)f_1 + (x_3^2 - x_1^2)f_2 + (x_1^2 - x_2^2)f_3}{(x_2 - x_3)f_1 + (x_3 - x_1)f_2 + (x_1 - x_2)f_3}\right] \tag{3.65}$$

To demonstrate the first stage of this search process, examine the four points in Fig. 3.10. We are interested to reduce the initial interval $[x_1, x_3]$. We assume that the function $f(x)$ is unimodal and has a minimum. Then by examining the values of $f(x)$, we can eliminate the interval from x_1 to x_2, and utilize the region $[x_2, x_3]$ as the new interval. This new interval contains three points, $(x_2, \tilde{x}, x_3)$ that can be utilized to estimate a x^* value from the Eq. (3.65), and so on. Usually, we evaluate $f(x^*)$ and remove from the set $\{x_1, x_2, x_3\}$ the point that corresponds to the highest value of the function $f(x)$, unless a bracket on the minimum of $f(x)$ is lost by doing this, in which case we remove the x so as to maintain the bracket.

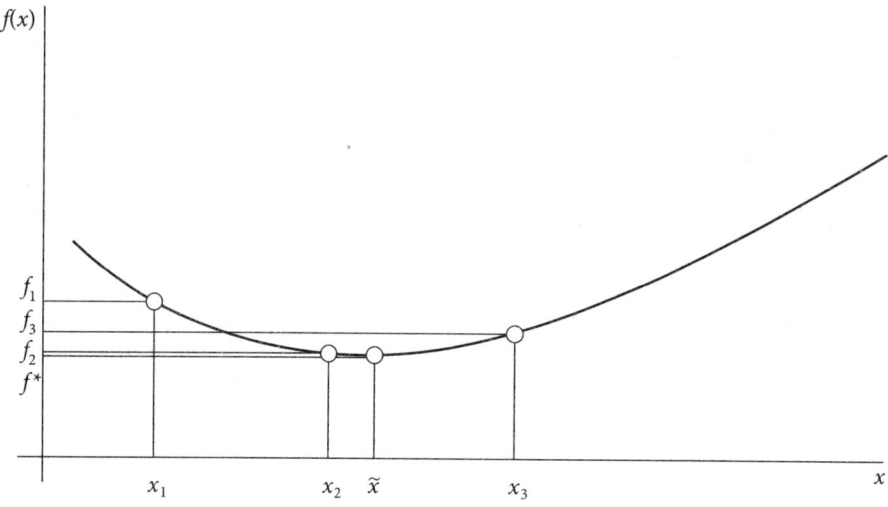

Fig. 3.10 Quadratic interpolation

3.5.2 Cubic interpolation

Cubic interpolation is used to find the minimum of $f(x)$ based on approximating the objective function by a polynomial of third degree within the interval of interest. After that, we determine the associated stationary point of the polynomial (3.66)

$$f(x) = a_1 x^3 + a_2 x^2 + a_3 x + a_4 \tag{3.66}$$

We need to compute four points (that bracket the minimum) for the estimation of the minimum, either four values of the polynomial $f(x)$, or the values of $f(x)$ and the derivative of $f(x)$, each at two points.

In the first case, we obtain four linear equations with the four unknown parameters (a_1, a_2, a_3, and a_4) that are the desired coefficients. Let the matrix X be represented as

$$X = \begin{bmatrix} x_1^3 & x_1^2 & x_1 & 1 \\ x_2^3 & x_2^2 & x_2 & 1 \\ x_3^3 & x_3^2 & x_3 & 1 \\ x_4^3 & x_4^2 & x_4 & 1 \end{bmatrix} \tag{3.67}$$

$$F^T = \begin{bmatrix} f(x_1) & f(x_2) & f(x_3) & f(x_4) \end{bmatrix} \tag{3.68}$$

$$A^T = \begin{bmatrix} a_1 & a_2 & a_3 & a_4 \end{bmatrix} \tag{3.69}$$

$$F = XA \tag{3.70}$$

The extremum of $f(x)$ is obtained by setting the derivative of $f(x)$ equal to zero as shown in Eq. (3.71) and then solving for $\tilde{x}$

$$\frac{df(x)}{dx} = 3a_1 x^2 + 2a_2 x + a_3 = 0 \tag{3.71}$$

so that,

$$\tilde{x} = \frac{-2a_2 \pm \sqrt{4a_2^2 - 12a_1 a_3}}{6a_1} \tag{3.72}$$

The sign of the second derivative of $f(\tilde{x})$ governs the sign to be used before the square root, that is to say, whether a minimum or maximum is sought. Equation (3.74) can be used to compute the vector A

$$F = XA \tag{3.73}$$

or $\quad A = X^{-1} F \tag{3.74}$

After predicting the optimum point $\tilde{x}$, it is employed as a new point for the next iteration and the point with the highest value of $f(x)$ [lowest value of $f(x)$ for maximization] is eliminated.

When, the first derivatives of $f(x)$ are available, only two points are required, and the cubic function can be fitted to the two pairs of the slope values and function values. The four coefficients can be uniquely related to these four pieces of information in the cubic equation, which can be optimized to predict the new, nearly optimal data point. If (x_1, f_1, f_1') and (x_2, f_2, f_2') are available, then the optimum $\tilde{x}$ is

$$\tilde{x} = x_2 - \left[\frac{f_2' + w - z}{f_2' - f_1' + 2w} \right] (x_2 - x_1) \tag{3.75}$$

where

$$z = \frac{3[f_1 - f_2]}{[x_2 - x_1]} + f_1' + f_2' \tag{3.76}$$

$$w = \left[z^2 - f_1' \cdot f_2' \right]^{1/2} \tag{3.77}$$

In a minimization problem, we require $x_1 < x_2$, $f_1' < 0$, and $f_2' > 0$ (the minimum is bracketed by x_1 and x_2). Calculate the function value $f(\tilde{x})$ at the new point $(\tilde{x})$ to determine which of the previous two points to replace. The application of this technique for solving nonlinear programming problem that use gradient information is straightforward and efficient.

Summary

- This cheaper discusses various methods for optimization of single-variable. The criteria and conditions for this optimization are given in this chapter. This section provides some idea about different search methods like direct search and gradient search methods. Bracketing of the optimization point is very crucial for unimodal function; this technique is given in this chapter. Here, we have presented unrestricted search method, exhaustive search, dichotomous search, interval halving method, Fibonacci method, and golden section method. Beside this, Newton method, Quasi-Newton method, secant method and polynomial approximation methods are also considered.

Exercise

3.1 Why single-variable unconstrained optimization is important for chemical engineers? Give some examples of the single-variable unconstrained optimization problem.

3.2 What is the difference between interpolation and elimination methods?

3.3 Describe the Fibonacci numbers.

3.4 What is a unimodal function? What are the different methods for optimizing the unimodal function?

3.5 Use the quadratic interpolation method to find the minimum of the function given in Example 3.3. Consider the initial step size of 0.1.

3.6 Use the cubic interpolation method to find the minimum of the function given in Example 3.3. Consider an initial step size of $t_0 = 0.1$.

3.7 Prove that a convex function is unimodal.

3.8 Determine the value of x within the interval (0, 1) that minimizes the function $f = (x-1)(x-2.5)$ to within ± 0.025 by (a) the Fibonacci method and (b) the golden section method.

3.9 Find the minimum value of the function $f = x^4 - x + 1$ using secant method.

3.10 Write an algorithm for finding the minimum resistance of heat transfer using Eq. (2.50). Use Newton method for the same.

3.11 Solve the Example 3.3 using Quasi-Newton method.

References

Biggs, M. B. 2008. *Nonlinear Optimization with Engineering Applications*, Springer Optimization and its Applications, vol. 19.

Edgar, T. F., Himmelblau, D. M., Lasdon, L. S. *Optimization of Chemical Processes* (2*nd ed.*), McGraw–Hill.

Rao, S. S. *Engineering Optimization Theory and Practice* (4*e*), New Jersey: John Wiley and Sons, Inc.

Ravindran, A., Ragsdell, K. M., Reklaitis, G. V. *Engineering Optimization Methods and Applications* (2*e*), John Wiley and Sons, Inc.

Snyman, J. A. *Practical Mathematical Optimization*, Springer.

Chapter 4

Trust-Region Method

4.1 Introduction

Iterative methods for optimization are categorized into two classes. One class is called line search methods and the other class as trust region algorithms. Trust-Region methods are iterative method in which a model (m_k) approximates the objective function (f) and this model is minimized in a neighborhood of the current iterate (the trust region). In case of a line-search method, the iterations are performed toward some particular directions; for example, the gradient directions are used to find the successive iterates in *steepest descent* [Liu and Chen, 2004]. However, in a Trust-Region algorithm, its iterates are derived by solving the corresponding optimization problem iteratively within an enclosed region. Therefore, we have more options to choose the iterates. Indeed, we can consider line-search methods as special cases of trust region methods [A. R. Conn *et al.*, (2000)]. Trust-Region methods first introduced by M. J. D. Powel in 1970 [M. J. D. Powell, (1970)]. Powell [M. J. D. Powell, (1975)] also established the convergence result of unconstrained Trust-Region method optimization. Fletcher [R. Fletcher, (1972)] first recommended Trust-Region algorithms to solve linearly constrained optimization problems and non-smooth optimization problems [R. Fletcher, (1982)]. Trust-Region methods are very essential and effective methods in the area of nonlinear optimization. These methods are also useful for non-convex optimization problems and non-smooth optimization problems [Sun (2004)].

As most of the research works on trust region algorithms are mostly started in the 80s, trust region algorithms are less mature compare to line search algorithms, and the applications of trust region algorithms are limited as compared to line search algorithms. However, trust region methods have two major advantages. One is that they are reliable and robust; another is that they have very strong convergence properties. The key contents of any trust region algorithm are how to calculate the trust region trial step and how a decision can be made if a trial step should be accepted or not. An iteration of a trust region algorithm has the following form; a trust region is available at the beginning of the iteration. This is possible by considering an initial guess value $X_0 \in \mathbb{R}^n$ and trust region radius $\Delta_0 > 0$. An approximate model (m_k) is constructed, and it is solved within the trust region, giving a solution s_k which is called the trial step. A merit function is selected, which is used for updating the next trust region and for deciding the new iterate point [Yuan, (1993)]. There are many applications of Trust-Region method like curve fitting (Helfrich and Zwick, 1996), optimization of pressure swing adsorption [Agarwal *et al.* 2009] etc.

4.2 Basic Trust-Region Method

4.2.1 Problem statement

The statement of Basic Trust-Region Method is given below

i. consider a local model m_k of the objective function f around X_k
ii. calculate a trial point $X_k + s_k$ that reduces the model m_k within the trust region $\|s_k\| \leq \Delta_k$
iii. calculate the reduction ratio

$$r_k = \frac{f(X_k) - f(X_k + s_k)}{m_k(X_k) - m_k(X_k + s_k)} \tag{4.1}$$

iv. if m_k and f agree at $X_k + s_k$, i.e., $r_k \geq \eta_1$
then
accept trial point: $X_{k+1} = X_k + s_k$ \hfill (4.2)
update the trust region radius:

$$\Delta_{k+1} \in \begin{cases} [\Delta_k, \infty] & \text{if } r_k \geq \eta_2 \\ [\gamma_2 \Delta_k, \Delta_k] & \text{if } r_k \in (\eta_1, \eta_2) \end{cases} \tag{4.3}$$

else
reject trial point: $X_{k+1} = X_k$
decrease trust region radius: $\Delta_{k+1} \in [\gamma_1 \Delta_k, \gamma_2 \Delta_k]$
where $0 < \eta_1 \leq \eta_2 < 1$ and $0 < \gamma_1 \leq \gamma_2 < 1$

The reduction ratio r_k plays two roles in BTR method

1. decides the acceptance of the trial point $X_k + s_k$
2. controls the update procedure of Trust-Region radius

Unlike the line search methods, Trust-Region methods will attempt to get the next approximate solution within a region of the current iterate. For trust region methods, three issues need to be emphasized: (i) Trust-Region radius, which determines the size of a trust region (ii) Trust-Region subproblem, which approximates a minimizer within the region and (iii) Trust-Region fidelity;

used to estimate the level of accuracy of an approximated solution [Liu and Chen, 2004]. The following sections describe Trust-Region radius, Trust-Region subproblem, and Trust-Region fidelity in depth.

4.2.2 Trust-Region radius

At kth iteration, the quadratic model of the function $f(X)$ around the current iterate X_k is represented by

$$m_k(X_k + d) = f(X_k) + g_k^T d + \frac{1}{2} d^T B_k d, \qquad (4.4)$$

The Trust-Region is characterized as the region where $\|d\| \leq \Delta_k$.

$$\Delta_{k+1} = \beta_k \Delta_k \qquad (4.5a)$$

$$\beta_k \in \begin{cases} [\gamma_1, 1] & \text{if } r_k < \mu_1 \\ [1, \gamma_2] & \text{if } r_k \geq \mu_2 \\ [\gamma_3, \gamma_4] & \text{otherwise} \end{cases} \qquad (4.5b)$$

where Δ_k is a trust region radius (Fig. 4.1). The constants are as follows $0 < \gamma_1 < 1 < \gamma_2$, $\gamma_1 \leq \gamma_3 < 1 < \gamma_4 \leq \gamma_2$ and $0 < \mu_1 < \mu_2 < 1$. A function called merit function is generally used to check whether the trial step is accepted, or any adjustment is needed for the trust region radius. Choosing a proper trust region radius (Δ_k) is crucial for any TR method algorithm. A trust region may show an improvement at very slow rate during the estimation of the solution, if the value of Δ_k is very small. On the other hand, the agreement between the model (m_k) and the objective function (f) will be very poor when the value of Δ_k is too large. At the initial step, selection of an initial trust region radius (Δ_0) is very important. During the implementation of a trust region method, we should be concerned regarding the issue of the proper selection of the initial trust region radius (ITRR) [M. J. D. Powell, (1970)]. A bad choice of Δ_0 can lead an increase in number of iteration consequently cost of optimization, particularly when the linear algebra required for each iteration is costly [A. SARTENAER, (1997)]. Now, we will discuss an algorithm for finding initial Trust-Region radius.

Initial Trust Region Radius (ITRR)

The initial trust region radius (Δ_0) can be estimated automatically by following a strategy that is introduced by Sartenaer [A. Sartenaer, (1997)]. The fundamental concept of this method is to find out a maximal initial radius by performing several repeated trials in the $-g_k$ direction that also guarantee a satisfactory agreement between the objective function and the model. The difficulty in estimating an ITRR (Δ_0) is to find a technique to check agreement between the objective function and the approximated model at the initial point X_0. The technique discussed in this chapter is based on the utilization of normally available information at this point, i.e., value of the function and its gradient. A reliable ITRR can be determined with the additional cost of some function evaluations.

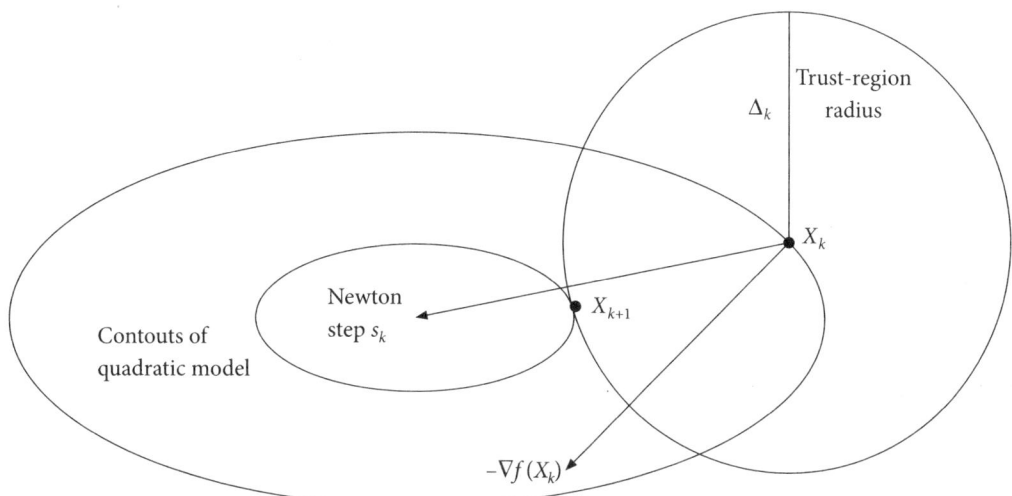

Fig. 4.1 Trust region

We need to perform some extra mathematical computations for this purpose. However, this technique will reduce the number of iterations required to obtain a solution.

This method determines the maximal radius, which shows a satisfactory agreement between the objective function and the model in the $-g_0$ direction, employing an iterative search procedure along this direction. At any iteration k of the search procedure, with a given radius estimation $\Delta_{k,0}$, the values of the model and objective function are calculated at the point $\left(X_0 - \Delta_{k,0}\hat{g}_0\right)$, is

$$m_{k,0} = m_0\left(X_0 - \Delta_{k,0}\hat{g}_0\right) \tag{4.6a}$$

and
$$f_{k,0} = f_0\left(X_0 - \Delta_{k,0}\hat{g}_0\right) \tag{4.6b}$$

where
$$\hat{g}_0 = \left(\frac{g_0}{\|g_0\|}\right) \tag{4.6c}$$

And the ratio,
$$r_{k,0} = \frac{f_0 - f_{k,0}}{f_0 - m_{k,0}} \tag{4.7}$$

is calculated, then the algorithm stores the maximum value among the estimates of $\Delta_{k,0}$, whose related $r_{k,0}$ is "close enough to one". Finally, it updates the current estimate $\Delta_{k,0}$. The updating phase for $\Delta_{k,0}$ follows the framework in Eq. (4.5a)–(4.5b), however, includes a more general test on $r_{k,0}$ because the predicted change in Eq. (4.7) is not guaranteed to be positive. Therefore, we set

$$\Delta_{k+1,0} = \beta_{k,0}\Delta_{k,0}$$

where

$$\beta_{k,0} \in \begin{cases} [\gamma_1, 1] & \text{if } |r_{k,0} - 1| > \mu_1 \\ [1, \gamma_2] & \text{if } |r_{k,0} - 1| \leq \mu_2 \\ [\gamma_3, \gamma_4] & \text{otherwise} \end{cases} \quad (4.8)$$

for some $0 \leq \mu_2 < \mu_1$. Note that updated Eq. (4.8) only considers the tolerability between the objective function and its approximated model, without taking care of the minimization of the objective function f.

4.2.3 Trust-Region subproblem

Trust region subproblems are one of the vital parts of trust region algorithms. Since, each iteration of a trust region algorithm requires to solve (exactly or inexactly) a trust region subproblem, finding efficient solver for trust region subproblems is very important. We consider subproblem (4.4) which has been studied by many authors. The following lemma is well known for solving Trust-Region subproblem.

Lemma 4.1

A vector $d^* \in \Re^n$ is a solution of the problem

$$\min_{d \in \Re^n} m_k(d) = g_k^T d + \frac{1}{2} d^T B_k d, \quad (4.9)$$

subject to $\|d\| \leq \Delta_k$ (4.10)

where $g_k \in \Re^n$, $B_k \in \Re^{n \times n}$ is a symmetric matrix, and $\Delta_k > 0$, if and only if there exists $\lambda^* \geq 0$ such that

$$(B_k + \lambda^* I) d^* = -g_k \quad (4.11)$$

and that $B_k + \lambda^* I$ is positive semi-definite, $\|d^*\| \leq \Delta_k$ and $\lambda^* (\Delta_k - \|d^*\|) = 0$

Similar to other algorithms, performance of Trust-Region methods depend on the selection of a set of parameters.

4.2.4 Trust-Region fidelity

When a subproblem is solved, the trial point $X_k + s_k$ will be checked to understand whether it is a good candidate for the next iterate. This is calculated explicitly using the equation

$$r_k = \frac{f(X_k) - f(X_k + s_k)}{m_k(X_k) - m_k(X_k + s_k)} \quad (4.12)$$

The trial point is accepted whenever $r_k \geq \eta_1$, i.e., $X_{k+1} = X_k + s_k$. If not, then $X_{k+1} = X_k$. As the parameter η_1 is a small positive number, the aforesaid rule favors a trial point only if the value of the objective function f is also decreased. The radius of the Trust-Region will be expanded for the next iteration while m_k approximates f nicely and gives a large value of r_k. Conversely, when the value of r_k is smaller than η_1 or r_k is negative, it implies that within this present trust region the objective function f is not approximated properly by the model m_k. As a result, the iterate remains unchanged. Then, the Trust-Region radius will be reduced in size to develop more suitable model and subproblem for the next iteration.

4.3 Trust-Region Methods for Unconstrained Optimization

During the minimization of an unconstrained problem, a step to a new iterate is estimated by minimizing a local model of the objective function within a bounded region that has the center at the current iterate. At every iteration of the trust region method, the nonlinear objective function is substituted by a simple model keeping its center at the current iterate. For this purpose, the first and probably second-order information available at this present iterate is used to construct this simple model. Therefore, this model is usually appropriate only within a certain restricted region neighboring this point. Thus, a trust region is defined where the approximated model is said to agree satisfactorily with the actual objective function. Then the trust region algorithms are composed of solving a series of subproblems in which the model is minimized approximately within the trust region, which gives a candidate for the next iterate [M. J. D. Powell, (1970)].

In this section, we consider trust region algorithms for unconstrained optimization problem:

$$\min_{X \in \Re^n} f(X) \tag{4.13}$$

where $f(X)$ is a nonlinear continuous differentiable function in $\Re^n$. At each iteration, a trial step is evaluated by solving the subproblem

$$\min_{d \in \Re^n} m_k(d) = g_k^T d + \frac{1}{2} d^T B_k d \tag{4.14}$$

subject to $\|d\| \leq \Delta_k$ (4.15)

where $g_k = \nabla f(X)$ is the gradient at the current approximate solution, B_k is an $n \times n$ symmetric matrix that approximates the Hessian matrix of $f(X)$ and $\Delta_k > 0$ is called the trust region radius. Let s_k be a solution of (4.14)–(4.15). The predicted reduction is defined by the reduction in the approximate model, that is

$$P_{red_k} = m_k(0) - m_k(s_k) \tag{4.16}$$

Unless the current point X_k is a stationary point and B_k is positive semi-definite, the predicted reduction P_{red_k} is always positive. The actual reduction is the reduction in the objective function:

$$A_{red_k} = f(X_k) - f(X_k + s_k) \qquad (4.17)$$

And we define the ratio between the actual reduction and the predicted reduction by

$$r_k = \frac{A_{red_k}}{P_{red_k}} \qquad (4.18)$$

which is employed to make a decision if the trial step is acceptable and to adjust the new trust region radius.

A general algorithm for unconstrained trust region optimization can be given as follows.

Step 1 Given $X_1 \in \Re^n$, $\Delta_1 > 0$, $\varepsilon \geq 0$, $B_1 \in \Re^{n \times n}$ symmetric; $0 < \tau_3 < \tau_4 < 1 < \tau_1$, $0 \leq \tau_0 \leq \tau_2 < 1$, $\tau_2 > 0$, $k = 1$

Step 2 If $\|g_k\| \leq \varepsilon$ then stop; Solve a trust region subproblem, giving s_k.

Step 3 Compute $r_k = \dfrac{P_{red_k}}{A_{red_k}}$; set $X_{k+1} = \begin{cases} X_k & \text{if } r_k \leq \tau_0 \\ X_k + s_k & \text{otherwise} \end{cases}$

choose Δ_{k+1} from the equation $\Delta_{k+1} = \begin{cases} [\tau_3 \|s_k\|, \ \tau_4 \Delta_k] & \text{if } r_k < \tau_2 \\ [\Delta_k, \ \tau_1 \Delta_k] & \text{otherwise} \end{cases}$

Step 4 Update B_{k+1}; $k = k + 1$; go to step 2

The constants $\tau_i \ (i = 0, 1, \ldots, 4)$ can be selected by users. Typical values are $\tau_0 = 0$, $\tau_1 = 2$, $\tau_2 = \tau_3 = 0.25$, $\tau_4 = 0.5$.

4.4 Trust-Region Methods for Constrained Optimization

For constrained problems, most trust region subproblems can be regarded as some kind of modification of the SQP subproblem of line search algorithm, which has the following form:

$$\min_{d \in \Re^n} m_k(d) = g_k^T d + \frac{1}{2} d^T B_k d, \qquad (4.19)$$

subject to

$$c_i(X) + d^T \nabla c_i(X) = 0; \quad i = 1, 2 \ldots, m_e \qquad (4.20)$$

$$c_i(X) + d^T \nabla c_i(X) \geq 0; \quad i = m_e + 1, \ldots, m \tag{4.21}$$

where $g_k = g(X_k) = \nabla f(X_k)$ and B_k is an approximate Hessian of the Lagrange function. The first type of trust region subproblems, being a slightly modification of SQP subproblem (4.19)–(4.21), have the following form:

$$\min_{d \in \Re^n} m_k(d) = g_k^T d + \frac{1}{2} d^T B_k d, \tag{4.22}$$

subject to

$$\theta_k c_i(X) + d^T \nabla c_i(X) = 0; \quad i = 1, 2 \ldots, m_e \tag{4.23}$$

$$\theta_k c_i(X) + d^T \nabla c_i(X) \geq 0; \quad i = m_e + 1, \ldots, m \tag{4.24}$$

$$\|d\| \leq \Delta_k \tag{4.25}$$

where $\theta_k \in (0,1)$ is a parameter that is introduce to overcome the possible non-feasibility of the linearized constraints (4.20) and (4.21) in the trust region (4.25). Another trust region subproblem is obtained by substituting the linearized constraints (4.20) and (4.21) by a single quadratic constraint. It can be written as:

$$\min_{d \in \Re^n} m_k(d) = g_k^T d + \frac{1}{2} d^T B_k d, \tag{4.26}$$

subject to

$$\left\|\left(c_k + A_k^T d\right)^-\right\| \leq \xi_k \tag{4.27}$$

$$\|d\| \leq \Delta_k \tag{4.28}$$

where $c_k = c(X) = (c_1(X), \ldots, c_m(X))^T$, $A_k = A(X) = \nabla c(X)^T$, $\xi_k \geq 0$ is a parameter and the superscript "−" means that $v_i^- = v_i (i = 1, 2, \ldots, m_e)$, $v_i^- = \min[0, v_i](i = m_e + 1, \ldots, m)$. This algorithm is given by Celis et al. [M. R. Celis et al., (1985)] and Powell and Yuan [Powell and Yuan, (1991)]

Trust region subproblems can also derived by using exact penalty functions. The following trust region subproblem is based on the L_∞ exact penalty function:

$$\min_{d \in \Re^n} m_k(d) = g_k^T d + \frac{1}{2} d^T B_k d + \sigma \left\|\left(c_k + A_k^T d\right)^-\right\|_\infty \tag{4.29}$$

subject to $\|d\| \leq \Delta_k$ (4.30)

Trust region subproblems based on exact penalty functions are closely related to subproblems of trust region algorithms for nonlinear systems of equations. Trust region algorithms that compute the trial step by solving (4.29)–(4.30) are also similar to trust region algorithms for nonsmooth optimization.

Once, a trial step s_k is computed by solving the trust region subproblem, the predicted reduction P_{red_k} is defined by the reduction of some approximate function $\bar{m}_k(d)$. It should be noted that in general $m_k(d) - \bar{m}_k(d)$. A merit function $P_k(X)$ is used to define the actual reduction A_{red_k}. $P_k(X)$ is normally some penalty function. And the functions $\bar{m}_k(d)$ and $P_k(X)$ are so constructed that

$$m_k(d) - \bar{m}_k(0) = P_k(X+d) - P_k(X) + O(\|d\|)$$ (4.31)

when $\|d\|$ is very small.

The algorithm can be stated as follows:

Algorihtm

Step 1 Given $X_1 \in \Re^n$, $\Delta_1 > 0$, $\varepsilon \geq 0$, $B_1 \in \Re^{n \times n}$ symmetric; $0 < \tau_3 < \tau_4 < 1 < \tau_1$, $0 \leq \tau_0 \leq \tau_2 < 1$, $\tau_2 > 0$, $k = 1$

Step 2 If $\|g_k\| \leq \varepsilon$ then stop; Solve a trust region subproblem, giving s_k.

Step 3 Compute $r_k = \dfrac{P_{\text{red}_k}}{A_{\text{red}_k}}$; set $X_{k+1} = \begin{cases} X_k & \text{if } r_k \leq \tau_0 \\ X_k + s_k & \text{otherwise} \end{cases}$

choose Δ_{k+1} from the equation $\Delta_{k+1} = \begin{cases} [\tau_3 \|s_k\|, \tau_4 \Delta_k] & \text{if } r_k < \tau_2 \\ [\Delta_k, \tau_1 \Delta_k] & \text{otherwise} \end{cases}$

Step 4 Update B_{k+1}; $k = k + 1$; go to step 2

The constants $\tau_i (i = 0,1,\ldots,4)$ can be chosen by users. Typical values are $\tau_0 = 0$, $\tau_1 = 2$, $\tau_2 = \tau_3 = 0.25$, $\tau_4 = 0.5$.

The parameter τ_0 is usually zero or a small positive constant. There is an advantage of using zero τ_0; whenever the objective function value is decreased, the trial step is accepted. Therefore, it would not discard a "good point", which is a desirable property particularly when the evaluations of functions are very costly.

4.5 Combining with Other Techniques

Combination of Trust-Region method with other techniques is also possible in constructing algorithms. Nocedal and Yuan [Nocedal and Yuan, (1998)] observed that the trial step of trust region algorithms for unconstrained optimization is also a descent direction. Therefore, when the trial step is unacceptable, it is still possible to carry out a line search along the direction of the trial step. An algorithm, which combines backtracking and trust region has been discussed by Nocedal and Yuan.

4.6 Termination Criteria

Convergence of a Trust-Region algorithm relies on the accuracy of the approximate solution to the Trust-Region subproblem (4.9). For an efficient algorithm, it is important that the Trust-Region subproblem not be solved exactly. Broadly speaking, it is desirable to compute a solution with as little effort as possible, subject to the requirement that the computed solution does not interfere with the overall convergence of the method.

Fletcher (1987) discussed some termination criteria. The conventional termination criteria are to terminate when any of $f(x_k) - f(x_{k+1})$, $x_k - x_{k+1}$, or $\nabla f(x_k)$ are small. All are used in practice. Fletcher also suggested

$$(x_k - x_{k+1})^T \nabla f(x_k) \tag{4.32}$$

which is invariant when the step is a Newton step.

It is also possible for $\Delta_k \to 0$ as $k \to \infty$. Thus, it is also necessary to consider termination when Δ_k is small

i. The change in the objective function value $|f(x) - f(x_{k+1})|$ is less than the termination criterion ε_1 in any step

ii. The change in the second-order Taylor series model of the objective function value $\left|(x_k - x_{k+1})^T \left(g_k + \frac{1}{2} B_k (x_k - x_{k+1})\right)\right|$ is less than the termination criterion ε_3 in any step

iii. The trust region radius Δ_k has shrunk to less than the termination criterion ε_4 in any step.

4.7 Comparison of Trust-Region and Line-Search

For trust region method, we have employed an approximated quadratic model m_k for the implementation. If we consider first two terms in RHS of Eq. (4.4), it will be reduced to a linear model. This shows that a Trust-Region method with an approximated linear model is similar to gradient descent, however, often it attains better performances due to its capability to regulate trust regions adaptively all through the iterations. For this reason, the line-search methods can be regarded as special cases of Trust-Region [Liu and Chen, 2004]. Both the Trust-Region and the line-search methods are guaranteed to converge to a local minimum. However, for real application, all these local minima are not significant. A typical line-search method (e.g., steepest descent or even Trust-Region with a linear approximation) may often converge to a local minimum, which is substandard compared to a nearby one.

Summary

- Trust-Region methods have been discussed for both unconstrained and constrained optimization problems. Various features of Trust-Region methods such as trust region radius, trust region subproblem, trust region fidelity are given in this chapter. Termination criteria are an important factor for any algorithm; it has been considered in this chapter. Various researches confirm that the trust region methods can be combined with other optimization methods. A comparison between line search and trust region methods is made to explain the advantages and disadvantages of the methods.

Exercise

4.1 Find the quadratic approximation of the function around the point (1, 1)
$R(C, t) = 77.92 + 9.41C + 3.86t - 3.53C^2 - 7.33t^2$

4.2 What are the advantages of trust region methods over line search methods?

4.3 Discuss the effect of initial trust region radius (ITRR) on the performance of TR algorithm.

4.4 Define trust region fidelity.

4.5 Find the maximum of the function using Trust region method
$f = 15x_1 + 8x_1x_2 + 5x_2$ subject to $x_1 + x_2 \leq 10$

4.6 Calculate the reduction ratio $r_k = \dfrac{f(X_k) - f(X_k + s_k)}{m_k(X_k) - m_k(X_k + s_k)}$ for the function given in problem 1.

4.7 Is the result of any Trust Region method depends on the initial guess value $X_0 \in \mathbb{R}^n$? Justify your answer.

4.8 Discuss the convergence criteria of Trust Region method, show how the other parameters (τ_1, τ_2, τ_3 and τ_4) affect the convergence.

4.9 When combination of other methods with Trust region method is advantageous?

4.10 Write an algorithm for problem 5, which combines Trust region and backtracking method.

References

Agarwal, A., Biegler, L. T. and Zitney, S. E. 2009. *Simulation and Optimization of Pressure Swing Adsorption Systems Using Reduced-Order Modeling;* Ind. Eng. Chem. Res., 48, 2327–43.

Celis, M. R., Dennis, J. E. and Tapia, R. A. *A Trust Region Algorithm for Nonlinear Equality Constrained Optimization,* in: P. T. Boggs, R. H. Byrd and R. B. Schnabel, eds., Numerical Optimization (SIAM, Philadelphia, 1985), 71–82.

Conn, A. R., Gould, N. I. M. and Toint, P. L. 2000. *Trust-Region Methods,* SIAM.

Fletcher, R. An *Algorithm for Solving Linearly Constrained Optimization Problems,* Math. Program. 2(1972) 133–65.

Fletcher, R. *A Model Algorithm for Composite Non-differentiable Optimization Problems,* Mathematical Programming Study, 17(1982): 67–76.

Fletcher, R. 1987. *Practical Methods of Optimization* (2e), John Wiley and Sons.

Helfrich, H. P. and Zwick, D. *A Trust Region Algorithm for Parametric Curve and Surface Fitting,* Journal of Computational and Applied Mathematics, 73(1996): 119–34.

Liu, T-L, Chen, H-T. 2004. *Real-Time Tracking Using Trust-Region Methods,* IEEE Transactions on Pattern Analysis and Machine Intelligence, vol. 26, No. 3.

Nocedal, J. and Yuan, Y. *Combining Trust Region and Line Search Techniques,* in: Y. Yuan, ed. Advances in Nonlinear Programming, (Kluwer, 1998), 153–75.

Powell, M. J. D. 1970. *A New Algorithm for Unconstrained Optimization, in Nonlinear Programming,* J. B. Rosen, O. L. Mangasarian, and K. Ritter, eds., New York: Academic Press.

Powell, M. J. D. 1975. *Convergence Properties of a Class of Minimization Algorithms,* in: O. L. Mangasarian, R. R. Meyer, S. M. Robinson (Eds.), Nonlinear Programming 2, New York: Academic Press.

Powell, M. J. D. and Yuan, Y. *A Trust Region Algorithm for Equality Constrained Optimization,* Math. Prog. 49(1991): 189–211.

Sartenaer, A. 1997. *Automatic Determination of an Initial Trust Region in Nonlinear Programming,* SIAM J. Sci. Comput., 18(6): 1788–1803.

Sun, W.; Appl. Math. Comput., 156(2004): 159–74.

Yuan, Y., *A New Trust Region Algorithm for Nonlinear Optimization,* in: D. Bainov and V. Covachev, eds., Proceedings of the First International Colloquium on Numerical Analysis (VSP, Zeist, 1993), pp. 141–52.

Chapter 5

Optimization of Unconstrained Multivariable Functions

5.1 Introduction

When the optimization of an objective function is required without any additional correlation, then this optimization is called unconstrained optimization. Unconstrained optimization problem appears in some cases in chemical engineering. It is the simplest multivariable optimization problem. Parameter estimation is a significant application in engineering and science, where, multivariable unconstrained optimization methods are required. Some optimization problems are inherently unconstrained; there is no additional function (section 2.7, 2.8). When there are some usual constraints on the variables, it is better to ignore these constraints and to consider that they do not have any impact on the optimal solution. Unconstrained problems also formed due to reconstructions of constrained optimization problems, in which the penalization terms are used to replace the constraints in the objective function that have the effect of discouraging constraint violations.

Rarely do we get any unconstrained problem as a practical design problem, the knowledge on this type of optimization problems is essential for the following purposes:

1. The constraints hold very less influence in some design problems.
2. To get basic idea about constrained optimization, the study of unconstrained optimization techniques is necessary.
3. Solving the unconstrained optimization problem is quite easy compared to constrained optimization.

Some robust and efficient methods are required for the numerical optimization of any nonlinear multivariable objective functions. The efficiency of the algorithm is very significant because these optimization problems require an iterative solution process, and this trial and error becomes unfeasible when number of variables is more than three. Generally, it is very difficult to predict the behavior of nonlinear function; there may exist local minima or maxima, saddle points, regions of convexity, and concavity. Therefore, robustness (the capability to get a desired solution) is desirable for these methods. In some regions, the optimization algorithm may proceed quite slowly toward the optimum, demanding excessive computational time. In this chapter, we will discuss various nonlinear programming algorithms for unconstrained optimization.

5.2 Formulation of Unconstrained Optimization

For an unconstrained optimization problem, we minimize the objective function that is constructed with real variables, without any limitations on the values of these variables. The mathematical representation of this minimization problem is:

$$\min_{x} f(x) \tag{5.1}$$

where $x \in \mathbb{R}^n$ is a real vector with $n \geq 1$ components and $f : \mathbb{R}^n \to \mathbb{R}$ is a smooth function.

Generally, it is desirable to find the global minimum of $f(x)$, a point where the objective function reaches its least value. We may face difficulty to find the global minima, since our knowledge on $f(x)$ is typically local only. We do not have an adequate idea on the overall shape of $f(x)$ as our algorithm does not visit thoroughly many points. Therefore, we can never be confirmed whether the function does not have a sharp valley or peak in some area that is not examined thoroughly by the algorithm. The majority optimization algorithms are capable of finding only a local minimum. Local minimum is a point that produces the least value of $f(x)$ in neighborhood of the starting point. The methods discussed in the following sections are local minimization method.

5.3 Direct Search Method

Hook and Jeeves coined the term "direct search" in 1961 [Hooke and Jeeves, (1961)]. Sometimes the classical methods fail or are not feasible to find the maximum or minimum value of a complicated function. 'Direct search' is a method for solving these problems by using a computer that follows a simple search strategy [Hooke and Jeeves, (1961)]. The phrase "direct search" has been used to describe sequential analysis of trial solutions by performing the comparison of each trial solution with the "best" obtained up to that time. This method also finds a strategy for determining what the next trial solution will be (as a function of earlier results) [Lewis *et al.* (2000)]. Although, most direct search methods have been developed by heuristic approaches, some of them have proved extremely effective in practice, particularly in applications where the objective function was non-differentiable, had discontinuous first derivatives, or was subject to random error [Swann (1972)].

5.3.1 Random search methods

A large class of optimization problems can be handled by random search techniques. Random search techniques were first introduced by Anderson [Anderson, (1953)] and later by Rastrigin [Rastrigin, (1963)] and Karnopp [Karnopp, (1963)]. The particular type of problem is presented here is called an optimization problem and is characterized by a search process, carried out in a

88 Optimization in Chemical Engineering

multidimensional space, $X = x_1, x_2, ..., x_n$. The search process is conducted for a set of values of the parameters (X) that gives an absolute extreme (minimum or maximum) of a criterion or reward function,

$$f(X) = f(x_1, x_2, ..., x_n) \tag{5.2}$$

Random numbers are utilized in these methods for determining the minimum point. These methods can be used quite efficiently as random number generator is available with most of the computer libraries. Here, we have presented some of the well-known random search methods.

5.3.1.1 Random jumping method

This method randomly generates points $x^{(k)}$ within a fixed region; selecting the one that provides the best value of the function over a huge number of trials. The Fig. 5.1 represents the random jumping method. The contour is given for an objective function. Minimum value of the function has been found using random jumping method. The algorithm of this method is given below:

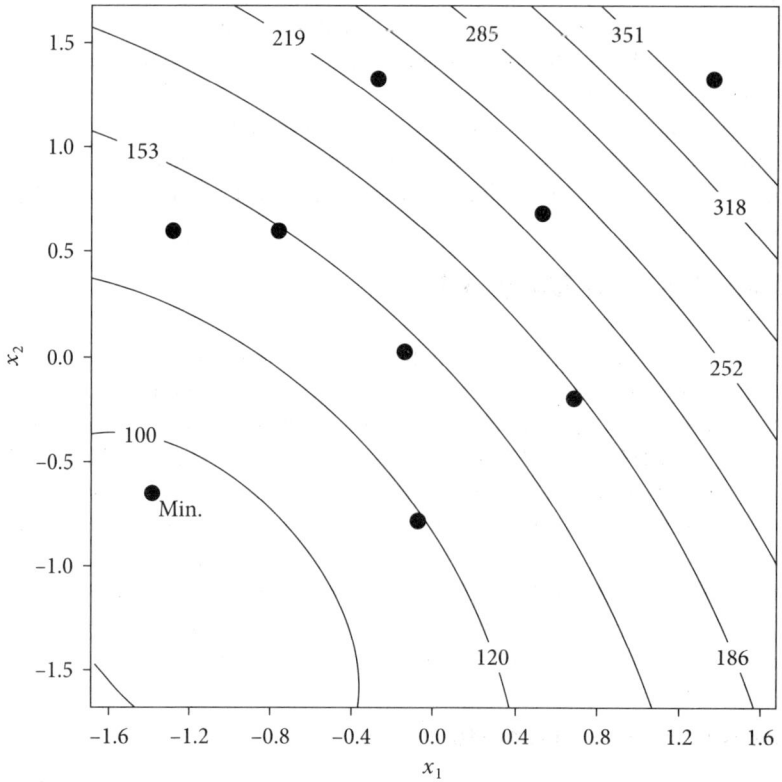

Fig. 5.1 Contour representation for random jumping method

Algorithm

Step 1 Formulate the optimization problem with a lower (l_i) and upper (u_i) bounds for each design variable x_i, $i = 1, 2, \ldots, n$ to generate the random values of x_i.
where $l_i \leq x_i < u_i$, $i = 1, 2, \ldots, n$

Step 2 Sets of n random numbers (r_1, r_2, $\ldots$, r_n) were generated. These random numbers are distributed uniformly between 0 and 1. Each set of these numbers is utilized to locate a point X, which is defined as

$$X = \begin{Bmatrix} x_1 \\ x_2 \\ \vdots \\ x_n \end{Bmatrix} = \begin{Bmatrix} l_1 + r_1(u_1 - l_1) \\ l_2 + r_2(u_2 - l_2) \\ \vdots \\ l_n + r_n(u_n - l_n) \end{Bmatrix} \tag{5.3}$$

At point X, the value of the function is calculated.

Step 3 Generate a huge number of random points X and evaluate the value of objective function at each of these points.

Step 4 For a minimization problem, select the smallest value of $f(X)$ as it is preferred. Conversely, select the highest value of $f(X)$ for a maximization problem.

5.3.1.2 Random walk method

Random Walk Optimization (RWO) is an optimization process, which is a zero order method. Zero order implies that no derivatives, only the function values are used to found the search vector. This method is suitable for non-smooth, discrete objective functions as it does not require the derivative of the functions. During each iteration, the direction of search process is a random direction. Random Walk optimization algorithm is a very simple and therefore, the optimization process is controlled by very few parameters.

The random walk method works on the principle of generating a series of approximations that improves gradually toward the minimum. Each of these approximations is determined from the previous approximation. Therefore, if X_i is the approximation to the minimum attained in the $(i - 1)$th iteration (or stage or step), the improved or new approximation in the ith iteration is estimated by the relation

$$X_{i+1} = X_i + \lambda u_i \tag{5.4}$$

where λ is a specified scalar step length and u_i denote a unit random vector generated in the ith iteration. The following steps [Fox, (1971)] describe the detailed operation of this method:

1. Start with the following parameters: X_1, initial point; an adequately large initial step size of λ; ε, a minimum permissible step size, and N, the maximum allowable number of iterations.

2. Calculate the value of the function $f_1 = f(X_1)$.

3. Set the iteration number as $i = 1$.

4. Create a set of n random numbers $r_1, r_2, \ldots, r_n$; all are enclosed in the interval $[-1, 1]$ and express the unit vector u as

$$u = \frac{1}{\left(r_1^2 + r_2^2 + \cdots + r_n^2\right)^{1/2}} \begin{Bmatrix} r_1 \\ r_2 \\ \vdots \\ r_n \end{Bmatrix} \quad (5.5)$$

It is expected that the directions created using Eq. (5.5) have a bias toward the diagonals of the unit hypercube [Fox, (1971)]. To keep away from such a bias, the length of the vector (R) is calculated as

$$R = \left(r_1^2 + r_2^2 + \cdots + r_n^2\right)^{1/2} \quad (5.6)$$

and the acceptance of the random numbers generated ($r_1, r_2, \ldots, r_n$) depend on the value of the vector (R). They are accepted only when $R \leq 1$ but are rejected if $R > 1$. When we accept the random numbers, then the unbiased random vector u_i is represented by Eq. (5.5).

5. Calculate the new vector as $X = X_1 + \lambda u$ and the corresponding value of the function $f = f(X)$.
6. Then the values of f_1 and f are compared. When $f_1 > f$, set the new values as $X_1 = X$ and $f_1 = f$ and go to step 3. If $f_1 \leq f$, move to step 7.
7. If $i \leq N$, update the new iteration number as $i = i + 1$ and go to step 4. Conversely, if $i > N$, move to step 8.
8. Calculate the reduced new step length as $\lambda = \lambda/2$. Whenever the new step length $\lambda \leq \varepsilon$, move to step 9; if not (i.e., if the new step length $\lambda > \varepsilon$), go to step 4.
9. Terminate the process by considering $X_{opt} = X_1$ and $f_{opt} = f_1$.

5.3.2 Grid search method

Grid search process includes mainly three steps: constructing a proper grid within the design space, estimating values of the objective function at all these grid points, and finally locating the grid point with the lowest function value (highest function value for maximization problem).

Consider for the ith design variable, the lower and upper bounds are represented by l_i and u_i, respectively. Split the entire range $[l_i, u_i]$ into $p_i - 1$ equal parts so that $x_i^{(1)}, x_i^{(2)}, \ldots x_i^{(p_i)}$ indicate the grid points along the x_i axis ($i = 1, 2, \ldots, n$), where n is the number of variables. This gives us a total $p_1 \times p_2 \times \cdots \times p_n$ number of grid points within the design space. Figure 5.2 shows a grid with $p_i = 7$ for a two-dimensional design space.

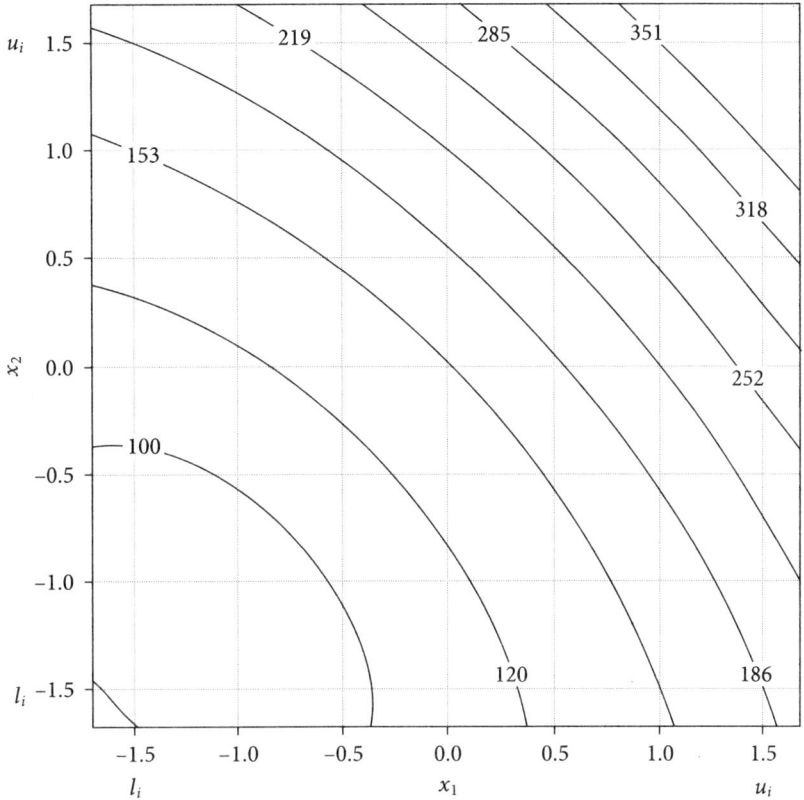

Fig. 5.2 Contour representation for grid search method

We can select the grid points based on methods of experimental design [Montgomery, (2008)]. It can be noticed that in most practical problems the grid search method needs a huge amount of function evaluations. For instance, consider a problem having 8 design variables ($n = 8$), the number of grid points will be $4^8 = 65536$ with $p_i = 4$ and $5^8 = 390625$ with $p_i = 5$. However, the grid search method can be applied easily to locate an approximate minimum for problems that have small number of design variables. In addition, the grid search method can be employed to establish an appropriate starting point for other more effective methods such as Newton's method. Details of this method with examples have been discussed in chapter 11.

Example 5.1

Experimental results show that the Cr(VI) removal by Powder Activated Carbon (PAC) at pH 4, follows the relation

Cr(VI) removal(%) = $77.92 + 9.41C + 3.86t - 3.53C^2 - 7.33t^2$ [Dutta *et al.* (2011)]

Find the optimum value of time and PAC concentration which will show maximum Cr(VI) removal.

Here, we solve the problem using grid search method. This is a two dimensional problem (time and PAC concentration as shown in Fig. 5.3) with upper limit and lower limit as follows:

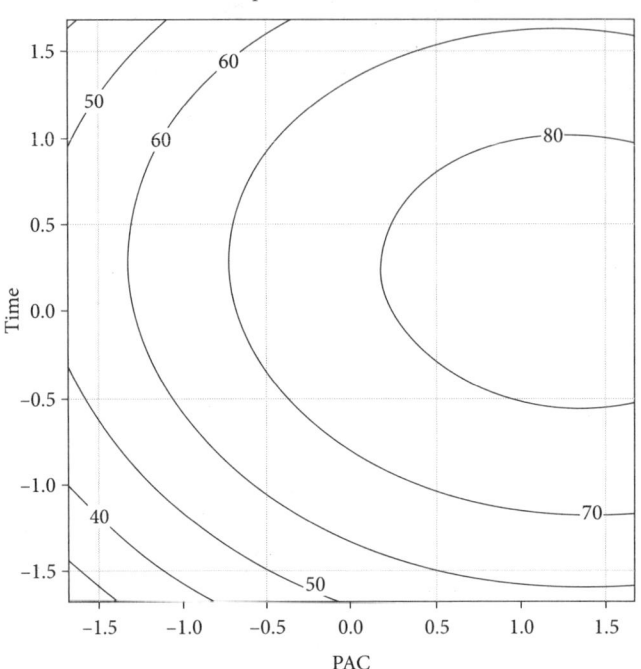

Fig. 5.3 Contour representation for Cr(VI) removal

Say PAC concentration (C) is denoted as variable 1 and time (t) is variable 2;

$l_1 = -1.5$, $u_1 = 1.5$ (Coded value)

$l_2 = -1.5$, $u_2 = 1.5$ (Coded value)

here, we consider $p_1 = 4$, $p_2 = 4$; therefore, we have $p_1 \times p_2 = 4 \times 4 = 16$ grid points.

from the contour plot it is clear that we have to find the function values of these 16 points.

The value of grid points are given in the Table 5.1. Study of various grid points show that the maximum function value is found at (1.5, 0.5) point. The corresponding function value is 84.19

Table 5.1 Value of objective function at different grid points

Grid point	Cr(VI) removal (%)	Grid point	Cr(VI) removal (%)	Grid point	Cr(VI) removal (%)	Grid point	Cr(VI) removal (%)
(−1.5, 1.5)	45.16	(−0.5, 1.5)	61.63	(0.5, 1.5)	71.04	(1.5, 1.5)	73.49
(−1.5, 0.5)	55.96	(−0.5, 0.5)	72.43	(0.5, 0.5)	81.84	(1.5, 0.5)	84.19
(−1.5, −0.5)	52.1	(−0.5, −0.5)	68.57	(0.5, −0.5)	77.98	(1.5, −0.5)	80.33
(−1.5, −1.5)	33.58	(−0.5, −1.5)	50.05	(0.5, −1.5)	59.46	(1.5, −1.5)	61.81

5.3.3 Univariate method

The univariate method generates trial solution for one decision variable keeping all others fixed. By this way, the best solution for each of the decision variables keeping others constant is obtained. The whole process is repeated iteratively until it converges. This is the easiest methods used to minimize functions of n variables where we find the minimum of the objective function by varying only one variable at a time, while holding all other variables fixed. Therefore, we are converting the process a one-dimensional minimization along each of the coordinate directions of an n-dimensional design space. This method is called the univariate search method.

Univariate method generates trial solution for one decision variable keeping all others fixed. Consider the ith iteration where we start at a base point X_i and vary that variable keeping the values of $n - 1$ variables fixed. The problem turns into a one-dimensional minimization problem as we are changing only one variable. Therefore, we can use any of the methods discussed in Chapter 3 to generate a new base point X_{i+1}. Best solutions for each of the variables keeping others constant are obtained. Hereafter, the search is extended in a new direction. This search in new direction is accomplished by varying any one of the remaining $n - 1$ variables that were fixed in the preceding iteration. In practice, this search process is continued by considering each coordinate direction one by one. The first cycle is completed after all the n directions are searched sequentially, and therefore, we repeat the whole process of sequential minimization [Rao, (2009)]. The whole process is repeated iteratively until convergence.

Theoretically, the univariate method is applicable for finding the minimum of any function that has continuous derivatives. However, the method may not even converge if the function possesses a steep valley. This process is very efficient for a quadratic function of the form

$$f(X) = \sum_{i=1}^{n} c_i x_i^2 \tag{5.7}$$

as the search directions line up with the principle axes as shown in Fig. 5.4a. However, this method does not work satisfactorily for more general form of quadratic objective functions as shown

$$f(X) = \sum_{i=1}^{n} \sum_{j=1}^{n} d_{ij} x_i x_j \tag{5.8}$$

For the later case as shown in Fig. (5.4b), the changes in X decreases gradually as it reaches close to the optimum point, so large number of iterations will be required to achieve high accuracy [Edgar et al. (2001)].

The univariate method is quite simple and can be performed easily. However, it does not converge quickly to an optimum solution because it has an oscillating tendency that progressively declines toward the optimum point. Therefore, it will be excellent if we stop the calculations at some point close to the optimum point instead of trying to get the exact optimum point [Rao, (2009)].

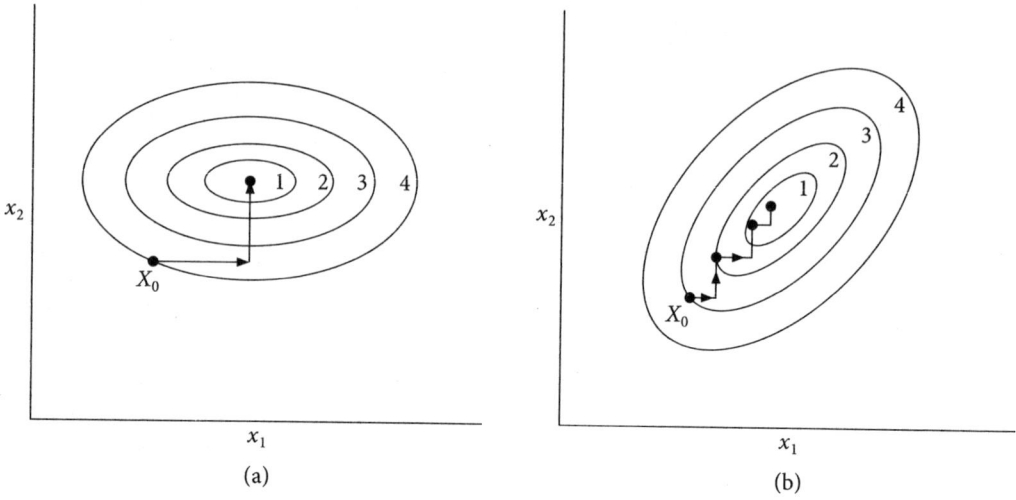

Fig. 5.4 Contour for quadratic function

5.3.4 Pattern search methods

In the univariate method, the search process is carried out for the minimum point along directions parallel to the coordinate axes. It is observed that this process might not converge in some circumstances and that even if the process converges, it will become very sluggish as it move toward the optimum point. We can avoid these difficulties by changing the search directions in a convenient way rather than keeping them constantly parallel to the coordinate axes. These changing directions are known as pattern directions. This process of going from a given point to the subsequent point is called '*move*'. A *move* is considered as success if the value of $f(X)$ decreases; otherwise, it is a

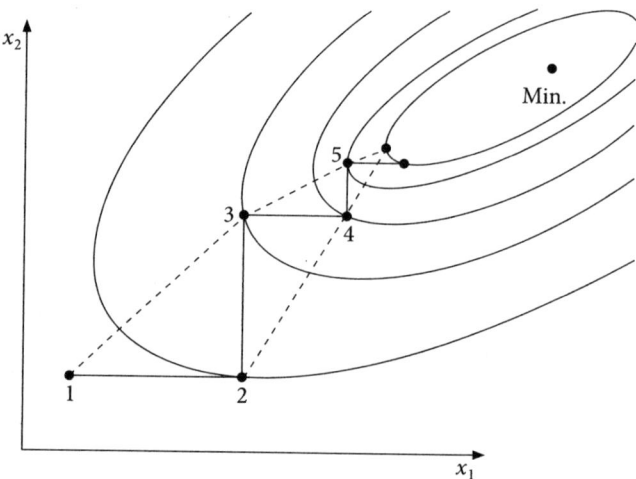

Fig. 5.5 Pattern search method

failure [Hooke and Jeeve (1960)]. It can be shown that for a quadratic objective function with two variables, all of these lines pass through a minimum. Unfortunately, this feature will not be true for functions with multiple variables, even if they are quadratics in nature. However, this concept can still be employed to accomplish a rapid convergence during the minimization of function with n-variable.

In Fig. 5.5, the move direction $1 \to 2 \to 3 \to 4 \to 5$ is univariate move whereas move direction $1 \to 3 \to 5$ (dotted line) is called pattern direction. Techniques that employ pattern directions as search directions are called the pattern search methods.

5.3.4.1 Powell's method

In the year 1964, Powell [Powell (1964)] delineated a method for finding out the minimizers without calculating derivatives. This method depends upon the characteristics of conjugate directions defined by a quadratic function. Powell's method can be explained as an extension of the original pattern search technique. This is one of the most extensively used direct search method and can be shown to be a method of conjugate directions. A conjugate directions method takes a finite number of steps to minimize a quadratic function. A conjugate directions method is supposed to accelerate the convergence of even general nonlinear objective functions because, near its minimum point, a general nonlinear function can be fairly well approximated by a quadratic function. The process of generating the conjugate directions, and the property of quadratic convergence are given in the following section.

To understand the Powell's pattern search method, we should first be well aware of the two important concepts, the conjugate directions and quadratic convergence.

A particular way of achieving quadratic termination is to call upon the idea of conjugacy of a set of non-zero vectors $s_1, s_2, \ldots, s_n$ to a specified positive definite matrix H. The property can be represented as

$$(s_i)^T H s_j = 0 \quad \forall i \neq j \tag{5.9}$$

A conjugate direction method is one that produces this kind of directions when it is employed to a quadratic function with Hessian H.

Theorem 5.1

A conjugate direction method terminates for a quadratic function in at most n exact line searches, and each X_{k+1} is the minimizer in the subspace generated by X_1 and the directions $s_1, s_2, \ldots, s_n$ (that is the set of points $\left\{ X \mid X = X_1 + \sum_{j=1}^{k} \alpha_j s_j \forall \alpha_j \right\}$).

The proof of this theorem is beyond the scope of this book. Readers may follow [Fletcher, 1987] for proof of this theorem.

Example 5.2

Consider the objective function:

$$f(X) = 2x_1^2 + x_2^2 - 5 \tag{5.10}$$

For minimizing the objective function, find the conjugate direction to the initial direction s_0. The starting point is $X_0 = [1 \ 1]^T$ with the initial direction being $s_0 = [-4 \ -2]^T$

Solution

The given values are $X_0 = [1 \ 1]^T$ and $s_0 = [-4 \ -2]^T$; therefore,

$$s_0 = -\begin{bmatrix} 4 \\ 2 \end{bmatrix} \text{ and } H_0 = \begin{bmatrix} 4 & 0 \\ 0 & 2 \end{bmatrix}$$

we have to solve Eq. (5.9) for $s_1 = \begin{bmatrix} s_1^1 & s_1^2 \end{bmatrix}^T$ with $s_0 = \begin{bmatrix} -4 & -2 \end{bmatrix}^T$. The equation is given below

$$(s_i)^T H s_j = 0 \quad 0 \leq i \neq j \leq n-1$$

substituting, the corresponding values we have

$$\begin{bmatrix} -4 & -2 \end{bmatrix} \begin{bmatrix} 4 & 0 \\ 0 & 2 \end{bmatrix} \begin{bmatrix} s_1^1 \\ s_1^2 \end{bmatrix} = 0$$

Because s_1^1 is not unique, we can choose $s_1^1 = 1$ and determine s_1^2 from above correlation

$$\begin{bmatrix} -16 & -4 \end{bmatrix} \begin{bmatrix} s_1^1 \\ s_1^2 \end{bmatrix} = 0$$

Thus, $s_1 = [1 \ -4]^T$ is a direction conjugate to $s_0 = [-4 \ -2]^T$.

We can achieve the minimum of $f(X)$ in two stages by utilizing first s_0 and then s_1.

5.3.4.2 Hooke–Jeeves method

Hooke–Jeeves method finds the minimum of a multivariable, unconstrained, nonlinear function. This is simple and fast optimizing method, which follows the 'hill climbing' technique. The algorithm takes a step in different 'directions' from the initial starting point, and performs a new model run. Then, the algorithm uses the new point as its best guess if the new likelihood score is better than the old one. When it is worse then, the algorithm remains unchanged to the old point. The search process continues in a series of these steps, each step slightly smaller than the previous one. When the algorithm finds a point, and any improvement is not possible with a small step in any direction, then this point is accepted as being the 'solution', and exits.

Hooke–Jeeves method consists of two main routines namely the "exploratory search" routine and the "pattern move" routine. The exploratory routine search the local proximity in the directions parallel to the coordinate axes for an improved objective function value. The pattern routine accelerates the search by moving to a new improved position in the direction of the previous optimal point obtained by the exploratory routine. A set of search direction is generated iteratively in the pattern search method. The main criterion for generating search directions is that they should traverse the search space completely. The search direction should be such that starting from one point in the search space, we can reach any other point in the search space by traversing along these search directions only. For any N-dimensional problem, as a minimum N number of linearly independent search directions are required. For instance, at least two search directions are necessary to move from any one point to any other point in a two-variable function. There are many probable combinations of N search directions, a few of them may be able to attain the goal faster (with lesser number of iterations), and some of them may need more iterations. An exploratory move is conducted systematically in the neighborhood of the current point to get the best point around the current point. Subsequently, two such points are used to make a pattern move [Deb, (2009)].

Exploratory move:

The base point or the current solution is represented by X^c. Consider that the variable X_i^c is perturbed by Δ_i. Set $i = 1$ and $X = X^c$.

Step 1   Compute $f = f(X)$, $f^+ = f(X_i + \Delta_i)$ and $f^- = f(X_i - \Delta_i)$.

Step 2   Find $f_{min} = \min(f, f^+, f^-)$. Then, set X corresponds to f_{min}.

Step 3   If $i = N$ go to step 4; else set $i = i + 1$ and go to Step 1.

Step 4   If $X \neq X^c$, success; Else failure.

During the exploratory move, the current point is perturbed in both negative and positive directions along each variable one at a time and the best point is recorded. At the end of each variable perturbation, the current position is changed to the best point. When the point obtained at the end of all variable perturbations is different from the original point then the exploratory move is a successful one; otherwise, the exploratory move is a failure. In any circumstance, the best point is believed to be the result of the exploratory move.

Pattern move:

A new point is obtained by jumping from the current best point X^c along a direction connecting the previous best point $X^{(k-1)}$ and the current base point $X^{(k)}$ as follows:

$$X_p^{(k-1)} = X^{(k)} + \left(X^{(k)} - X^{(k-1)}\right) \tag{5.11}$$

The Hooke–Jeeves method involves an iterative application of an exploratory move in the vicinity of the current point and a subsequent jump using the pattern move. If the pattern move does not take the solution to a better region, the pattern move is not accepted and the extent of the exploratory search is reduced. The algorithm works as follows:

Step 1   Choose the following parameters: the starting point $X^{(0)}$, variable increments $\Delta_i (i = 1, 2, ..., N)$, a step reduction factor $\alpha > 1$, and a termination parameter, ε. Set $k = 0$.

Step 2   Perform an exploratory move considering $X^{(k)}$ as the base point. Say X is the result of the exploratory move. If the exploratory move is success, set $X^{(k+1)} = X$ and move to Step 4;
   Else, move to Step 3.

Step 3   If $\|\Delta\| < \varepsilon$, Terminate;
   Else, set $\Delta_i = \dfrac{\Delta_i}{\alpha}$ for $i = 1, 2, ..., N$ and go to Step 2.

Step 4 Set $k = k + 1$ and conduct the pattern move $X_p^{(k-1)} = X^{(k)} + \left(X^{(k)} - X^{(k-1)}\right)$.

Step 5 Conduct another exploratory move with $X_p^{(k-1)}$ as the base point. Let the result be $X^{(k-1)}$.

Step: Is $f\left(X^{(k+1)}\right) < f\left(X^{(k)}\right)$? If yes, go to Step 4;
Else go to Step 3.

The search strategy is simple and straightforward. The algorithm requires less storage for variables; only two points ($X^{(k)}$ and $X^{(k-1)}$) need to be stored at any iteration. The numerical calculations involved in the process are also simple. However, since the search largely depends on the moves along the coordinate directions (X_1, X_2, and so on) during the exploratory move, the algorithm may prematurely converge to a wrong solution, especially in the case of functions with highly nonlinear interactions among variables. The algorithm may also get stuck in the loop of generating exploratory moves either between Steps 5 and 6 or between Steps 2 and 3. Another feature of this algorithm is that it terminates only by exhaustively searching the vicinity of the converged point. This requires a large number of function evaluations for convergence to a solution with a reasonable degree of accuracy. The convergence to the optimum point depends on the parameter α. A value $\alpha = 2$ is recommended.

The pattern search method is clearly a simple strategy which is easily programmed and which require very little computer storage, and it has been found to be extremely useful in a wide variety of applications ranging from curve fitting to the on-line performance optimization of chemical processes [Swann, (1972)].

The working principle of the algorithm can be better understood through the following hand-simulation on a numerical exercise problem.

Example 5.3

Show the exploratory move for the function given below

$$\min F(X) = 1200\left(2x_1x_3 + 2x_2x_3\right) + 2500x_1x_2 + 500\left(\frac{1000}{x_1x_2x_3}\right) + 100\left(\frac{1000}{10x_1x_2x_3}\right) \quad (5.12)$$

Solution

The value represented here is $f(X) = F(X)/1000$

Set iteration number $i = 1$

Consider the current point $X^{(1)} = [1 \ 1 \ 1]^T$ and $f^{(1)} = 517.3$

The increment vector $\Delta_1 = [0.2, 0.2, 0.2]^T$

now $f^+ = f(1.2, 1.2, 1.2) = 305.651$

and $f^- = f(0.8, 0.8, 0.8) = 1000.766$

$f_{\min} = \min(f^+, f, f^-) = 305.3$

Set $X^{(2)} = [1.2 \ 1.2 \ 1.2]^T$ for next iteration $i = 2$

Set iteration number $i = 2$

for this iteration $f = f(1.2, 1.2, 1.2) = 305.651$

Calculate $\quad f^+ = f(1.4, 1.4, 1.4) = 200.167$

$\quad\quad\quad\quad\quad f^- = f(1.0, 1.0, 1.0) = 517.3$

Therefore, $\quad f_{min} = (f^+, f, f^-) = 200.167$

Note For each iteration, we need to calculate the objective function value only at one point; other two values are stored in the previous iteration.

5.4 Gradient Search Method

Gradient search methods require the gradient vector of the objective function. Consider a function $f(X)$ where $X = [x_1, x_2, \ldots x_n]^T$. The gradient vector of $f(X)$ is given by the partial derivatives with respect to each of the independent variables

$$\nabla f(X) \equiv g(X) \equiv \left[\frac{\partial f}{\partial x_1} \ \frac{\partial f}{\partial x_2} \ \cdots \ \frac{\partial f}{\partial x_n} \right]^T \tag{5.13}$$

The gradient possess a very useful property. In an n-dimensional space if we move from any point along the gradient direction, the value of $f(X)$ increases at the fastest rate. Therefore, the gradient direction is known as the direction of steepest ascent. Unfortunately, this steepest ascent direction is not a global property and it is a local one as the shape of the contours is not uniform.

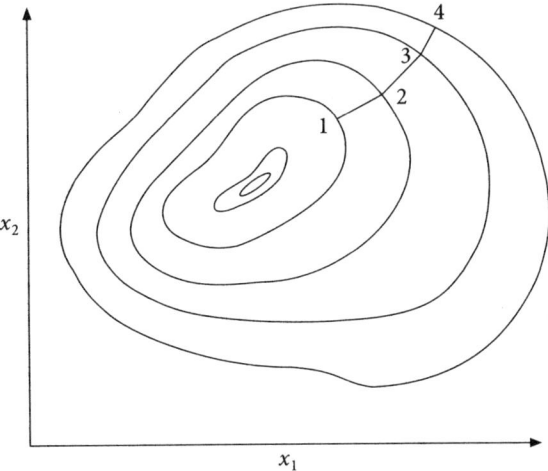

Fig. 5.6 Gradient search method

Figure 5.6 shows that the direction of search changes from point to point. At point 1, steepest direction is 1–2. Similarly, at point 2 and 3, steepest directions are 2–3 and 3–4 respectively.

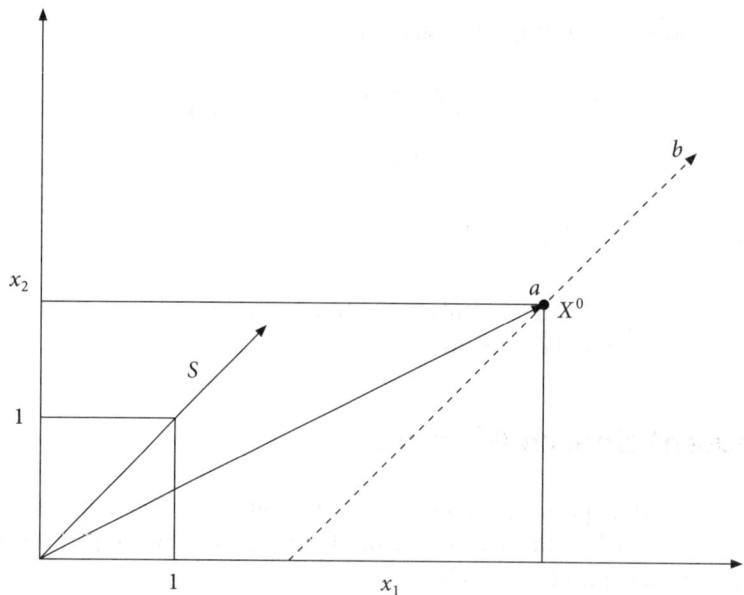

Fig. 5.7 Direction of movement of any point

Figure 5.7 explains how any point moves from a particular point X^0 with a direction s. Here, the line "$a - b$" starts from point a (X^0) and moves with a direction $s = [1, 1]$.

5.4.1 Steepest descent (Cauchy) method

Cauchy in 1847 [Cauchy, 1847] first utilized the negative of the gradient vector ($-g(X)$) as a the direction for minimization problem. The gradient is the vector at any point X, which gives the direction (local) of the greatest rate of change of $f(X)$. It is orthogonal to the contour of $f(X)$ at X. Steepest descent method starts from an initial trial point X_1 and move iteratively along the directions of steepest descent until the optimum point is reached. The steepest descent method can be described by the following steps:

Let $f : \mathbb{R}^n \to \mathbb{R}$ be real-valued function of n variables:

1. Start from an arbitrary initial point X_1. Set the iteration number as $i = 1$.
2. Find the direction of search s_i as

$$s_i = -\nabla f_i = -\nabla f(X_i) \qquad (5.14)$$

3. Calculate λ_i^*, the optimal step length in the direction s_i and set

$$X_{i+1} = X_i + \lambda_i^* s_i = X_i - \lambda_i^* \nabla f_i \qquad (5.15)$$

here, λ_i^* determines how far to go in the given direction

4. Check the new point, X_{i+1}, for optimality. If X_{i+1} is optimum, stop the search process. If not, move to step 5.
5. Set the new iteration number $i = i + 1$ and go to step 2.

This process may seem to be the best unconstrained minimization method as all one-dimensional search start in the "best" direction. However, this process is not really effective in most problems as the steepest descent direction is a local property [Rao, (2009)]. Another drawback of this method is the convergence of this method largely depends on the nature of the objective function. The iterate can oscillate back and forth in a zigzag manner, making the process very slow.

Convergence Criteria:

The following criteria can be applied to terminate the iterative process.

1. If the change of the function value in two successive iterations is very small:

$$\left|\frac{f(X_{i+1}) - f(X_i)}{f(X_i)}\right| \leq \varepsilon_1 \tag{5.16}$$

2. If the values of the partial derivatives (components of the gradient) of the function f are very small:

$$\left|\frac{\partial f}{\partial X_i}\right| \leq \varepsilon_2 \quad i = 1, 2, \ldots, n \tag{5.17}$$

3. If the change in the design vector in two successive iterations is very small:

$$|X_{i+1} - X_i| \leq \varepsilon_3 \tag{5.18}$$

Here, $\varepsilon_1, \varepsilon_2$ and ε_3 are very small numbers.

Example 5.4

Find the optimum of the following function using Cauchy's method.

$$f(X) = 9x_1^2 + 4x_1x_2 + 7x_2^2$$

Solution

The elements of the gradient are

$$\frac{\partial f}{\partial x_1} = 18x_1 + 4x_2 \quad \text{and} \quad \frac{\partial f}{\partial x_2} = 4x_1 + 14x_2$$

Now we will employ the steepest descent method to get the solution

Consider $X_0 = [1\ 1]^T$, then

$$f(X_0) = 20 \text{ and}$$

$$\nabla f(X_0) = \begin{bmatrix} 22 & 18 \end{bmatrix}^T$$

from Eq. (5.15), we have

$$X_1 = X_0 - \lambda_0^* \nabla f(X_0)$$

or $\begin{bmatrix} x_1 \\ x_2 \end{bmatrix} = \begin{bmatrix} 1 \\ 1 \end{bmatrix} - \lambda_0^* \begin{bmatrix} 22 \\ 18 \end{bmatrix}$

now we have to find λ_0^* such that $f(X_1)$ will be minimum.

If we substitute value of X_1 in equation

$$f(X) = 9x_1^2 + 4x_1 x_2 + 7x_2^2$$

we have $f(\lambda_0^*) = 20 - 808\lambda_0^* + 8208(\lambda_0^*)^2$

The optimum value of λ_0^* is 0.05 when $\dfrac{df(\lambda_0^*)}{d\lambda_0^*} = 0$

therefore, $X_1 = \begin{bmatrix} x_1 \\ x_2 \end{bmatrix} = \begin{bmatrix} 1 \\ 1 \end{bmatrix} - 0.05 \begin{bmatrix} 22 \\ 18 \end{bmatrix} = \begin{bmatrix} -0.1 \\ 0.1 \end{bmatrix}$ and $f(X_1) = 0.12$

we will reach the optimum point following this iterative method.

5.4.2 Conjugate gradient (Fletcher–Reeves) method

In view of Theorem 5.1, which equates conjugacy and exact line searches with quadratic termination, it is attractive to try to associate conjugacy properties with the steepest descent method in an attempt to achieve both efficiency and reliability. This is the aim of conjugate gradient method [R. Fletcher, (1987)].

The convergence characteristics of the steepest descent method can be greatly improved by modifying this into a conjugate gradient method. We can also consider this method as a conjugate directions method including the application of the gradient of the function [Rao, (2009)]. In the previous discussion, we have seen that a minimization method is quadratically convergent when it uses the conjugate directions. This characteristic of quadratic convergence is quite helpful since it gives assurance that this process will minimize a quadratic function in n steps or less. Every quadratically convergent method is supposed to achieve the optimum point in a finite number of iterations because near the optimum point any general function can be well approximated by a quadratic function. We have noticed that Powell's conjugate direction method performs n single-variable minimizations per iteration and at the end of each iteration it establishes a new conjugate direction. Therefore, usually, n^2 single-variable minimizations are required to determine the minimum of a quadratic function. In contrast, if we are able to estimate the gradients of the objective function, we can establish a new conjugate direction after each one-dimensional minimization, and consequently a faster convergence can be achieved. The development of conjugate directions and formulation of the Fletcher–Reeves method are explained in the following section.

Development of the Fletcher–Reeves Method

The Fletcher–Reeves method is formulated by performing some modification of the steepest descent method in such a way that it becomes quadratically convergent. The search process starts from an arbitrary point X_1, the quadratic function

$$f(X) = \frac{1}{2} X^T [A] X + B^T X + C \tag{5.19}$$

can be minimized by using the search procedure along the steepest descent search direction $s_1 = -\nabla f_1$ with the step length

$$\lambda_1^* = -\frac{s_1^T \nabla f_1}{s_1^T A s_1} \tag{5.20}$$

A linear combination (Eq. (5.21)) of s_1 and $-\nabla f_2$ is used to find the direction of second search (s_2)

$$s_2 = -\nabla f_2 + \beta_2 s_1 \tag{5.21}$$

where the constant β_2 can be estimated by making s_1 and s_2 conjugate with respect to $[A]$. This lead to the following equation.

$$\beta_2 = -\frac{\nabla f_2^T \nabla f_2}{\nabla f_1^T s_1} = \frac{\nabla f_2^T \nabla f_2}{\nabla f_1^T \nabla f_1} \tag{5.22}$$

This process has been continued to derive the general formula for the ith search direction as

$$s_i = -\nabla f_i + \beta_i s_{i-1} \tag{5.23}$$

where

$$\beta_i = \frac{\nabla f_i^T \nabla f_i}{\nabla f_{i-1}^T \nabla f_{i-1}} \tag{5.24}$$

Therefore, the algorithm of Fletcher–Reeves method can be written as follows.

Algorithm

The algorithm of Fletcher–Reeves method is given below:

1. Start from X_1, an arbitrary initial point.
2. Calculate the first search direction $s_1 = -\nabla f(X_1) = -\nabla f_1$.
3. Find the point X_2 using the relation

$$X_2 = X_1 + \lambda_1^* s_1 \tag{5.25}$$

   where λ_1^* is the optimal step length in the direction s_1. Set $i = 2$ and move to the next step.

4. Determine $\nabla f_i = \nabla f(X_i)$, and set

$$s_i = -\nabla f_i + \frac{|\nabla f_i|^2}{|\nabla f_{i-1}|^2} s_{i-1} \tag{5.26}$$

5. Calculate λ_i^*, the optimum step length in the direction s_i, and locate the new point

$$X_{i+1} = X_i + \lambda_i^* s_i \tag{5.27}$$

6. Check for the optimality of the point X_{i+1}. Stop the process if X_{i+1} is optimum; or else, set $i = i + 1$ and go to step 4.

Remark

This process should converge in n cycles or less for a quadratic function as the directions s_i used in this method are A-conjugate. However, the method may utilize much more than n cycles for convergence when quadratics are ill-conditioned (when contours are highly distorted and eccentric). The cumulative effect of rounding errors has been considered as the main reason for this. Since, s_i is represented by Eq. (5.26), any error emanating from the inaccuracies occurred during the estimation of λ_i^*, and from the round-off error in the process of the accumulation of successive $|\nabla f_i|^2 s_{i-1}/|\nabla f_{i-1}|^2$ terms, is carried forward through the vector s_i. Therefore, these errors progressively contaminate the search directions s_i. Thus in practice, it is required to restart the process periodically after some number of steps; say, m number of steps by considering the new search direction as the steepest descent direction. Therefore, after every m steps, s_{m+1} is replaced by $-\nabla f_{m+1}$ rather than using its usual form (Eq. (5.26)). A value of $m = n + 1$ is suggested by Fletcher and Reeves, where n denotes the number of design variables.

5.4.3 Newton's method

Newton's method can be used efficiently to minimize the multivariable functions. For this purpose, we have to consider a quadratic approximation of the function $f(X)$ at $X = X_i$ by means of the Taylor's series expansion

$$f(X) \approx f(X_i) + \nabla f_i^T (X - X_i) + \frac{1}{2}(X - X_i)^T [H_i](X - X_i) \tag{5.28}$$

where $[H_i] = [H]|_{X_i}$ is called the Hessian matrix (matrix of second partial derivatives) of f estimated at the point X_i. For the minimum of $f(X)$, the partial derivatives of Eq. (5.28) have been set equal to zero

$$\frac{\partial f(X)}{\partial X_j} = 0, \quad j = 1, 2, \ldots, n \tag{5.29}$$

Form Eqs (5.28) and (5.29), we get

$$\nabla f = \nabla f_i + [H_i](X - X_i) = 0 \tag{5.30}$$

If $[H_i]$ is nonsingular, Eq. (5.30) can be solved to get an improved approximation ($X = X_{i+1}$) as

$$X_{i+1} = X_i - [H_i]^{-1} \nabla f_i \tag{5.31}$$

Since, the terms of higher-order have been neglected in Eq. (5.28). The Eq. (5.31) should be employed iteratively to get the optimum solution X^*. Provided that $[H_1]$ is nonsingular, it can be shown that the series of points $X_1, X_2, \ldots, X_{i+1}$ converge to the actual solution X^* starting from any initial point X_1 that is satisfactorily close to the solution X^*. It can be noticed that Newton's method is a second-order method since the second partial derivatives of the objective function is utilized in the form of the matrix $[H_i]$.

Example 5.5

Minimize the function using Newton's method

$$f(X) = 5x_1^2 + x_2^2 - 3x_1 x_2$$

Starting point $X_0 = [1 \ 1]^T$

Solution

Calculate the value of

$$\nabla f(X) = \begin{bmatrix} 10x_1 - 3x_2 \\ 2x_2 - 3x_1 \end{bmatrix} \quad \nabla f(X_0) = \begin{bmatrix} 7 \\ -1 \end{bmatrix}$$

and

$$H(X) = \begin{bmatrix} 10 & -3 \\ -3 & 2 \end{bmatrix} \quad H^{-1}(X) = \begin{bmatrix} 0.1818 & 0.2727 \\ 0.2727 & 0.9091 \end{bmatrix}$$

from Eq. (5.31)

$$X_1 = X_0 - [H_0]^{-1} \nabla f_0$$

$$X_1 = \begin{bmatrix} 1 & 1 \end{bmatrix}^T - \begin{bmatrix} 0.1818 & 0.2727 \\ 0.2727 & 0.9091 \end{bmatrix} \begin{bmatrix} 7 \\ -1 \end{bmatrix}$$

$$X_1 = \begin{bmatrix} 1 & 1 \end{bmatrix}^T - \begin{bmatrix} 1 \\ 1 \end{bmatrix}$$

$$X_1 = \begin{bmatrix} 0 \\ 0 \end{bmatrix}$$

calculate the value of the function at the point X_1

$$f(X_1) = f(X^*) = 0$$

Note The solution is found in a single step using Newton's method

5.4.4 Marquardt method

When the design vector X_i is away from the optimum point X^*, the steepest descent method adequately reduces the value of the function. On the other hand, the Newton method, converges fast when the design vector X_i is very close to the optimum point X^*. The Marquardt method [Marquardt, 1963] tries to take benefit of both the Newton's method and steepest descent. In this method, the diagonal elements of the Hessian matrix, $[H_i]$ have been modified by the equation

$$[\tilde{H}_i] = [H_i] + \alpha_i [I] \tag{5.32}$$

where $[I]$ and α_i represent an identity matrix and a positive constant respectively. The constant α_i ensures the positive definiteness of $[\tilde{H}_i]$ when $[H_i]$ is not positive definite. We can note that while α_i is sufficiently large (on the order of 10^4), the term $\alpha_i [I]$ dominates $[H_i]$ and the inverse of the matrix $[\tilde{H}_i]$ becomes

$$[\tilde{H}_i]^{-1} = [[H_i] + \alpha_i [I]]^{-1} \approx [\alpha_i [I]]^{-1} = \frac{1}{\alpha_i}[I] \tag{5.33}$$

Therefore, if the search direction s_i is calculated as

$$s_i = -[\tilde{H}_i]^{-1} \nabla f_i \tag{5.34}$$

Optimization of Unconstrained Multivariable Functions

For large values of α_i, the direction s_i converts to a direction of steepest descent. In the Marquardt method, the value of α_i is considered to be large during the starting and then decreased gradually to zero as the process proceeds iteratively. Therefore, the characteristics of this search technique change from those of a steepest descent method to those of the Newton method as the value of α_i decreases from a large value to zero. The iterative procedure of the modified version of Marquardt method can be illustrated as given bellow.

Algorithm

1. Start with the following parameters: X_1, an arbitrary initial point; a large constant α_1 (on the order of 10^4), a small constant ε (on the order of 10^{-2}), $c_1 (0 < c_1 < 1)$, and $c_2 (c_2 > 1)$. Set the iteration number as $i = 1$.
2. Calculate the gradient of the function, $\nabla f_i = \nabla f(X_i)$.
3. Check point X_i for optimality. If $\|\nabla f\|_i = \|\nabla f(X_i)\| \leq \varepsilon$, X_i is optimum and thus stop the process. If not, move to step 4.
4. Estimate the new vector X_{i+1} by using the equation

$$X_{i+1} = X_i + s_i = X_i - \left[[H_i] + \alpha_i[I]\right]^{-1} \nabla f_i \tag{5.35}$$

5. Compare the values of f_{i+1} and f_i. If $f_{i+1} < f_i$, move to step 6. If $f_{i+1} \geq f_i$, go to step 7.
6. Set $\alpha_{i+1} = c_1 \alpha_i$, $i = i + 1$, and go to step 2.
7. Set $\alpha_i = c_2 \alpha_i$ and go to step 4.

The main advantage of this technique is the nonexistence of the step size λ_i along the search direction s_i. In fact, the above algorithm can be customized by introducing an optimal step length in Eq. (5.35) as

$$X_{i+1} = X_i + \lambda_i^* s_i = X_i - \lambda_i^* \left[[H_i] + \alpha_i[I]\right]^{-1} \nabla f_i \tag{5.36}$$

where λ_i^* is calculated using any of the one-dimensional search processes explained in Chapter 3.

Example 5.6

Minimize the function $f(X) = x_1 - x_2 + 3x_1^2 + 2x_1 x_2 + x_2^2$

Consider the starting point $X_1 = \begin{bmatrix} 0 \\ 0 \end{bmatrix}$ using Marquardt method with $\alpha_1 = 10^4$, $c_1 = \dfrac{1}{4}$, $c_2 = 2$ and $\varepsilon = 10^{-2}$

Solution

We will start with the first iteration

For $i = 1$:

The value of the function $f_1 = f(X_1) = 0.0$ and

$$\nabla f_1 = \begin{bmatrix} \dfrac{\partial f}{\partial x_1} \\ \dfrac{\partial f}{\partial x_2} \end{bmatrix}_{(0,0)} = \begin{bmatrix} 1+6x_1+2x_2 \\ -1+2x_1+2x_2 \end{bmatrix}_{(0,0)} = \begin{bmatrix} 1 \\ -1 \end{bmatrix}$$

Since, $\|\nabla f_1\| = 1.4142 > \varepsilon$, we compute

$$[H_1] = \begin{bmatrix} \dfrac{\partial^2 f}{\partial x_1^2} & \dfrac{\partial^2 f}{\partial x_1 x_2} \\ \dfrac{\partial^2 f}{\partial x_1 x_2} & \dfrac{\partial^2 f}{\partial x_2^2} \end{bmatrix}_{(0,0)} = \begin{bmatrix} 6 & 2 \\ 2 & 2 \end{bmatrix}$$

$$X_2 = X_1 - \left[[H_1] + \alpha_1[I]\right]^{-1} \nabla f_1$$

$$= \begin{bmatrix} 0 \\ 0 \end{bmatrix} - \begin{bmatrix} 6+10^4 & 2 \\ 2 & 2+10^4 \end{bmatrix}^{-1} \begin{bmatrix} 1 \\ -1 \end{bmatrix} = \begin{bmatrix} -0.9996 \\ 1.0000 \end{bmatrix} \times 10^{-4}$$

as $f_2 = f(X_2) = -1.9994 \times 10^{-4} < f_1$, we set $\alpha_2 = c_1\alpha_1 = 2500$, $i = 2$, and proceed to the next iteration.

Iteration 2:

The gradient vector corresponding to X_2 is represented by

$$\nabla f_2 = \begin{bmatrix} \dfrac{\partial f}{\partial x_1} \\ \dfrac{\partial f}{\partial x_2} \end{bmatrix}_{(-0.9996\times 10^{-4},\, 1.0000\times 10^{-4})} = \begin{bmatrix} 1+6x_1+2x_2 \\ -1+2x_1+2x_2 \end{bmatrix}_{(-0.9996\times 10^{-4},\, 1.0000\times 10^{-4})} = \begin{bmatrix} 0.9996 \\ -1.0000 \end{bmatrix}$$

$\|\nabla f_2\| = 1.4139 > \varepsilon$, hence, we compute

$$X_3 = X_2 - \left[[H_2] + \alpha_2[I]\right]^{-1} \nabla f_2$$

$$= \begin{bmatrix} -0.9996 \times 10^{-4} \\ 1.0000 \times 10^{-4} \end{bmatrix} - \begin{bmatrix} 2506 & 2 \\ 2 & 2502 \end{bmatrix}^{-1} \begin{bmatrix} 0.9996 \\ -1.0000 \end{bmatrix}$$

$$= \begin{bmatrix} -4.992 \times 10^{-4} \\ 5.000 \times 10^{-4} \end{bmatrix}$$

since, $f_3 = f(X_3) = -0.9987 \times 10^{-3} < f_2$, we set $\alpha_3 = c_1 \alpha_2 = 625$, $i = 3$, and proceed to the next iteration. The iteration process is to be continued until it satisfy the convergence criteria $\|\nabla f_i\| < \varepsilon$.

5.4.5 Quasi-Newton method

The Newton's method uses the basic iterative process that is given by Eq. (5.31):

$$X_{i+1} = X_i - [H_i]^{-1} \nabla f(X_i) \tag{5.37}$$

where $[H_i]$ is the Hessian matrix that consist of the second partial derivatives of the function f and varies with the design vector X_i for a non-quadratic (common nonlinear) objective function f. The fundamental concept used in the variable metric or Quasi-Newton methods is to approximate either $[H_i]$ by an another matrix $[A_i]$ or $[H_i]^{-1}$ by an another matrix $[B_i]$, considering only the first partial derivatives of f. If $[H_i]^{-1}$ is approximated by $[B_i]$, Eq. (5.37) can be represented as

$$X_{i+1} = X_i - \lambda_i^* [B_i] \nabla f(X_i) \tag{5.38}$$

where λ_i^* is considered as the optimal step length along the direction

$$s_i = -[B_i] \nabla f(X_i) \tag{5.39}$$

It can be observed that the steepest descent direction method can be shown as a special case of Eq. (5.39) by adjusting $[B_i] = [I]$.

Calculation of $[B_i]$

To execute the iteration process given by Eq. (5.38), an approximate inverse of the Hessian matrix, $[B_i] \equiv [A_i]^{-1}$, is to be calculated. For this purpose, first we expand the gradient of the function f about an arbitrary reference point (X_0) using Taylor's series as

$$\nabla f(X) \approx \nabla f(X_0) + [H_0](X - X_0) \tag{5.40}$$

If we select any two points X_i and X_{i+1} and use $[A_i]$ to approximate $[H_0]$, Eq. (5.40) can be written as

$$\nabla f_{i+1} = \nabla f(X_0) + [A_i](X_{i+1} - X_0) \tag{5.41}$$

$$\nabla f_i = \nabla f(X_0) + [A_i](X_i - X_0) \tag{5.42}$$

Subtracting Eq. (5.42) from Eq. (5.41) yields

$$g_i = [A_i]d_i \qquad (5.43)$$

where

$$d_i = X_{i+1} - X_i \qquad (5.44)$$

$$g_i = \nabla f_{i+1} - \nabla f_i \qquad (5.45)$$

The solution of Eq. (5.43) for d_i can be given as

$$d_i = [B_i]g_i \qquad (5.46)$$

where the inverse of the Hessian matrix, $[H_0]^{-1}$ is approximated by $[B_i] \equiv [A_i]^{-1}$. It can be noticed that Eq. (5.46) represents a system of n equations with n^2 unknown elements of the matrix $[B_i]$. So for $n > 1$, the selection of $[B_i]$ is not unique and we have to select $[B_i]$ that is very close to $[H_0]^{-1}$, in some sense. Various number of procedures have been recommended in the literature for the calculation of $[B_i]$. Once $[B_i]$ is known, the value of $[B_{i+1}]$ is computed using the iterative process. The symmetry and positive definiteness of the matrix $[B_i]$ is to be maintained in addition to satisfying Eq. (5.46), which is a major concern for this method. If $[B_i]$ is symmetric and positive definite, the $[B_{i+1}]$ must remain symmetric and positive definite.

Rank 1 Updates

The matrix $[B_i]$ can be updated by using the general formula given as

$$[B_{i+1}] = [B_i] + [\Delta B_i] \qquad (5.47)$$

where $[\Delta B_i]$ is considered as the correction or update matrix added to $[B_i]$. Theoretically it can be considered that the matrix $[B_i]$ have the rank as high as n. However, in practical application, most updates $[\Delta B_i]$ are only of either rank 1 or rank 2. To develop a rank 1 update, simply we select a scaled outer product of a vector z for $[\Delta B_i]$ as

$$[\Delta B_i] = czz^T \qquad (5.48)$$

where the n-component vector z and constant c the are to be estimated. Combination of Eqs (5.47) and (5.48) gives us

$$[B_{i+1}] = [B_i] + czz^T \qquad (5.49)$$

If Eq. (5.49) is forced to satisfy the Quasi-Newton condition, we have

$$d_i = [B_{i+1}]g_i \qquad (5.50)$$

we get

$$d_i = \left([B_i] + czz^T\right)g_i = [B_i]g_i + cz\left(z^T g_i\right) \tag{5.51}$$

since, $\left(z^T g_i\right)$ in Eq. (5.51) is a scalar, Eq. (5.51) can rewritten as

$$cz = \frac{d_i - [B_i]g_i}{z^T g_i} \tag{5.52}$$

Therefore, an easy selection for z and c would be

$$z = d_i - [B_i]g_i \tag{5.53}$$

$$c = \frac{1}{z^T g_i} \tag{5.54}$$

This gives us the unique rank 1 update formula for $[B_{i+1}]$:

$$[B_{i+1}] = [B_i] + [\Delta B_i] \equiv [B_i] + \frac{\left(d_i - [B_i]g_i\right)\left(d_i - [B_i]g_i\right)^T}{\left(d_i - [B_i]g_i\right)^T g_i} \tag{5.55}$$

This formula has been credited to Broyden [Broyden, (1967)]. To employ Eq. (5.55), at the start of the algorithm an initial symmetric positive definite matrix is selected for $[B_i]$. After that, the next point X_2 is calculateed using Eq. (5.38). Then the new matrix $[B_2]$ is calculated using Eq. (5.55) and the new point X_3 is found out from Eq. (5.38). Until the convergence is achieved, this iterative process is continued. Equation (5.55) confirms that whenever $[B_i]$ is symmetric, $[B_{i+1}]$ is also symmetric. Though, there is no assurance that $[B_{i+1}]$ will remain positive definite even if $[B_i]$ is positive definite. This may cause a breakdown of the method, particularly when it is applied for the optimization of non-quadratic functions. It can be confirmed easily that the columns of the matrix $[\Delta B_i]$ described by Eq. (5.55) are multiples of each other. The updating matrix possesses only one independent column and therefore, the rank of the matrix will be 1. For this reason, the Eq. (5.55) is considered to be a rank 1 updating formula. Although the Broyden formula given in Eq. (5.55) is not robust, it has the characteristic of quadratic convergence [Broyden et al. (1975)]. The rank 2 update formulas discussed in the next section give the assurance of both symmetry and positive definiteness of the matrix $[B_{i+1}]$. These formulas are more robust in minimizing common nonlinear functions; therefore, they are favored in practical applications.

Theorem 5.2

If it is well defined, and if $d_1, d_2, ..., d_n$ are independent, then the rank 1 method terminates on a quadratic function in at most $n + 1$ searches, with $B_i + 1 = H^{-1}$.

Rank 2 Updates

In rank 2 updates we decide the update matrix $[\Delta B_i]$ as the sum of two rank 1 updates as

$$[\Delta B_i] = c_1 z_1 z_1^T + c_2 z_2 z_2^T \tag{5.56}$$

where c_1 and c_2 are constants and z_1 and z_2 are the n-component vectors. We have to determine these constants and vectors. Combining Eqs (5.47) and (5.56) we get

$$[B_{i+1}] = [B_i] + c_1 z_1 z_1^T + c_2 z_2 z_2^T \tag{5.57}$$

By Eq. (5.57) is forced to satisfy the Quasi-Newton condition, Eq. (5.50), we get

$$d_i = [B_i]g_i + c_1 z_1 \left(z_1^T g_i\right) + c_2 z_2 \left(z_2^T g_i\right) \tag{5.58}$$

where $\left(z_1^T g_i\right)$ and $\left(z_2^T g_i\right)$ can be identified as scalars. Although the vectors z_1 and z_2 in Eq. (5.58) are not unique, the choices can be made to satisfy Eq. (5.58) are as follows:

$$z_1 = d_i \tag{5.59}$$

$$z_2 = [B_i]g_i \tag{5.60}$$

$$c_1 = \frac{1}{z_1^T g_i} \tag{5.61}$$

$$c_2 = \frac{1}{z_2^T g_i} \tag{5.62}$$

Therefore, the formula for rank 2 update can be stated as

$$[B_{i+1}] = [B_i] + [\Delta B_i] \equiv [B_i] + \frac{d_i d_i^T}{d_i^T g_i} - \frac{([B_i]g_i)([B_i]g_i)^T}{([B_i]g_i)^T g_i} \tag{5.63}$$

This equation is known as the Davidon–Fletcher–Powell (DFP) formula [Davidon, (1959)] [Fletcher and Powell, (1963)]. Since,

$$X_{i+1} = X_i + \lambda_i^* s_i \tag{5.64}$$

where s_i is the search direction, we can rewrite $d_i = X_{i+1} - X_i$ as

$$d_i = \lambda_i^* s_i \tag{5.65}$$

Thus, Eq. (5.63) can be represented as

$$[B_{i+1}] = [B_i] + \frac{\lambda_i^* s_i s_i^T}{s_i^T g_i} - \frac{([B_i]g_i)([B_i]g_i)^T}{([B_i]g_i)^T g_i} \tag{5.66}$$

Equation (5.66) gives the updated value of matrix B.

5.4.6 Broydon–Fletcher–Goldfrab–Shanno method

As discussed previously, the major difference of the BFGS and the DFP methods is that in the BFGS method, the Hessian matrix is updated iteratively instead of the inverse of the Hessian matrix. The following steps can illustrate the BFGS method.

1. Start with X_1, an initial point and a $n \times n$ positive definite symmetric matrix $[B_1]$ as an initial estimation of the inverse Hessian matrix of the function f. We can take $[B_1]$ as the identity matrix $[I]$ without any additional information. Calculate the gradient vector $\nabla f_1 = \nabla f(X_1)$ and set the iteration number as $i = 1$.
2. Calculate ∇f_i, the gradient of the function f at point X_i; and set

$$s_i = -[B_i]\nabla f_i \tag{5.67}$$

3. Find, λ_i^*, the optimal step length in the direction s_i and compute

$$X_{i+1} = X_i + \lambda_i^* s_i \tag{5.68}$$

4. Check the point X_{i+1} for optimality. If $\|\nabla f_{i+1}\| \leq \varepsilon$, where the value of ε is very small, take $X^* \approx X_{i+1}$ and stop the process. Or else, move to step 5.
5. Update the Hessian matrix as

$$[B_{i+1}] = [B_i] + \left(1 + \frac{g_i^T[B_i]g_i}{d_i^T g_i}\right)\frac{d_i d_i^T}{d_i^T g_i} - \frac{d_i g_i^T[B_i]}{d_i^T g_i} - \frac{[B_i]g_i d_i^T}{d_i^T g_i} \tag{5.69}$$

where

$$d_i = X_{i+1} - X_i = \lambda_i^* s_i \tag{5.70}$$

$$g_i = \nabla f_{i+1} - \nabla f_i = \nabla f(X_{i+1}) - \nabla f(X_i) \tag{5.71}$$

6. Set the new iteration number as $i = i + 1$ and go to step 2.

Notes

1. We can consider the BFGS as a conjugate gradient, Quasi-Newton, and variable metric method.
2. The BFGS method can be termed an indirect update method, because the inverse of the Hessian matrix is approximated.
3. The matrix $[B_i]$ maintains its positive definiteness as the value of i increases, if the step lengths λ_i^* are found accurately. Therefore, for real life application, the matrix $[B_i]$ might become indefinite or even singular if λ_i^* are not computed correctly. For this reason, a resetting of the matrix $[B_i]$ to the identity matrix $[I]$ is required periodically. However, practical experience shows that the BFGS method is less sensitive to the errors in λ_i^* compared to the DFP method.
4. It has been found that the BFGS method has super-linear convergence near the point X^* [Dennis and More, (1977)].

5.5 Levenberg–Marquardt Algorithm

Newton's method with a line search algorithm converges quickly when $f(X)$ is convex. The matrix $H(X)$ may not be positive-definite in all places if $f(X)$ is not strictly convex (often it is the case in regions far from the optimum). Therefore, replacing $H(X)$ by another positive-definite matrix is one way to forcing convergence. The Marquardt–Levenberg method is one approach to accomplish this, as illustrated in the subsequent section.

Whenever the Hessian matrix $H(X_k)$ is not positive, the search direction $d_k = H(X_k)^{-1} g_k$ may not point in a descent direction. A simple procedure is followed to make sure that the search direction is a descent direction. The so-called Levenberg–Marquardt modification is introduced to Newton algorithm:

$$X_{k+1} = X_k - \left(H(X_k) + \mu_k I\right)^{-1} g_k \tag{5.72}$$

where $\mu \geq 0$

The fundamental concept of the Levenberg–Marquardt modification is given in this section. Consider H, a symmetric matrix that may not be positive definite. Let the eigenvalues of H are $\lambda_1, \ldots, \lambda_n$ and the corresponding eigenvectors are $v_1, \ldots, v_n$. The eigenvalues $\lambda_1, \ldots, \lambda_n$ are real, but may not all be positive. After that, consider the matrix $G = H + \mu I$, where $\mu \geq 0$. Note that the eigenvalues of G are $\lambda_1 + \mu, \ldots, \lambda_n + \mu$. Indeed,

$$Gv_i = (H + \mu I)v_i \tag{5.73}$$

$$= Hv_i + \mu I v_i \qquad (5.73a)$$

$$= \lambda_i v_i + \mu v_i \qquad (5.73b)$$

$$= (\lambda_i + \mu) v_i \qquad (5.73c)$$

which shows that v_i is also an eigenvector of G with eigenvalue $\lambda_i + \mu$, for all $i = 1,\ldots, n$. Therefore, when the value of μ is sufficiently large, all the eigenvalues of G are positive, and G is positive definite. Consequently, if the parameter μ_k is sufficiently large in the Levenberg–Marquardt modification of Newton's algorithm, then the search direction $d_k = -\left(H(X_k) + \mu_k I\right)^{-1} g_k$ always points in a descent direction. In this case, if we further introduce a step size α_k as illustrated in the preceding section,

$$X_{k+1} = X_k - \alpha_k \left(H(X_k) + \mu_k I\right)^{-1} g_k \qquad (5.74)$$

then it is assured that the descent property holds.

When we consider that $\mu_k \to 0$, the Levenberg–Marquardt modification of Newton's algorithm can be made to move toward the performance of the pure Newton's method. Conversely, by considering $\mu_k \to \infty$, the algorithm becomes a pure gradient method with small step size. In practice, we can start the process with a small μ_k, and then increase the value slowly till we observe that the iteration is descent, that is, $f(X_{k+1}) < f(X_k)$.

Levenberg–Marquardt method, first derived by Levenberg in 1944 [Levenberg, (1944)] and re-derived by Marquardt in 1963 [Marquardt, (1963)], is a method for solving nonlinear equations. This method is often mentioned when the history of trust region algorithms is discussed. The reason is that the technique of trust region is, in some sense, equivalent to that of the Levenberg–Marquardt method.

Consider a system of nonlinear equations

$$f_i(X) = 0, \quad i = 1, 2, \ldots, m \qquad (5.75)$$

where $f_i(X)$, $i = 1, 2, \ldots, m$ are continuous differentiable functions in $\Re^n$. We try to compute a least squre solution, which means that we need to solve the nonlinear least squares problem

$$\min_{X \in \Re^n} \|F(X)\|_2^2 \qquad (5.76)$$

where $F(X) = (f_1(X), \ldots, f_m(X))^T$. The Gauss–Newton method is iterative, and at current iterate X_k, the Gauss-Newton step is

$$d_k = -\left(A(X_k)^T\right) + F(X_k) \qquad (5.76)$$

Summary

- This chapter elucidates various techniques for optimization of multivariable unconstrained problem. Various search methods like direct search method (e.g., random search methods, grid search method, univariate method, and pattern search methods) and gradient search method (e.g., steepest descent (Cauchy) method, conjugate gradient (Fletcher-Reeves) method, Newton's method, Marquardt method, Quasi-Newton method, and BFGS method) has been discussed with examples. A modified form of Marquardt method that is Levenberg–Marquardt algorithm method is also discussed in this chapter.

Review Questions

5.1 Show that one iteration is sufficient to find the minimum of a quadratic function by using the Newton's method.

5.2 Give some examples of unconstrained optimization in the field of chemical engineering.

5.3 Find the of the function

$$f(X) = 2x_1^2 + x_2^2 - 2x_1 x_2$$

using Random Jumping method.

5.4 The dye removal by TiO_2 adsorption (at pH 5.5) is given by (Anupam K. *et al.* 2011)

dye removal = $13.08(TiO_2) + 15.77(Time) - 9.996(TiO_2)^2 - 12.347(Time)^2$

Plot the contour and find the maximum dye removal using Grid Search method.

5.5 What are basic differences between Exploratory move and Pattern move?

5.6 Consider the following minimization problem

$$f(X) = x_1^2 + 2x_1 x_2 + x_2^2 + 3x_1$$

Find the minimum using Steepest Descent method.

5.7 Solve the problem 6 using Newton's method.

5.8 What are advantages of Marquardt method over Steepest Descent and Newton's method.

5.9 The BFGS method can be considered as a Quasi-Newton, conjugate gradient, and variable metric method- Justify this statement with proper example.

5.10 In Quasi-Newton method, the matrix $[B_i]$ is updated using the formula

$$[B_{i+1}] = [B_i] + \frac{\lambda_i^* s_i s_i^T}{s_i^T g_i} - \frac{([B_i]g_i)([B_i]g_i)^T}{([B_i]g_i)^T g_i}$$

Discuss the effect of λ_i^* in this iteration process.

References

Anderson, R. L. *Recent Advances in Finding Best Operating Conditions*, J. Amer. Statist. Assoc., 48(1953): 789–98.

Anupam, K. *et al. The Canadian Journal of Chemical Engineering*, 89(2011): 1274–80.

Broyden, C. G. 1967. *Quasi-Newton Methods and their Application to Function Minimization*, Mathematics of Computation, vol. 21, pp. 368.

Broyden, C. G., Dennis, J. E. and More, J. J. *On the Local and Superlinear Convergence of Quasi-Newton Methods*, Journal of the Institute of Mathematics and Its Applications, 12(1975): 223.

Cauchy, A. L. 1847. *Méthode Générale Pour La Résolution Des Systèmes D'équations Simultanées*, Comptes Rendus de l'Academie des Sciences, Paris, vol. 25, pp. 536–38.

Davidon, W. C. 1959. *Variable Metric Method of Minimization*, Report ANL–5990, Argonne National Laboratory, Argonne, IL.

Deb, K. 2009. *Optimization for Engineering Design: Algorithms and Examples*, PHI Learning Pvt. Ltd.

Dennis, J. E. Jr. and More, J. J. 1977. *Quasi-Newton Methods, Motivation and Theory*, SIAM Review, vol. 19, No. 1, pp. 46–89.

Dutta, S. *et al. Adsorptive Removal of Chromium (VI) from Aqueous Solution Over Powdered Activated Carbon: Optimisation through Response Surface Methodology*, Chemical Engineering Journal, 173(2011): 135–43.

Edgar, T. F., Himmelblau, D. M., Lasdon, L. S. 2001. *Optimization of Chemical Processes*, McGraw-Hill.

Fletcher, R. and Powell, M. J. D. 1963. A Rapidly Convergent Descent Method for Minimization, Computer Journal, vol. 6, No. 2, pp. 163–68.

Fox, R. L. 1971. *Optimization Methods for Engineering Design*, Addison–Wesley, Reading, MA.

Hooke, R., Jeeves, T. A. *Direct Search Solution of Numerical and Statistical Problems*, Assoc. Computing Mechinery J., 8(1960): 212–29.

Hooke, R. and Jeeves T. A. 1961. *"Direct Search" Solution of Numerical and Statistical Problems*, J. Ass. Comput. Mach., 8: 212–29.

Karnopp, D. C., *Random Search Techniques for Optimization Problems*, Automatica, 1(1963): 111–21.

Levenberg, K. *"A Method for the Solution of Certain Nonlinear Problems in Least Squares"*, Qart. Appl. Math., 2(1944): 164–66.

Lewis, R. M., Torczon, V., Trosset, M. W. *Direct Search Methods: Then and Now;* Journal of Computational and Applied Mathematics, 124(2000): 191–207.

Marquardt, D. W. *An Algorithm for Least-Squares Estimation of Nonlinear Inequalities,* SIAM J. Appl. Math., 11(1963): 431–41.

Montgomery, D. C. 2008. *Design and Analysis of Experiments,* John Wiley and Sons, INC.

Powell, M. J. D. An Efficient Method for Finding the Minimum of a Function of Several Variables without Calculating Derivatives, The Computer Journal, 7(1964): pp. 155–62.

Rao, S. S. 2009. *Engineering Optimization: Theory and Practice* (4e), John Wiley and Sons.

Rastrigin, L. A. The *Convergence of the Random Search Method in the External Control of a Many-Parameter System,* Automat. Remote Control, 24(1963): 1337–42.

Swann, W. H. 1972. *Direct Search Methods, in: W. Murray (Ed.), Numerical Methods for Unconstrained Optimization,* New York: Academic Press, pp. 13–28.

Chapter 6

Multivariable Optimization with Constraints

6.1 Formulation of Constrained Optimization

The design variables in an optimization problem cannot be chosen arbitrarily. In some optimization problems, design variables should satisfy some additional specified functional and other requirements. These additional restrictions on design variables are collectively called design constraints.

Most of the constrained optimization problems contain objective function(s) with constraints. The constrained optimization problem can be represented as:

$$\text{Minimize } f(X) \quad X \in \mathbb{R}^n \tag{6.1a}$$

$$\text{subject to } c_i(X) = 0, \quad i \in E \tag{6.1b}$$

$$c_i(X) \geq 0, \quad i \in I \tag{6.1c}$$

where $f(X)$ is the objective function, $c_i(X)$ are constraint functions. E stands for the index set of equations or equality constraints in the problem, I stands for the set of inequality constraints, and both are finite sets. Generally, the constraint equations can be put into different forms for our convenience: such as $c_i(X) \leq b$ turns into $b - c_i(X) \geq 0$. If any point X' (point (2,20) in Fig. 6.1) satisfies all the constraints that point is called a feasible point. The set of all such feasible points is referred to as feasible region (shaded region in Fig. 6.1). Generally, we look for a local (relative) minimum rather than a global minimum. The definition of a constrained local minimizer X^* is that $f(X^*) \leq f(X)$ for all feasible X sufficiently close to X^* [Fletcher, (1986)].

120　Optimization in Chemical Engineering

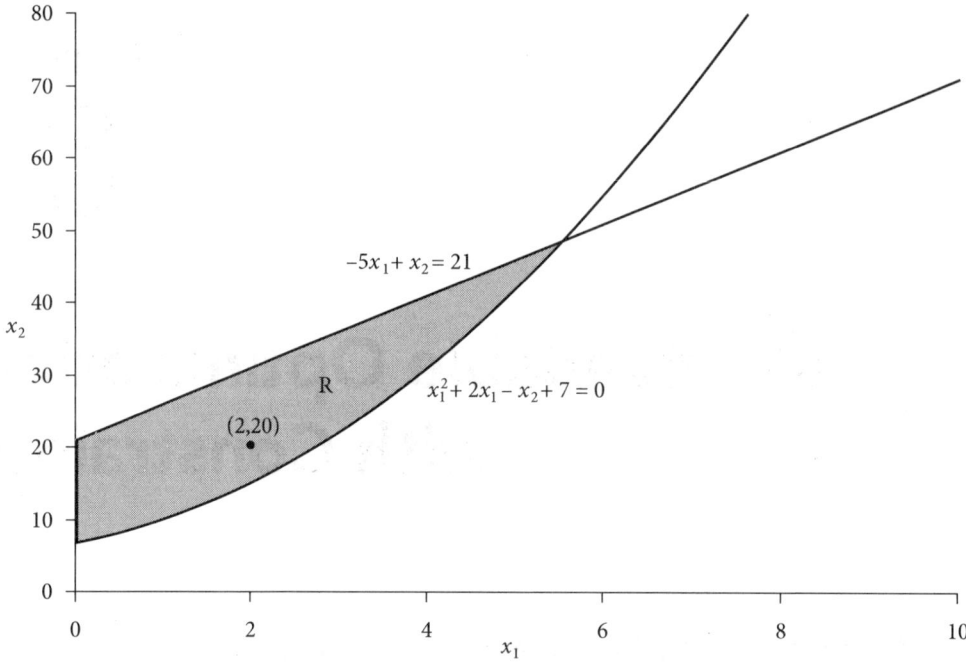

Fig. 6.1　Graphical representation of feasible region

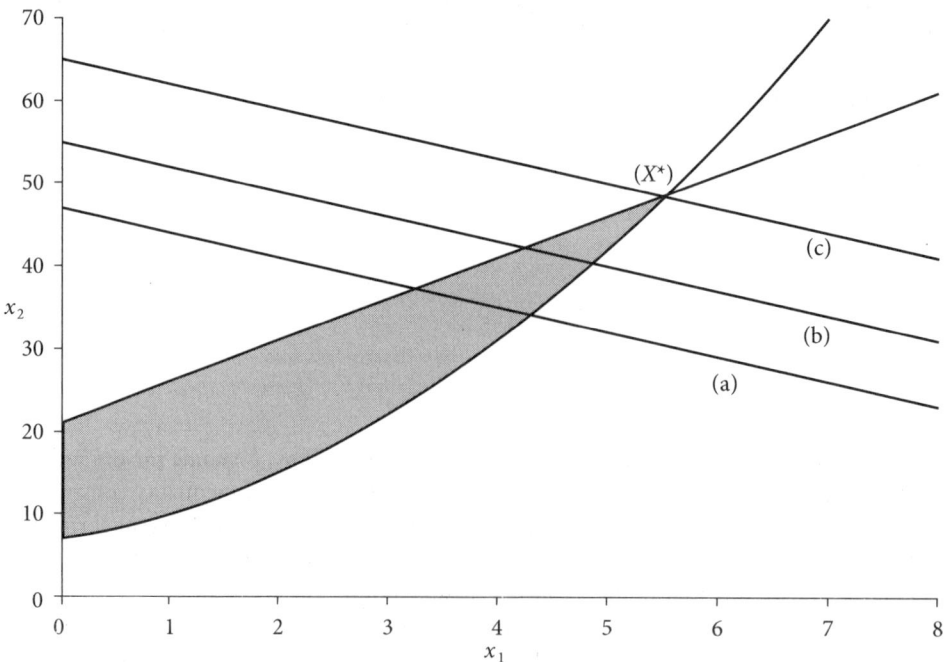

Fig. 6.2　Constrained optimization problem

Figure 6.1 is a graphical representation of a constrained optimization problem. Where R (shaded area) is the feasible region with inequality constraints.

$$x_2 - x_1^2 + 2x_1 \geq 7 \qquad (6.2a)$$

$$x_2 - 5x_1 \leq 21 \qquad (6.2b)$$

$$x_1 \geq 0 \qquad (6.2c)$$

Any objective function that satisfies the constraints (Eq. (6.2a)–(6.2c)) can be optimized by the graphical method. Figure 6.2 explains how maximum value of the objective function $f = 3x_1 + x_2$ (or minimum of $-f$) was obtained. When the line moves from (a) to (b); objective function value (f) increases from 47 to 55. Finally, it moves from (b) to (c) and reaches the maximum point within the feasible region. The maximum value of objective function is 65 at the point (5.5, 48.5).

Depending on the nature of constraints, feasible region may be bounded or unbounded. In Fig. 6.1, the feasible region R is bounded whereas in Fig. 6.3 feasible region is unbounded. If the feasible region is unbounded then the objective function $f(X) \to \mp\infty, X \in R$ or the function may not have a minimum (or maximum) point. The problem also does not provide any solution when the region R is empty; that is when the constraints are inconsistent.

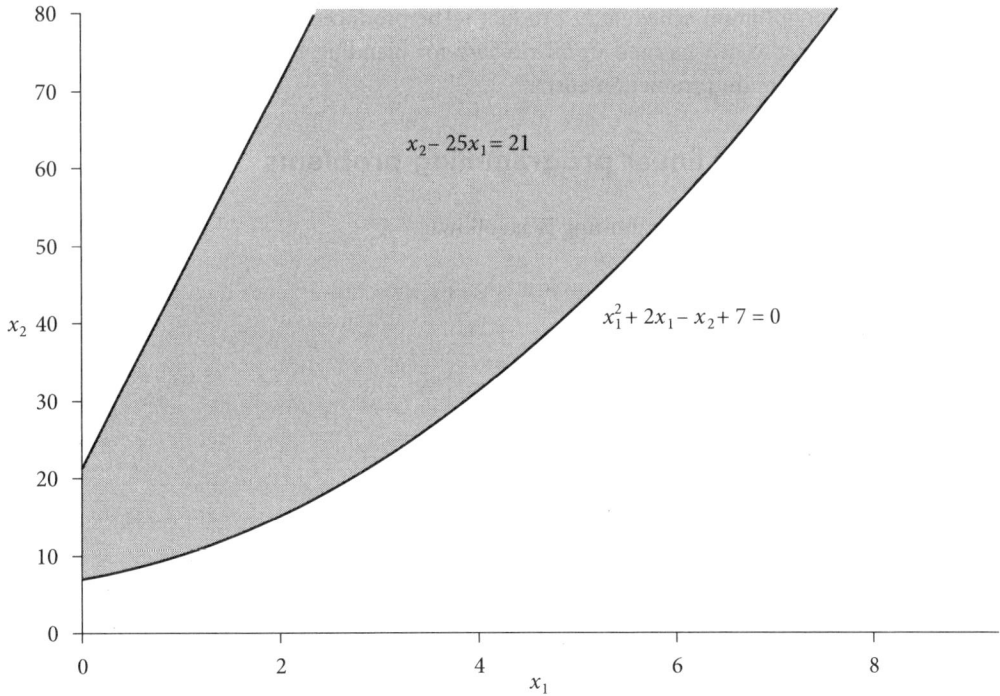

Fig. 6.3 Unbounded feasible region

Constrained multivariable optimization problem can be broadly classified in to two categories; linear programming and nonlinear programming.

6.2 Linear Programming

Linear programming can be viewed as part of great revolutionary development which has given mankind the ability to state general goals and to lay out a path of detailed decisions to take in order to "best" achieve its goals when faced with practical situations of great complexity [Dantzig, 2002]. Linear Programming (LP) is the simplest type of constrained optimization problem. In LPs the objective function $f(X)$ and the constraint functions $c_i(X)$ are all linear functions of decision variables X. The aim of linear programming (LP) is to find out the values of X that minimize a linear objective function subject to a finite number of linear constraints. The linear constraints may be equality and inequality functions of decision variables. The main purpose is to obtain a point that minimizes the objective function and simultaneously satisfies all the constraints. The points that satisfy the constraints are referred as a feasible point. [Chong and Zak (2001)]. This type of optimization problem using linear programming was first developed in 1930s for the optimal allocation of resources. In 1939, L.V. Kantorovich presented several solutions for production and transportation planning problems. T.C. Koopmans contributed significantly during World War II to solve the transportation problems. In 1975, Koopmans and Kantorovich were jointly awarded Nobel Prize in economics for their contribution on the theory of optimal allocation of resources. There are many applications of LP in the field of chemical engineering. Many oil companies are using LP for determining the optimum schedule of product to be produced from the crude oils available. Linear programming also can be used in oil refinery for blending purpose, which optimizes the composition with minimum production cost.

6.2.1 Formulation of linear programming problems

The standard form of a linear programming is as follows:

Scalar form: Minimize $f(X) = c_1 x_1 + c_2 x_2 + \ldots + c_n x_n$ (6.3a)

subject to
$$a_{11} x_1 + a_{12} x_2 + \ldots + a_{1n} x_n = b_1$$
$$a_{21} x_1 + a_{22} x_2 + \ldots + a_{2n} x_n = b_2$$
$$\vdots$$
$$a_{m1} x_1 + a_{m2} x_2 + \ldots + a_{mn} x_n = b_m$$
(6.3b)

$$x_1 \geq 0$$
$$x_2 \geq 0$$
$$\vdots$$
$$x_n \geq 0$$
(6.3c)

where a_{ij} ($i = 1, 2,\ldots,m$; $j = 1, 2,\ldots,n$), b_j, and c_j are known constants, and x_j are called decision variables.

Matrix form: Minimize $f(X) = c^T X$ (6.4a)

subject to $aX = b$ (6.4b)

$X \geq 0$ (6.4c)

where $X = \begin{Bmatrix} x_1 \\ x_2 \\ \vdots \\ x_n \end{Bmatrix}, b = \begin{Bmatrix} b_1 \\ b_2 \\ \vdots \\ b_m \end{Bmatrix}, c = \begin{Bmatrix} c_1 \\ c_2 \\ \vdots \\ c_n \end{Bmatrix}$ and $a = \begin{bmatrix} a_{11} & a_{12} & \cdots & a_{1n} \\ a_{21} & a_{22} & \cdots & a_{2n} \\ \vdots & \vdots & \ddots & \vdots \\ a_{m1} & a_{m2} & \cdots & a_{mn} \end{bmatrix}$

Here, a is an $m \times n$ matrix consists of real entries. The value of all b_i should be positive; if any element of b is negative, say the ith element, we have to multiply the ith constraint with -1 to achieve a positive right-hand side. All theorems and solving methods for linear programs are usually written in standard form (Eq. (6.3a)–(6.3c) or Eq. (6.4a)–(6.4c)). All other forms of linear programs can be transformed to the standard form. If a linear programming problem is in the form

Minimize $f(X) = c^T X$ (6.5a)

subject to $aX \geq b$ (6.5b)

$l \leq X \leq u$ (6.5c)

Here, l and u are lower and upper limit of the variable X. The original problem can be converted into the standard form by introducing so-called slack and surplus variables y_j. Algorithms for solving LP problems are divided into two categories (i) simplex method and (ii) non-simplex method

6.2.1.1 Basic LP definitions and results

Now we will simplify the ideas demonstrated in the previous section for problems with 2 to n dimensions. Proofs of the following theorems are not in the scope of this book, it can be found in Dantzig (1963). At the beginning of LPP discussion some standard definitions are given.

Definition 6.1

In an n-dimensional space, a point X has been characterized by an ordered set of n values or coordinates $(x_1, x_2, \ldots, x_n)$. The coordinates of X can also termed as the components of X.

Definition 6.2

If the coordinates of any two points A and B are presented by $x_j^{(1)}$ and $x_j^{(2)}$ ($j = 1, 2, \ldots, n$), the line segment (L) connecting these points is the group of points $X(\lambda)$ whose coordinates are expressed by the equation $x_j = \lambda x_j^{(1)} + (1 - \lambda) x_j^{(2)}$, $j = 1, 2, \ldots, n$, with $0 \leq \lambda \leq 1$.

Thus, we can write

$$L = \{X | X = \lambda X^{(1)} + (1-\lambda) X^{(2)}\} \tag{6.6}$$

Definition 6.3
A hyperplane in an n-dimensional space is defined as the set of points whose coordinates satisfy a linear equation

$$a_1 x_1 + a_2 x_2 + \ldots + a_n x_n = \mathbf{a}^T X = \mathbf{b} \tag{6.7}$$

A hyperplane H can be described as

$$H(\mathbf{a}, b) = \{X | \mathbf{a}^T X = b\} \tag{6.8}$$

Definition 6.4
A feasible solution of the linear programming problem is a vector $X = (x_1, x_2, \ldots, x_n)$, which satisfies Eqs (6.5b) and the bounds (6.5c).

Definition 6.5
A basis matrix is an $m \times n$ nonsingular matrix created from some m columns of the constraint matrix A (since the rank of matrix A rank(A) = m, A possesses at least one basis matrix).

Definition 6.6
A basic solution of a linear program is the unique vector found out by taking a basis matrix, setting each of the $n-m$ variables associated with columns of A not in the basis matrix equal to either l_j or u_j, and solving the resulting square, nonsingular system of equations for the m remaining variables.

Definition 6.7
A basic solution in which all variables satisfy their bounds is called a basic feasible solution (Eq. (6.5c)).

Definition 6.8
A basic feasible solution in which all basic variables x_j are strictly between their bounds, that is, $l_j < x_j < u_j$ is known as a non-degenerate basic feasible solution.

Definition 6.9
An optimal solution is a feasible solution that also minimizes $f(X)$ in Eq. (6.5a).

These definitions give us the following result:

Result 1
The minimum of the objective function $f(X)$ is reachable at a vertex of the feasible region. If minimum value is available at more than one vertex, then the value of the objective function is same at every point of the line segment joining any two optimal vertices.

Result 1 confirms that we need only look at vertices to search for a solution. Therefore, it is our prime concern to know how to characterize vertices algebraically for multi-dimension problems. The next result gives this information.

6.2.1.2 Duality of linear programming

One of the most interesting concepts in linear programming is the duality theory. Every linear programming problem is associated with duality, where there is another linear programming problem with the same data and closely related optimal solutions. These two problems are called to be duals of each other. While one of the problems is called the primal, the other is called dual.

The duality concept is very important due to two main reasons. Firstly, when the primal problem contains a large number of constraints and a smaller number of variables, they can be converted to a dual problem that reduces the computation effort considerably while solving. Secondly, during any decision making in future, the interpretation of the dual variables from the cost or economic point of view seems to be extremely useful.

Primal problem

$$\text{Maximize } f = \sum_{i=1}^{n} c_j x_j \tag{6.9a}$$

$$\text{subject to } \sum_{j=1}^{n} a_{ij} x_j \leq b_i, \; (i = 1, 2, \ldots, m) \tag{6.9b}$$

$$x_j \geq 0, \; (j = 1, 2, \ldots, n) \tag{6.9c}$$

The corresponding dual problem can be given by

Dual problem

$$\text{Minimize } z = \sum_{i=1}^{m} b_i y_i \tag{6.10a}$$

$$\text{subject to } \sum_{i=1}^{m} a_{ij} y_i \leq c_j, \; (j = 1, 2, \ldots, n) \tag{6.10b}$$

$$y_i \geq 0, \; (i = 1, 2, \ldots, m) \tag{6.10c}$$

Example 6.1

A chemical company produces two chemical products that produced through two different parallel reactions as shown below

$$A + B \xrightarrow{k_1} P_1 \tag{6.11a}$$

$$A + B \xrightarrow{k_2} P_2 \tag{6.11b}$$

The raw materials A and B have limited supply of 36 kg and 14 kg per day respectively. The reaction 6.11a takes 3 kg A and 1kg B to produce 1 kg P_1, and the reaction 6.11b takes 2 kg A and 1 kg B to produce 1 kg P_2. The profit of the company from these products is \$14 per kg P_1 and \$11 per kg P_2. Formulate a linear programming problem and maximize the daily profit of the company.

Solution

For maximizing the daily profit of the company, we have to formulate the LPP. The objective function (daily profit) can be written as

$$F = 14x_{P_1} + 11x_{P_2} \tag{6.12a}$$

where F is the daily profit, production rates (kg/day) of P_1 and P_2 are x_{P_1} and x_{P_2} respectively.

Constraint functions are formulated based on the limited supply of A and B.

Supply of A: $3x_{P_1} + 2x_{P_2} \leq 36$ \hfill (6.12b)

Supply of B: $x_{P_1} + x_{P_2} \leq 14$ \hfill (6.12c)

The production rates (decision variables) can never be negative.

$$x_{P_1}, x_{P_2} \geq 0 \tag{6.12d}$$

We can write this problem in the similar form of Eq. (6.5a)–(6.5c)

Max $F = 14x_{P_1} + 11x_{P_2}$ \hfill (6.12a)

subject to

$3x_{P_1} + 2x_{P_2} \leq 36$ \hfill (6.12b)

$x_{P_1} + x_{P_2} \leq 14$ \hfill (6.12c)

$$x_{P1}, x_{P2} \geq 0 \qquad (6.12d)$$

This LPP can be solved easily by using graphical method. Figure 6.4 gives us an idea about the graphical method. The shaded area is the feasible region, which satisfies the inequality constraints (Eq. (6.12b)–(6.12d)). The objective function F touches the extreme point (8,6) with a value of 178. The solution of the problem is given below:

$$F^* = 178, \; x_{P1}^* = 8, \text{ and } x_{P2}^* = 6$$

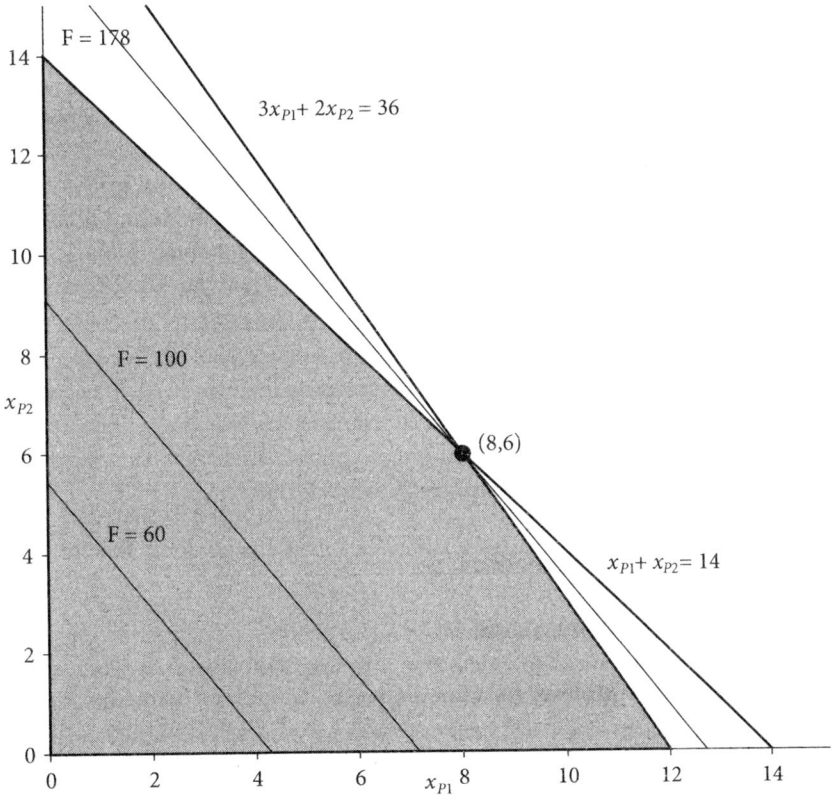

Fig. 6.4 Graphical representation of problem 6.12a–6.12d

6.2.2 Simplex method

For solving linear programming problems, Dantzig [Dantzig, (1963)] developed the simplex method in 1947. It is the perceptive idea behind simplex algorithm that there always exists a vertex of the feasible region, which is an optimal solution to the LP. A vertex is an optimal solution if there is no better neighboring vertex. The terminologies of "vertex", "neighboring vertex" have very clear geometric interpretations. However, these terms are not useful for computer programming. The simplex algorithm is always initiated with an algorithm whose equations are in canonical form. In the first step of the simplex method, certain artificial variables (slack/surplus) are introduced into the standard form. The resulting auxiliary problem is in canonical form. The simplex algorithm consists of a sequence of pivot operations referred to as Phase I that creates a series of different canonical forms. The objective is to find a feasible solution if one exists. If the final canonical form yields such a solution, the simplex algorithm is again applied in a second succession of pivot operations referred to as Phase II.

Algorithm

The simplex method for solving a linear programming problem in standard form produces a series of feasible points $X^{(1)}, X^{(2)}, \ldots$ that terminates to the solution. Each iterate $X^{(k)}$ represents an extreme point and there exists an extreme point at which the solution comes about. Therefore, $n - m$ number of the variables have zero value at $X^{(k)}$ and are referred to as non-basic variables ($N^{(k)}$). The remaining m variables possess a non-negative value, are mentioned to as basic variables ($B^{(k)}$). The simplex method systematically modifies to these sets after each iteration, in order to find the alternative that offers the optimal solution [Fletcher, (1986)]. At each iteration it is convenient to assume that the variables are permuted such that the basic variables are in first m elements of X. Then, we can write $\mathbf{X}^T = (\mathbf{X}_B^T, \mathbf{X}_N^T)$, where $\mathbf{X}_B^T$ and $\mathbf{X}_N^T$ refer to the basic and non-basic variables respectively. The matrix a in Eq. (6.4b) can also be partitioned similarly into a = [a_B : a_N] where a_B is $m \times m$ basic matrix and a_N is $m \times (n-m)$ non basic matrix. Then, Eq. (6.4b) can be written as

$$[\mathbf{a}_B : \mathbf{a}_N]\begin{pmatrix}\mathbf{X}_B\\ \mathbf{X}_N\end{pmatrix} = \mathbf{a}_B\mathbf{X}_B + \mathbf{a}_N\mathbf{X}_N = \mathbf{b} \tag{6.13}$$

Since, $\mathbf{X}_N^{(k)} = 0$, it is possible to write

$$\mathbf{X}^{(k)} = \begin{pmatrix}\mathbf{X}_B\\ \mathbf{X}_N\end{pmatrix}^{(k)} = \begin{pmatrix}\hat{\mathbf{b}}\\ 0\end{pmatrix} \tag{6.14}$$

where $\hat{\mathbf{b}} = \mathbf{a}_B^{-1}\mathbf{b}$. Since the basic variables must take non-negative values it is expected that $\hat{\mathbf{b}} \geq 0$.

The partitioning $B^{(k)}$ and $N^{(k)}$ and the extreme point (vertex) $X^{(k)}$ with the above properties ($\mathbf{X}_B^{(k)} = \hat{\mathbf{b}} \geq \mathbf{0}, \mathbf{X}_N^{(k)} = \mathbf{0}$, $\mathbf{a}_B$ non-singular) is referred to as a Basic Feasible Solution (BFS).

From above discussion, an LP in canonical form with m linear constraints and n decision variables, may have a basic feasible solution for every choice of $n-m$ non-basic variables (or equivalently, m basic variables). Therefore, the number of vertices (or BFS) of the feasible region might as well be the same as that of the choice of $n-m$ non-basic variables. How many such choices? From combinatorics, the number of choices are

$$\binom{n}{n-m} = \binom{n}{m} = \frac{n!}{m!(n-m)!} \tag{6.15}$$

Even though, this is a finite number, it could be really large, even if the n, m are relatively small. For example, with $n = 20$, $m = 10$, the number of choices are 184756. In practice, the simplex algorithm usually finds the optimal solution after scanning some $4m$ to $6m$ BFS, and very rarely beyond $10m$.

Example 6.2

The steps for a simplex method can be explained through this example

Maximize: $Z = 3x_1 + 2x_2$

subject to constraints

$2x_1 + x_2 \leq 10$

$x_1 + x_2 \leq 8$

$x_1 \leq 4$

and

$x_1 \geq 0, x_2 \geq 0$

Solution

We should first write the problem in canonical form by introducing slack variables.

$2x_1 + x_2 + s_1 = 10$

$x_1 + x_2 + s_2 = 8$

$x_1 + s_3 = 4$

in this case, $n = 5$ and $m = 3$ and degree of freedom $n - m = 2$

initialization (finding a starting BFS or a starting vertex): It is easy in this case-just set

NBV = (x_1, x_2) = (0, 0) and BV = (s_1, s_2, s_3) = (10, 8, 4)

Optimality test (is the current BFS or vertex optimal?): The current BFS will be optimal if and only if it is better than every neighboring vertex (or every BFS share all but one basic variables). To do this, we try to determine whether there is any way Z can be increased by increasing one of the non-basic variables from its current value zero while all other non-basic variables remain zero (while we adjust the values of the basic variables to continue satisfying the system of equations).

In this case the objective function is $Z = 3x_1 + 2x_2$, and Z take value 0 at (0, 0). It is easy to see that no matter we increase x_1 (while holding $x_2 = 0$) or increase x_2 (while holding $x_1 = 0$), we are going to increase Z since all the coefficients are positive. We conclude that the current BFS is not optimal.

Moving to the neighboring BFS (or vertex): Two neighboring BFS share all but one basic variables. In other wards, one of the variables (x_1, x_2) is going to become a basic variable (entering basic variable), and one of (s_1, s_2, s_3) is going to become a non-basic variable (leaving basic variable).

a. Determining the entering basic variable: Choosing an entering basic variable amounts to choosing a non-basic variable to increase from zero. Note $Z = x_1 + 2x_2$. The value of Z is increased by 3 if we increase x_1 by 1, and by 2 if we increase x_2 by 1. Therefore, we choose x_1 as the entering basic variable.

b. Determining how large the entering basic variable can be: We cannot increase the entering variable x_1 arbitrarily, since it may cause some variables to become negative. What is the largest possible value that x_1 can attain? Note x_2 is held at zero. Hence,

$s_1 = 10 - 2x_1 \geq 0$; x_1 cannot exceed 5

$s_2 = 8 - x_1 \geq 0$; x_1 cannot exceed 8

$s_3 = 4 - x_1 \geq 0$; x_1 cannot exceed 4

it follows that the largest x_1 can be is the 4

c. Determine the leaving basic variable: When x_1 takes value 4, s_3 becomes zero. Therefore, s_3 is the leaving basic variable.

Therefore, the neighboring vertex we select is

NBV = (s_3, x_2) BV = (s_1, s_2, x_1)

Pivoting (solving for the new BFS): Recall that we have

$$\begin{array}{llllllll} Z & -3x_1 & -2x_2 & & & & = 0 & (0) \\ & 2x_1 & +x_2 & +s_1 & & & = 10 & (1) \\ & x_1 & +x_2 & & +s_2 & & = 8 & (2) \\ & x_1 & & & & +s_3 & = 4 & (3) \end{array}$$

The goal is to solve for the BFS, and it is going to be achieved by Gaussian elimination. We end up with

$$\begin{array}{llllllll} Z & & -2x_2 & & & +3s_3 & = 12 & (0) \\ & & x_2 & +s_1 & & -2s_3 & = 2 & (1) \\ & & x_2 & & +s_2 & -s_3 & = 4 & (2) \\ & x_1 & & & & +s_3 & = 4 & (3) \end{array}$$

In other wards, each basic variable has been eliminated from all but one row (*i*th row) and has coefficient +1 in that row. The Gaussian elimination always starts with the row of the leaving basic variable (or the row that achieve the minimal ratio in the preceding step), or the entering basic variable's row is always the row of the leaving basic variable, or the entering basic variable's row is always the row that achieves the minimal ratio in the preceding step.

What we have is that the BFS is

NBV = (s_3, x_2) = $(0, 0)$ BV = (s_1, s_2, x_1) = $(2, 4, 4)$

and

$Z = 12 + 2x_2 - 3s_3$

While, taking value 12 at this BFS.

Iteration The above BFS is not optimal, since we can increase x_2, which increases Z. We do not want to increase S_3, which decreases the value of Z. So the entering basic variable is x_2. How large can x_2 be? Note $s_3 = 0$, we have

$$x_2 + s_1 = 2 \; ; \; x_2 \text{ cannot exceed } 2$$

$$x_1 + s_2 = 4 \; ; \; x_2 \text{ cannot exceed } 4$$

$$x_1 = 4 \; ; \text{ no upper bound for } x_2$$

The maximum of x_2 is therefore, 2 achieved at row (1), and the leaving basic variable is the (original) basic variable in row (1), i.e., s_1. In other wards

NBV $= (s_1, s_3)$, BV $= (s_2, x_1, x_2)$ Gaussian elimination yields, starting from row (1) yield,

$$
\begin{array}{lrrrrl}
Z & & 2s_1 & & -s_3 & = 16 \quad (0) \\
& x_2 & +s_1 & & -2s_3 & = 2 \quad (1) \\
& & -s_1 & +s_2 & +s_3 & = 2 \quad (2) \\
& x_1 & & & +s_3 & = 4 \quad (3)
\end{array}
$$

or the new BFS is

$$\text{NBV} = (s_1, s_3) = (0, 0), \quad \text{BV} = (s_2, x_1, x_2) = (2, 4, 2).$$

The value of Z is

$$Z = 16 - 2s_1 + s_3$$

and it attains value 16 at this BFS.

The BFS is still not optimal, and clearly s_3 will be the entering basic variable. Note $s_1 = 0$, we have

$$x_2 - 2s_3 = 2 \; ; \text{ no upper bound for } s_3$$

$$s_2 + s_3 = 2 \; ; \; s_3 \text{ cannot exceed } 2$$

$$x_1 + s_3 = 4 \; ; \; s_3 \text{ cannot exceed } 4$$

The maximum of s_3 is therefore, 2 achieved at row (2), and the leaving basic variable is the (original) basic variable in row (1), i.e., s_2. In other wards

$$\text{NBV} = (s_1, s_2), \quad \text{BV} = (x_1, x_2, s_3).$$

the Gaussian elimination yields, starting from row (1) yields,

$$
\begin{array}{lrrrrl}
Z & & +s_1 & +s_2 & = 18 & (0) \\
 & x_2 & -s_1 & +2s_2 & = 6 & (1) \\
 & & -s_1 & +s_2 & +s_3 = 2 & (2) \\
x_1 & & +s_1 & -s_2 & = 2 & (3)
\end{array}
$$

or the new BFS is

$$\text{NBV} = (s_1, s_2) = (0,0), \quad \text{BV} = (x_1, x_2, s_3) = (2, 6, 2).$$

The value of Z is

$$Z = 18 - s_1 - s_3$$

and it attains value 18 at this BFS.

The BFS turns out to be optimal; any increase in the non-basic variable will decrease the value of Z. Hence,

Max $Z = 18$, achieved at $(x_1^*, x_2^*) = (2, 6)$.

Degeneracy in Linear Programming

A Linear Programming is degenerate if in a basic feasible solution, one of the basic variables takes on a zero value. Degeneracy is caused by redundant constraint(s) and could cost simplex method extra iterations.

Constraints are either redundant or necessary; redundant constraints are constraints that may be deleted from the set without changing the region defined by the set. It is observed that redundancies exist in most practical problems. The importance of detecting and removing redundancy in a set of linear constraints is the avoidance of all calculations associated with those constraints when solving an associated mathematical programming problem.

Example 6.3

$$\text{Max } Z = 2x_1 + x_2 + x_3 \tag{6.16a}$$

$$\text{subject to: } x_1 + 2x_2 \le 1 \tag{6.16b}$$

$$-x_2 + x_3 \le 0 \tag{6.16c}$$

$$x_1, x_2, x_3 \ge 0 \tag{6.16d}$$

this example shows that $x_2 \ge 0$, follows from constraints $-x_2 + x_3 \le 0$ and $x_3 \ge 0$. Therefore, $x_3 \ge 0$ is redundant constraint.

6.2.3 Nonsimplex methods

Simplex method is extensively used in practice to solve LP problems. However, the computational time required to get a solution using the simplex method increases rapidly when the number of components n of the variable $x \in \mathbb{R}^n$ increases. The major drawback of simplex algorithm is exponential complexity. As the number of component n of the variable x increases, the required time also increases exponentially. This correlation is also called the complexity of the algorithm. Therefore, we can say that the simplex algorithms possesses an exponential complexity. The complexity of the simplex algorithm is often written as $O(2^n - 1)$. For a number of years, exponential complexity and polynomial complexity have been distinguished by the computer scientists. If any algorithm used to solve linear programming problems has polynomial complexity, in that case the time required to find the solution is bounded by a polynomial in n. It is obvious that the polynomial complexity is more desirable than exponential complexity. This concept led to a concern in developing algorithms to solve linear programming problems, which have polynomial complexity. These algorithms are able to find the solution in an amount of time that is restricted by a polynomial in the number of variables. Therefore, the existence of an algorithm to solve linear programming problems with polynomial complexity is an important issue. This issue was somewhat resolved by Khachiyan [L. G. Khachiyan, (1979)], this method is also called the ellipsoid algorithm. Then, Karmarkar's interior point method [N. K. Karmarkar. (1984)] was developed in 1984. In this chapter we will consider two non-simplex methods namely Khachiyan's ellipsoid method and Karmarkar's interior point method.

6.2.3.1 Khachiyan's ellipsoid method

The first polynomial time algorithm for linear programming was developed in 1979 by the Russian mathematician Khachiyan [L. G. Khachiyan, (1979)]. He has shown that linear programs (LPs) can be solved efficiently; more accurately for LP problems that are polynomially solvable. Khachiyan's approach was developed based on ideas which are analogous to the Ellipsoid Method arising from convex optimization [Rebennack (2008)]. The ellipsoid method is an algorithm that finds an optimal solution of LPP in a finite number of steps. This method generates a sequence of ellipsoids; volume of those ellipsoids decreases uniformly at each step, thus, enclosing the minimum of a convex function (see Fig. 6.5).

The Basic Ellipsoid Algorithm

The linear programming problem

$$\text{Optimize } z = c^T x \tag{6.17a}$$

$$\text{subject to } Ax \leq b \tag{6.17b}$$

$$x \geq 0 \tag{6.17c}$$

A is an $m \times n$ matrix i.e., $A \in R^{m \times n}$, c is an n-vector, i.e., $c \in R^n$, x is an n-tuple, $x \in R^n$ and $b \in R^m$ suppose we are interested to find an n-vector x, satisfying

$$A^T x \leq b \tag{6.18}$$

The column of A corresponding to outward drawn normals to the constraints are denoted by $\alpha_1, \alpha_2, \ldots, \alpha_n$ and the components of b denoted by $\beta_1, \beta_2, \ldots, \beta_n$. Thus, Eq. (6.18) can be restated as

$$\alpha_i^T x \leq \beta_i, \ i = 1, 2, \ldots, n \quad (6.19)$$

throughout, the calculation we assume that n is greater than one.

The basic ellipsoid iteration

The ellipsoid method construct a sequence of ellipsoids; $E_0, E_1, E_2, \ldots E_k, \ldots$ each of which contains a point satisfying Eq. (6.18), if one exist. E_k is defined as follows:

$$E_k = \left\{ x \in R^n \,\middle|\, (x - x_k)^T B_k^{-1} (x - x_k) \leq 1 \right\} \quad (6.20)$$

Where

$$x_{k+1} = x_k - \tau \frac{B_k \alpha}{\sqrt{\alpha^T B_k \alpha}} \quad (6.21)$$

$$B_{k+1} = \delta \left(B_k - \frac{\sigma (B_k \alpha)(B_k \alpha)^T}{\alpha^T B_k \alpha} \right) \quad (6.22)$$

$$\tau = \frac{1}{n+1}, \sigma = \frac{2}{n+1}, \text{ and } \delta = \frac{n}{\sqrt{n^2 - 1}} \quad (6.23)$$

τ is known as the step parameter, while δ and σ are the expansion and dilation parameters respectively. On the $(k + 1)$th iteration, the algorithm verifies whether the centre x_k of the current ellipsoid E_k satisfies the constraints Eq. (6.18). If so, the iteration stops. If not some constraints violated by x_k, say

$$\alpha_i^T x \leq B_i \quad (6.24)$$

are chosen and the ellipsoid of minimum volume that contains the half ellipsoid:

$$\left\{ x \in E_k \,\middle|\, \alpha^T x \leq \alpha^T x_k \right\} \quad (6.25)$$

constructed.

The new ellipsoid and its centre are represented by E_{k+1} and x_{k+1} respectively as defined above. The iteration is again repeated. Note that the initial ellipsoid, E_0 is arbitrarily assumed.

Apart from the initial ellipsoid, these steps give a possibly infinite iterative algorithm for determining the feasibility of (6.18).

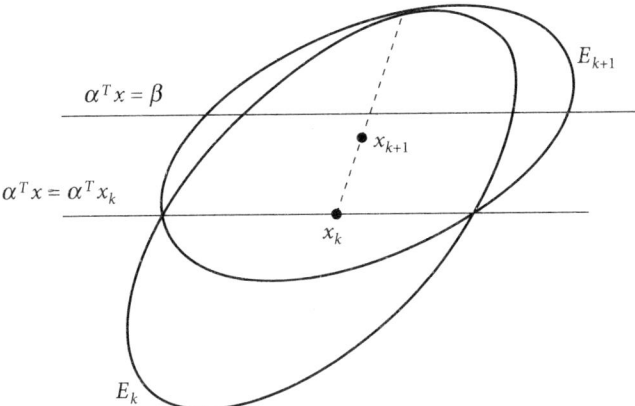

Fig. 6.5 Ellipsoid method

In essence, Khachiyan showed that it is possible to determine whether Eq. 6.18 is feasible or not within a pre-specified (polynomial) number of iterations (i.e., $6n^2L$ iteration) by

a) revising this algorithm to account for finite precision arithmetic

b) applying it to a appropriate perturbation of system

c) selecting E_0 appropriately

System (6.18) is feasible if and only if it terminates with feasible solution of the perturbed system within the stipulated number of iteration [Bland, (1981)].

6.2.3.2 Karmarkar's interior point method

In 1984, Karmarkar [N. K. Karmarkar, (1984)] proposed a new linear programming algorithm that has polynomial complexity. This algorithm is able solve some complex real-world problems such as scheduling, routing and planning more efficiently than the simplex method. This important work of Karmarkar led to the development of many other non-simplex techniques usually referred to as interior point methods.

Unlike the simplex method, the Karmarkar's interior point methods traverse the interior (internal space) of the feasible region as shown in Fig. 6.6. In exterior point method, the search algorithm follows a path A, B, C and reaches the optimum point. Whereas the interior point methods follow the path like 1, 2, 3, 4; then reaches the optimum. The main difficulty with interior point methods is that it needs to identify the best among all feasible directions at a specified solution. The main objective is to reduce the number of iterations to improve computational effort [Ravindran *et al.* (2006)]. It is found that, for a large problem, Karmarkar's method is 50 times faster than the simplex method.

Algorithm

Karmarkar's method is developed based on two observations:

i. When the current solution is close to the centre of the feasible region, we can move along the steepest descent direction that decreases the value of the objective function f by the maximum amount.

ii. Transformation of the solution space is always possible without changing the characteristic of the problem in order to keep the current solution near the centre of the feasible region.

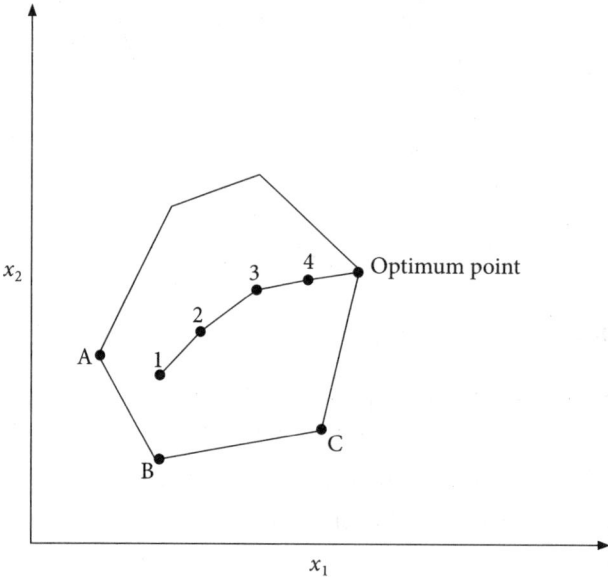

Fig. 6.6 Interior point method

In many numerical problems, the numerical instability can be reduced by changing the units of data or rescaling. Karmarkar noticed that the variables could be transformed in such a way that straight lines remain straight lines whereas distances and angles change for the feasible space. Karmarkar's projective scaling algorithm begins with an interior solution, by transforming the feasible region in a way so that the current solution is positioned at the center of the transformed feasible region. Figure 6.7(a) is the feasible region with the point $X^{(k)}$, far from the center. This region was converted to region 6.7(b), where the point $X'^{(k)}$ is near the center of the region. Where

$$X^{(k)} = \begin{bmatrix} x_1^{(k)} \\ x_2^{(k)} \\ \vdots \\ x_n^{(k)} \end{bmatrix}, \; X'^{(k)} = \begin{bmatrix} \dfrac{1}{n} \\ \dfrac{1}{n} \\ \vdots \\ \dfrac{1}{n} \end{bmatrix} = \dfrac{e}{n} \; \text{ and } e = \begin{bmatrix} 1 & 1 & \cdots & 1 \end{bmatrix}^T \qquad (6.26)$$

After transformation, the searching route follows the steepest descent direction with a step that stops very close to the boundary of the transformed feasible region. Afterward, the improved solution was mapped to the original feasible region by using an inverse transformation. This procedure is repeated until an optimum is achieved with the desired accuracy.

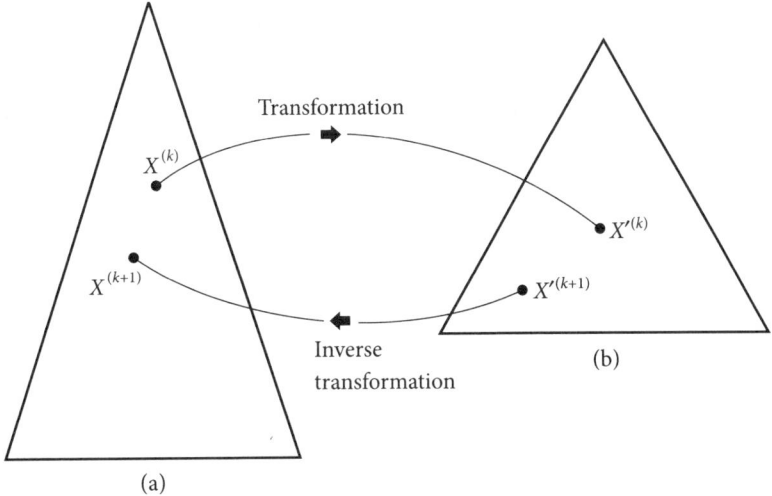

Fig. 6.7 Karmarkar's region inversion

Karmarkar's method employs the LPP in the following form

$$\text{Minimize } f = c^T X \tag{6.27a}$$

subject to

$$[a]X = 0 \tag{6.27b}$$

$$e^T X = 1 \tag{6.27c}$$

$$X \geq 0 \tag{6.27d}$$

where $X = \{x_1 \ x_1 \ ... \ x_n\}^T$, $c = \{c_1 \ c_2 \ ... \ c_n\}^T$, $e = \{1 \ 1 \ ...1\}^T$, and $[a]$ is an $m \times n$ matrix. Beside this, an interior feasible starting solution to Eq. (6.27b)–(6.27d) must be known. Usually, $X = \left\{\dfrac{1}{n} \ \dfrac{1}{n} \ ... \ \dfrac{1}{n}\right\}^T$ is preferred as the starting point. The optimum value of the objective function f must be zero for the problem. Therefore,

$$X^{(1)} = \left\{\dfrac{1}{n} \ \dfrac{1}{n} \ ... \ \dfrac{1}{n}\right\}^T = \text{interior feasible} \tag{6.28}$$

$$f_{\min} = 0 \tag{6.29}$$

Although, most LPPs may not be available in the form of Eq. (6.27b)–(6.27d) while satisfying the conditions of Eq. (6.28), it is possible to put any LPP in a form that satisfies Eq. (6.27b)–(6.27d) and (6.28) as indicated below. The process of this transformation is discussed below. Karmarkar used "Projective Transformations" which are non-linear transformations under which lines and subspaces are preserved but distances and angles are distorted [Lemire, (1989)].

The algorithm creates a sequence of points $X^{(1)}$, $X^{(2)}$, ...,$X^{(k)}$. The whole process can be divided into 4 steps:

Step 1 Initialization $X^{(1)}$ = center of the simplex

Step 2 Computation of the next point in the sequence

$$X^{(k+1)} = \varphi\left(X^{(k)}\right) \quad (6.30)$$

Step 3 Checking for infeasibility

Step 4 Checking for optimality
go to Step 1.

Now we will discuss the steps in detail

1. The function $b = j(a)$ can be defined by the following sequence of operations.
   Let $D = \text{diag}\{a_1, a_2 \ldots a_n\}$ be a diagonal matrix whose ith diagonal entry is a_i.
   Let

$$B = \begin{bmatrix} AD \\ e^T \end{bmatrix} \quad (6.31)$$

   i.e., augment the matrix AD with a row of all 1's. This guarantees that $KerB$

2. Compute the orthogonal projection of Dc into the null space of B.

$$c_p = \left[I - B^T\left(BB^T\right)^{-1}B\right]Dc \quad (6.32)$$

3. $\hat{c} = \dfrac{c_p}{|c_p|}$ i.e., $\hat{c}$ is the unit vector in the direction of c_p.

4. $b' = a_0 - \alpha r\hat{c}$ i.e., take a step of length αr in the direction $\hat{c}$, where r is the radius of largest inscribed sphere. By the Euclidean distance formula:

$$r = \frac{1}{\sqrt{n(n-1)}} \quad (6.33)$$

   and $\alpha \in (0,1)$ is a parameter which can be set equal to 1/4. The value of α $(0 < \alpha < 1)$ ensures that all iterates are interior points.

5. Apply inverse projective transformation to b'

$$b = \frac{Db'}{e^T Db'} \quad (6.34)$$

Return *b*.

Step 3. Check for in feasibility.

A "potential" function can be defined by

$$f(X) = \sum_i \ln \frac{c^T X}{x_i} \tag{6.35}$$

At each step, a certain improvement δ in the potential function is expected. The value of δ depends on how the parameter α is selected in the above Step. For example, if value of $\alpha = 1/4$ then $\delta = 1/8$. If the expected improvement is not found, i.e., if $f(X^{(k+1)}) > f(X^{(k)}) - \delta$ then we stop and conclude that the minimum value of the objective function must be strictly positive. The standard linear program problem can be transformed to the canonical form, then this condition corresponds to the situation that the original problem does not possess a finite optimum i.e., it is either unbounded or infeasible.

Step 4. Check for optimality

The optimality check should be done periodically. It includes moving from the current interior point to an extreme point without increasing the objective function value and checking the extreme point for optimality. This process is followed only when the time spent since the last check exceeds the time required for checking [Karmarkar, (1984)].

6.2.4 Integer linear programming

The linear programming problems that have been discussed so far are all continuous, in the sense that all decision variables are permitted to be fractional. For instance, we might optimize rate of production, flow rate of reactant and time of operation for batch process. However, in some cases fractional solutions are unrealistic. We have to consider the integer number for those decision variables. For example, number of worker, number of batches per day or month and number of heat exchanger in any HEN.

The optimization problem can be written for these systems

$$\text{Minimize } f = \sum_{j=1}^{n} c_j x_j \tag{6.36a}$$

subject to

$$\sum_{j=1}^{n} a_{ij} x_j = b_i, \; (i=1,2,\ldots,m; \; j=1,2,\ldots,n) \tag{6.36b}$$

$$x \in Z_+^n \tag{6.36c}$$

where Z_+^n denotes the set of dimensional vectors n having integer non-negative components.

The problem of Eq. (6.36a)–(6.36c) is called linear integer programming problem. When, a few variables but not all are restricted to be an integer, it is said to be a mixed integer program and is termed a pure integer program whilst all decision variables must be integers. Shaded area in Fig. 6.8 is the feasible region when variables (x_1, x_2) are real numbers, whereas red dots represents the feasible points for integer linear programming. Now, we will discuss two useful Mixed Integer Linear Programming (MILP) models.

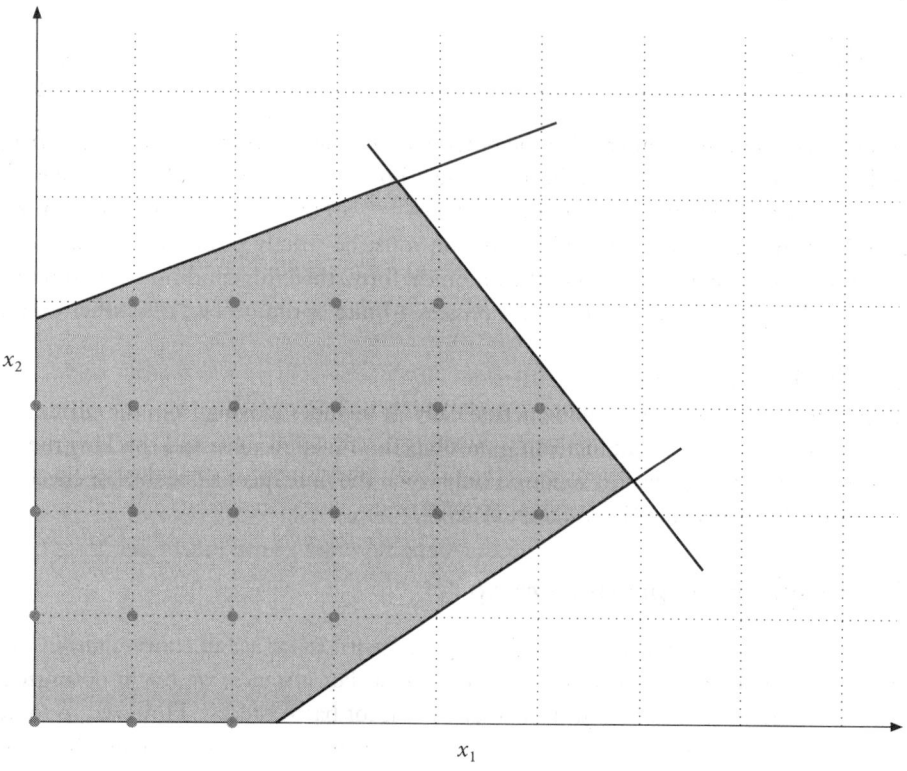

Fig. 6.8 Integer linear programming

Example 6.4 Warehouse Location During modeling any distribution system, a trade off between cost of transportation and operating cost of distribution centers is required for taking any decision. Say for example, a marketing manager required to take decision on which of n warehouses to be used that will meet the demands of m customers for a substance. The decisions to be made are which warehouses to be open and how much to ship from any warehouse to any customer. Let,

$$\text{Minimize } f = \sum_{i=1}^{m}\sum_{j=1}^{n} c_{ij} x_{ij} + \sum_{i=1}^{m} f_i y_i \tag{6.37a}$$

subject to

$$\sum_{i=1}^{m} x_{ij} = d_j \ (j = 1, 2, \ldots, n) \tag{6.37b}$$

$$\sum_{j=1}^{n} x_{ij} - y_i \left(\sum_{j=1}^{n} d_j \right) \leq 0 \ (i = 1, 2, \ldots, m) \tag{6.37c}$$

$$x_{ij} \geq 0 \quad (.., j = 1, 2, \ldots, n) \tag{6.37d}$$

$$y_i = 0 \text{ or } 1 \quad (i = 1, 2, \ldots, m) \tag{6.37e}$$

where

x_{ij} = Amount to be sent from warehouse i to customer j.

$$y_i = \begin{cases} 1 & \text{if warehouse } i \text{ is opened} \\ 0 & \text{if warehouse } i \text{ is not opened} \end{cases}$$

f_i = Fixed operating cost for warehouse i, if opened (for example, a cost to lease the warehouse),
c_{ij} = Per-unit operating cost at warehouse i plus the transportation cost for shipping from warehouse i to customer j.

Here, we have considered

i. the warehouses must fulfill the demand d_j of each customer; and
ii. shipment of goods from a warehouse is possible only if it is opened.

Example 6.5 Blending problem We have to prepare a blend from a given list of ingredients. The list gives us various information such as weight, value, cost, and analysis of each ingredient [Edger et al. (2001)].

Our objective is to prepare a blend of some specified total weight with satisfactory analysis by selecting a set of ingredients from the list with a minimum cost for a blend. Let, x_j is the amount of ingredient j available (continuous amounts) and y_k indicate ingredients to be utilized in discrete quantities v_k (if it is used y_k = 1 and y_k = 0 if not used). The quantities c_j and d_k be the respective costs of the ingredients and a_{ij} be the fraction of the component i in ingredients j. The problem can be stated is

$$\text{Minimize } f = \sum_j c_j x_j + \sum_k d_k v_k y_k \tag{6.38a}$$

subject to:

$$W^l \leq \sum_j x_j + \sum_k v_k y_k \leq W^u \tag{6.38b}$$

$$A_i^l \leq \sum_j a_{ij} x_j + \sum_k a_{ik} v_k y_k \leq A_i^u \tag{6.38c}$$

$$0 \leq x_j \leq u_j \text{ for all } j \tag{6.38d}$$

$$y_k = (0, 1) \text{ for all } k \tag{6.38e}$$

where, u_j = upper limit of the jth ingredient,

W^u and W^l = the upper and lower bounds on the weights respectively
A_i^u and A_i^l = the upper and lower bounds on the analysis for component i respectively
The following section elucidates the Branch and Bound method for solving MILP.

Branch-and-Bound method

Branch-and-Bound method has been developed by Land and Doig [Land and Doig (1960)]. This method is very efficient for solving mixed-integer linear and nonlinear programming. Feroymson and Ray used this method for plant location problem [Feroymson and Ray, (1966)]. This method consists of two basic operations. Branching, which divide the whole set of solution into subsets and bounding that consists of establishing bounds on the value of the objective function over the subsets. The branch-and-bound method involves recursive application of the branching and bounding operations, with a provision for eliminating subsets that do not contain any optimal solution [L. G. Mitten (1970)].

The integer problem is not directly solved in branch-and-bound method. Whereas, it relaxes the integer restrictions on the variables and solves as a continuous problem [Rao (2001)]. Sometimes the solution of that continuous problem may be an integer solution; on that case, we can consider it as the optimum solution of the integer problem. Otherwise, we have to assume at least one of the integer variables, say x_j, must have a nonintegral value. Whenever x_i is not an integer, always we can find an integer $[x_i]$ such that,

$$[x_i] < x_i < [x_i] + 1 \tag{6.39}$$

Then, we have to formulate two subproblems, one with the additional upper bound constraint

$$x_i \leq [x_i] \tag{6.40}$$

and another with lower bound as the additional constraint

$$x_i \geq [x_i] + 1 \tag{6.41}$$

This method of establishing the subproblems is called branching.

The branching operation removes some part of the continuous space, which is not feasible for the integer problem. At the same time, we have to ensure that any of the integer feasible solutions is not discarded. Then we have to solve these two subproblems as continuous problems. It is observed that the solution to the continuous problem forms a node and two branches may originate from this node.

This practice of branching and solving a series of continuous problems is continued until an integer feasible solution is obtained for one of the two continuous problems. Whenever such a feasible integer solution is obtained, the objective function value at that point turns into an upper bound on the minimum value of the objective function. At this instant, further consideration is not required for all these continuous solutions (nodes) that have objective function values larger than the upper bound. The eliminated nodes are said to have been fathomed since it is impossible to get a better integer solution from these nodes (solution spaces) than whatever we have at the present. Whenever a better bound is obtained, then the value of the upper bound on the objective function is updated [Rao (2001)]. This process can be illustrated by the example 6.6.

Example 6.6

Solve the problem using branch and bound method.

Maximize: $Z = 5x_1 + 7x_2$

subject to constraints

$x_1 + x_2 \leq 6$

$5x_1 + 9x_2 \leq 43$

$x_1, x_2 \geq 0$ and Integer

Solution

At the first step, we solve the problem as a continuous problem. The corresponding result is

$x_1 = 2.75$, $x_2 = 3.25$ and $Z = 36.50$

As x_1, x_2 are not integer, we have to form two subproblems (Eqs (6.40), (6.41))

one with additional upper bound $x_1 \leq 2$ (subproblem 1) and another is with lower bound $x_1 \geq 3$ (subproblem 2)

Table 6.1 Subproblems for branch and bound method

Subproblem 1	Subproblem 2
Maximize: $Z = 5x_1 + 7x_2$ subject to constraints	Maximize: $Z = 5x_1 + 7x_2$ subject to constraints
$x_1 + x_2 \leq 6$	$x_1 + x_2 \leq 6$
$5x_1 + 9x_2 \leq 43$	$5x_1 + 9x_2 \leq 43$
$x_1 \leq 2$	$x_1 \geq 3$
$x_1, x_2 \geq 0$	$x_1, x_2 \geq 0$

By solving these subproblems, we get

Table 6.2 Results of the subproblems in Table 6.1

Solution of subproblem 1	Solution of subproblem 2
$x_1 = 2$	$x_1 = 3$
$x_2 = 3.667$	$x_2 = 3$
$Z_1^* = 35.667$	$Z_2^* = 36$

Here, subproblem 2 is fathomed as x_1, x_2 are integer.

As $Z_1^* \leq Z_2^*$, we can stop branching. Then, the final solution is

$x_1^* = 3$

$x_2^* = 3$

$Z^* = 36$

Note The code for solving this problem using LINGO is given in chapter 12.

6.3 Nonlinear Programming with Constraints

In chapter 3 and 5, we have discussed various methods of nonlinear optimization without any constraint. Optimization with constraints is very common in chemical engineering. Most of the chemical process plant faces the limitations for raw materials, manpower, utilities, and space. We can include these constraints during optimization of the process variables. Another constraints might appear due to environmental considerations. The following sections elucidate the constraints nonlinear optimization

6.3.1 Problems with equality constraints

The constrained optimization problems with equality constraints could be represented as the following:

$$\text{Minimize } f(X) \quad X \in \mathbb{R}^n \tag{6.42a}$$

$$\text{subject to } c_i(X) = 0 \tag{6.42b}$$

$$X \geq 0$$

where $f(X)$ is the objective function, $c_i(X)$ are constraint functions.

6.3.1.1 Direct substitution

For a problem with m equality constraints consist of n variables, theoretically we are able to solve these m equality constraints simultaneously and represent any set of m variables in terms of the remaining $n - m$ variables. The substitution of these expressions into the original objective function gives us a new objective function that involves only $n - m$ variables. This new objective function becomes an unconstrained problem that is not subjected to any constraint. Therefore, optimization of this objective function can be performed by using the unconstrained optimization techniques discussed in Chapter 5.

$$\text{Minimize: } f(X) = 2x_1^2 + 2x_2^2 + x_3^2 - 2x_1x_2 - 4x_1 - 6x_2 \tag{6.43a}$$

subject to: $x_1 + x_2 + x_3 = 2$ (6.43b)

$x_1^2 + 5x_2 = 5$ (6.43c)

We will consider this problem to elucidate the "Direct Substitution Method". In the first step, Eq. (6.43c) is rearranged as

$$x_2 = \frac{5 - x_1^2}{5}$$ (6.44)

Now, replace x_2 from Eq. (6.43a) and (6.43b) using Eq. (6.44).
after rearrangement the final form is

Minimize: $f(X) = \frac{2}{25}x_1^4 + \frac{12}{5}x_1^2 + x_1^3 + \frac{2}{5}x_1^2 x_3^2 - 6x_1 - 4$ (6.45a)

subject to: $-x_1^2 + 5x_1 + 5x_3 - 5 = 0$ (6.45b)

this problem has 2 variables and 1 equality constraint
Again, we have to substitute x_3 from Eq. (6.45b) to Eq. (6.45a). Then, the final equation will be

Minimize: $f(X) = \frac{2}{25}x_1^4 + \frac{12}{5}x_1^2 + x_1^3 + \frac{2}{125}x_1^2 \left(5 - 5x_1 + x_1^2\right)^2 - 6x_1 - 4$ (6.46)

Equation (6.46) is an unconstrained optimization problem. We will discuss the problem formulated in section 2.3.1.

Example 6.7

The price per unit side area is $5 and price per unit area of top and bottom is $8. Find the optimum diameter and height of the for a 1000 liter tank.

$$\min_{L,D} f = c_s \pi DL + c_t \left(\pi/2\right) D^2$$ (2.1)

subject to $V = \left(\pi/4\right) D^2 L$ (2.2)

Solution

Incorporating the values $c_s = 5$ and $c_t = 8$ in Eq. (2.1), we get

$$\min_{L,D} f = 15.7 DL + 12.56 D^2$$ (6.47a)

subject to $D^2 L = 1273.9$ (6.47b)

and $V = 1000$ \hfill (6.47c)

We can use "Direct substitution method" to solve this problem. For this purpose Eq. (6.47b) can be written as

$$L = \frac{1273.9}{D^2} \hfill (6.48)$$

Now substitute L to the Eq. (6.47a), to get

$$\min_D f = \frac{20000.23}{D} + 12.56D^2 \hfill (6.49)$$

The Eq. (6.49) is an unconstrained problem with $2 - 1 = 1$ variable. Now this problem can be solved by using any method described in chapter 3.

6.3.1.2 Lagrange multiplier method

Lagrange multiplier method can be used for problems with equality constraints. The Lagrange multiplier method is discussed by the Example 6.8 of two variables with one constraint. The form of a general problem with n variables and m constraints is given later.

Consider a problem as follows:

$$\text{Minimize } f(x_1, x_2) \hfill (6.50a)$$

$$\text{subject to } g(x_1, x_2) = 0 \hfill (6.50b)$$

The necessary condition for the existence of an extreme point at $X = X^*$

$$\left(\frac{\partial f}{\partial x_1} - \frac{\partial f / \partial x_2}{\partial g / \partial x_2} \frac{\partial g}{\partial x_1} \right)\bigg|_{(x_1^*, x_2^*)} = 0 \hfill (6.51)$$

The Lagrange multiplier (λ) can be defined as

$$\lambda = -\left(\frac{\partial f / \partial x_2}{\partial g / \partial x_2} \right)\bigg|_{(x_1^*, x_2^*)} \hfill (6.52)$$

The Eq. (6.51) can be expressed as

$$\left(\frac{\partial f}{\partial x_1} + \lambda \frac{\partial g}{\partial x_1} \right)\bigg|_{(x_1^*, x_2^*)} = 0 \hfill (6.53)$$

and Eq. (6.52) can be expressed as

$$\left(\frac{\partial f}{\partial x_2} + \lambda \frac{\partial g}{\partial x_2}\right)\bigg|_{(x_1^*, x_2^*)} = 0 \tag{6.54}$$

In addition, the constraint equation should satisfy the extreme point, that is,

$$g(x_1, x_2)\big|_{(x_1^*, x_2^*)} = 0 \tag{6.55}$$

These are the necessary conditions for the point (x_1^*, x_2^*) to be an extreme point.

Notice that the partial derivative $(\partial g / \partial x_2)\big|_{(x_1^*, x_2^*)}$ has to be nonzero to be able to define λ by Eq. (6.52). This is because the variation dx_2 was expressed in terms of dx_1. On the other hand, if dx_1 is expressed in terms of dx_2, we would have obtained the requirement that $(\partial g / \partial x_1)\big|_{(x_1^*, x_2^*)}$ be nonzero to define λ. Therefore, the derivation of the necessary conditions using the method of Lagrange multipliers requires that at least one of the partial derivatives of $g(x_1, x_2)$ be nonzero at an extreme point.

Construct a function L, known as Lagrange function

$$L(x_1, x_2, \lambda) = f(x_1, x_2) + \lambda g(x_1, x_2) \tag{6.56}$$

by treating L as a function of the three variables x_1, x_2 and λ, the necessary conditions for its extremum are given by

$$\frac{\partial L}{\partial x_1}(x_1, x_2, \lambda) = \frac{\partial f}{\partial x_1}(x_1, x_2) + \lambda \frac{\partial g}{\partial x_1}(x_1, x_2) = 0 \tag{6.57}$$

$$\frac{\partial L}{\partial x_2}(x_1, x_2, \lambda) = \frac{\partial f}{\partial x_2}(x_1, x_2) + \lambda \frac{\partial g}{\partial x_2}(x_1, x_2) = 0 \tag{6.58}$$

$$\frac{\partial L}{\partial \lambda}(x_1, x_2, \lambda) = g(x_1, x_2) = 0 \tag{6.59}$$

Sufficiency conditions: A sufficient condition for $f(X)$ to have a constrained relative minimum at X^* is given by the following theorem.

Theorem 6.1

A sufficient condition for $f(X)$ to have relative minimum at X^* is that the quadratic, Q, defined by

$$Q = \sum_{i=1}^{n} \sum_{j=1}^{n} \frac{\partial^2 L}{\partial x_i \partial x_j} dx_i dx_j \tag{6.60}$$

evaluate at $X = X^*$ must be positive definite for all values of dX for which the constraints are satisfied.

It has been shown by Hancock [Hancock, (1960)] that a necessary condition for the quadratic form Q in Eq. (6.60), to be positive (negative) definite for all admissible variations dX is that each root of the polynomial z_j, defined by the following determinantal equation, be positive (negative):

$$\begin{vmatrix} L_{11}-z & L_{12} & L_{13} & \cdots & L_{1n} & g_{11} & g_{21} & \cdots & g_{m1} \\ L_{21} & L_{22}-z & L_{21} & \cdots & L_{21} & g_{12} & g_{22} & \cdots & g_{m2} \\ \vdots & & & & & & & & \\ L_{n1} & L_{n2} & L_{n3} & \cdots & L_{nn}-z & g_{1n} & g_{2n} & \cdots & g_{mn} \\ g_{11} & g_{12} & g_{13} & \cdots & g_{1n} & 0 & 0 & \cdots & 0 \\ g_{21} & g_{22} & g_{23} & \cdots & g_{2n} & 0 & 0 & \cdots & 0 \\ \vdots & & & & & & & & \\ g_{m1} & g_{m2} & g_{m3} & \cdots & g_{mn} & 0 & 0 & \cdots & 0 \end{vmatrix} = 0 \qquad (6.61)$$

where

$$L_{ij} = \frac{\partial^2 L}{\partial x_i \partial x_j}(X^*, \lambda^*) \qquad (6.62)$$

$$g_{ij} = \frac{\partial g_i}{\partial x_j}(X^*) \qquad (6.63)$$

Equation (6.61) is an $(n-m)$th order polynomial in z. The point X^* is not an extreme point when some of the roots of this polynomial are positive whereas the others are negative.

Example 6.8

Minimize the function using Lagrange multiplier method

Minimize $f(x_1, x_2) = 25 + x_1 - 3x_2 + x_1^2 + 2x_2^2 - 5x_1 x_2$

subject to $x_1^2 + x_2^2 = 7$

Solution

The Lagrange function $L(x_1, x_2, \lambda) = \left(25 + x_1 - 3x_2 + x_1^2 + 2x_2^2 - 5x_1 x_2\right) + \lambda\left(x_1^2 + x_2^2 - 7\right)$

The necessary conditions

$$\frac{\partial L}{\partial x_1} = 1 + 2x_1 - 5x_2 + \lambda(2x_1) = 0$$

$$\frac{\partial L}{\partial x_2} = -3 + 4x_2 - 5x_1 + \lambda(2x_2) = 0$$

$$\frac{\partial L}{\partial \lambda} = x_1^2 + x_2^2 - 7 = 0$$

Solving these equations, we get

$x_1^* = 1.8585$, $x_2^* = 1.8831$, and $\lambda^* = 1.2641$

which gives

$f^* = 14.25676$

To check whether this solution really corresponds to the minimum of f, we can apply the sufficiency condition.

$$L_{11} = \left.\frac{\partial^2 L}{\partial x_1^2}\right|_{(X^*,\lambda^*)} = 2 + 2\lambda^* = 4.5282$$

$$L_{12} = \left.\frac{\partial^2 L}{\partial x_1 \partial x_2}\right|_{(X^*,\lambda^*)} = -5$$

$$L_{22} = \left.\frac{\partial^2 L}{\partial x_2^2}\right|_{(X^*,\lambda^*)} = 4 + 2\lambda^* = 6.5282$$

$$g_{11} = \left.\frac{\partial g_1}{\partial x_1}\right|_{(X^*,\lambda^*)} = 2x_1^* = 3.717$$

$$g_{12} = \left.\frac{\partial g_1}{\partial x_2}\right|_{(X^*,\lambda^*)} = 2x_2^* = 3.7662$$

equation (6.61) can be written as

$$\begin{vmatrix} 4.5282 - z & -5 & 3.717 \\ -5 & 6.5282 - z & 3.7662 \\ 3.717 & 3.7662 & 0 \end{vmatrix} = 0$$

which gives $z = 10.51462$. This result confirms that the f^* is the minimum value.

6.3.2 Problems with inequality constraints

The general form of an optimization with inequality constraints is given below

Minimize $f(X)$ $X \in \mathbb{R}^n$ (6.64a)

subject to $c_j(X) \geq 0, \ j = 1, 2, \ldots m$ (6.64b)

$X \geq 0$ (6.64c)

where $f(X)$ is the objective function and $c_j(X) \geq 0$ are inequality constraints. Lagrange multiplier method can be used to solve these problems.

An inequality constraint $c_j(X) \geq 0$ is said to be active at X^* if $c_j(X^*) \geq 0$. It is inactive at X^* if $c_j(X^*) > 0$ [Chong and Zak, (2001)].

6.3.2.1 Kuhn–Tucker condition

If the problem has inequality constraints, the Kuhn–Tucker condition can be used to identify the optimum point. However, these methods produce a set of nonlinear simultaneous equations that may be difficult to solve.

The nonlinear programming (NLP) problem with one objective function $f(X)$ and m constraint functions c_j which are continuously differentiable, can be represented as follows:

Minimize $f(X), \ X \in \Re^n$ (6.65a)

subject to $c_j(X) \leq 0, \ j = 1, 2, \ldots m$ (6.65b)

In the preceding notation, n is the dimension of the function $f(X)$, and m is the number of inequality constraints.

$$L(X, \lambda) = f(X) - \sum_{j=1}^{m} \lambda_j c_j(X) \tag{6.66}$$

is the Lagrange function, and the coefficients λ_j are the Lagrange multipliers.

If the functions $f(X)$ and c_j are twice differentiable, the point X^* is an isolated local minimizer of the NLP problem, if there exists a vector $\lambda^* = (\lambda_1^*, \lambda_2^*, \ldots, \lambda_m^*)$ that meets the following conditions:

$$\frac{\partial f}{\partial x_i} + \sum_{j \in J_1} \lambda_j \frac{\partial c_j}{\partial x_i} = 0, \ i = 1, 2, \ldots n \tag{6.67a}$$

$\lambda_j > 0, \ j \in J_1$ (6.67b)

For convex programming, the Kuhn–Tucker conditions are necessary and sufficient for a global minimum. The Kuhn–Tucker conditions can be stated as follows:

$$\frac{\partial f}{\partial x_i} + \sum_{j=1}^{m} \lambda_j \frac{\partial g_j}{\partial x_i} = 0, \ i = 1, 2, \ldots n \tag{6.68a}$$

$$\lambda_j g_j = 0, \quad j = 1, 2, \ldots m \tag{6.68b}$$

$$g_j \leq 0, \quad j = 1, 2, \ldots m \tag{6.68c}$$

$$\lambda_j \geq 0, \quad j = 1, 2, \ldots m \tag{6.68d}$$

6.3.2.2 Logarithmic barrier method

During optimization process, the algorithm may cross the boundary of feasible region. We can follow a technique to prevent the optimization algorithm from crossing the boundary by assigning a penalty to approaching it. The most accepted way of doing this is to augment the objective function by a logarithmic barrier term:

$$B(x,\mu) = f(x) - \mu \sum_{i=1}^{p} \log(h_i(x)) \tag{6.69}$$

Here, log denotes the natural logarithm. Since,

$$-\log(t) \to \infty \text{ as } t \to 0, \tag{6.70}$$

The term $B(x,\mu)$ "blows up" at the boundary and consequently presents an optimization algorithm with a "barrier" to crossing the boundary. Certainly, the solution to an inequality-constrained is expected to lie on the boundary of the feasible set, so the barrier should be eliminated gradually by decreasing the value of μ toward zero. The following strategy has been suggested:

Choose $\mu_0 > 0$ and a strictly feasible point $x^{(0)}$.
For $k = 1, 2, 3, \ldots$
Choose $\mu_k \in (0, \mu_{k-1})$ (perhaps $\mu_k = \beta \mu_{k-1}$ for some constant $\beta \in (0,1)$.
Using $x^{(k-1)}$ as the starting point, solve

$$\min B(x, \mu_k) \tag{6.71}$$

to get $x^{(k)}$

Under certain conditions, $B(x,\mu)$ has a unique minimizer x_μ^* in a neighbourhood of x^* and that $x_\mu^* \to x^*$ as $\mu \to 0$.
Then for all μ sufficiently small

$$\nabla B(x_\mu^*, \mu) = 0 \tag{6.72}$$

6.3.3 Convex optimization problems

Problems involving the optimization of convex functions (objective as well as constraints functions) are called convex optimization problems. This class of problems has some advantages over other form of optimization problems; local optimizer is the global optimizer for them.

The standard form of a convex optimization problem

$$\text{Minimize } f_0(X) \tag{6.73a}$$

$$\text{subject to } f_i(X) \leq 0, \ i = 1, 2, \ldots m \tag{6.73b}$$

$$h_i(X) = 0, \ i = 1, 2, \ldots p \tag{6.73c}$$

where

$X \in R^n$ is the optimization variable (convex)

$f_0 : R^n \to R$ is the objective or cost function (convex function)

$f_i : R^n \to R, \ i = 1, 2, \ldots m$, are the inequality constraint functions (convex function)

$h_i : R^n \to R$ are equality constraint functions (affine functions)
optimal value of the problem (6.73a–6.73c) are as follows:

$$p^* = \inf \{ f_0(X) | f_i(X) \leq 0, \ i = 1, \ldots, m, \ h_i(X) = 0, \ i = 1, \ldots, p \} \tag{6.74a}$$

$$p^* = \infty \text{ if problem is infeasible (no } X \text{ satisfies the constraints)} \tag{6.74b}$$

$$p^* = -\infty \text{ if problem is unbounded below} \tag{6.74c}$$

i. X is a feasible if $X \in \text{dom } f_0$ and it satisfies the constraints and a feasible X is optimal if $f_0(X) = p^*$; X_{opt} is the set of optimal points.
ii. X is locally optimal if there is an $R > 0$ such that X is optimal for

$$\text{Minimize (over } z) \ f_0(z) \tag{6.75a}$$

$$\text{subject to } f_i(z) \leq 0, \ i = 1, 2, \ldots m; \ h_i(z) = 0, \ i = 1, 2, \ldots p \tag{6.75b}$$

$$\|z - X\|_2 \leq R \tag{6.75c}$$

The problem is quasiconvex when $f(X)$ is quasiconvex (and $g_i(X)$ are convex) often written as

$$\text{Minimize } f(X) \tag{6.76a}$$

$$\text{subject to } g_i(X) \leq 0, \ i = 1, 2, \ldots m \tag{6.76b}$$

$$AX = b, \ i = 1, 2, \ldots p \tag{6.76c}$$

important property: feasble set of a convex optimization problem is convex

Optimality condition

Considering the optimization problem above, which we write as

$$\min_{x \in X} f_0(x) \tag{6.77}$$

where X is the feasible set.

When f_0 is differentiable, then we know that for every $x, y \in \operatorname{dom} f_0$,

$$f_0(y) \geq f_0(x) + \nabla f_0(x)^T (y - x) \tag{6.78}$$

Then, x is optimal if and only if

$$x \in X \text{ and } \forall y \in X : \nabla f_0(x)^T (y - x) \geq 0 \tag{6.79}$$

If $\nabla f_0(x) \neq 0$, then it defines a supporting hyperplane to the feasible set at x.

When, the problem is unconstrained, we obtain the optimality condition:

$$\nabla f_0(x) = 0 \tag{6.80}$$

Note that these conditions are not always feasible, since the problem may not have any minimizer. This can happen for example when the optimal value is only attained in the limit; or, in constrained problems, when the feasible set is empty.

Theorem 6.2

Any locally optimal point of a convex problem is (globally) optimal

Proof

Let x^* be a local minimizer of f_0 on the set X, and let $y \in X$. By definition, $x^* \in \operatorname{dom} f_0$. We need to prove that $f_0(y) \geq f_0(x^*) = p^*$. There is nothing to prove if $f_0(y) = +\infty$, so let us assume that $y \in \operatorname{dom} f_0$. By convexity of f_0 and X, we have $x_\theta := \theta y + (1 - \theta) x^* \in X$, and:

$$f_0(x_\theta) - f_0(x^*) \leq \theta \big(f_0(y) - f_0(x^*) \big) \tag{6.81}$$

as x^* is a local minima, the left-hand side in this inequality is nonnegative for all small enough values of $\theta > 0$. We conclude that the right-hand side is nonnegative, i.e., $f_0(y) \geq f_0(x^*)$, as claimed.

Also, the optimal set is convex, since it can be written

$$X^{\text{opt}} = \{ x \in R^n : f_0(x) \leq p^*, x \in X \} \tag{6.82}$$

This proofs the statement of the theorem.

Summary

- Formulation and solution methods for multivariable and constrained optimization have been considered in this chapter. Both the methods of linear and nonlinear programming algorithms are discussed with proper examples. Integer linear programming that is extensively applicable for distribution of chemical products from different warehouses to different distributors and the blending process are considered. This chapter also enlightens the theories and examples of nonlinear constrained optimization processes.

    Further studies Real-time optimization of the pulp mill benchmark problem; Mehmet Mercangöz, Francis J. Doyle III; Computers and Chemical Engineering 32 (2008) 789–804

Review Questions

6.1 Is it possible to convert a constrained optimization problem to an unconstrained optimization problem? Explain your answer with proper example.

6.2 Maximize $f = 5x_1 + 3x_2$

subject to the constraints

$2x_1 + x_2 \leq 1000$

$x_1 \leq 400$

$x_2 \leq 700$

$x_1, x_2 \geq 0$

6.3 Convert the following problem to its dual form

Maximize $f = 400x_1 + 200x_2$

subject to the constraints

$18x_1 + 3x_2 \leq 800$

$9x_1 + 4x_2 \leq 600$

$x_1, x_2 \geq 0$

6.4 A refinery distills two crude petroleum, A and B, into three main products: jet fuel, gasoline and lubricants. The two crudes differ in chemical composition and thus, yield different product mixes (the remaining 10 per cent of each barrel is lost to refining): Each barrel of crude A yields 0.4 barrel of jet fuel, 0.3 barrel of gasoline, and 0.2 barrel of lubricants; Each barrel of crude B yields 0.2 barrel of jet fuel, 0.4 barrel of gasoline, and 0.3 barrel of lubricants. The crudes also differ in cost and availability: Up to 5,000 barrels per day of crude A are available at the cost $25 per barrel; Up to 3,000 barrels per day of Saudi crude are also available at the lower cost $18 per barrel.

Contracts with independent distributors require that the refinery produce 2,000 barrels per day of gasoline, 1,500 barrels per day of jet fuel, and 500 barrels per day of lubricants.

6.5 What are the advantages of Khachiyan's ellipsoid method over simplex method?

6.6 Minimize $f = -3x_1 + 2x_2$

Subject to $x_1 + x_2 \leq 9$

$x_2 \leq 6$

$x_1, x_2 \geq 0$

make the "off-center" point to an "equidistant" from the coordinate axes in a transformed feasible region by "variable scaling".

6.7 Solve the following problem by branch and bound method.

Maximize $f = x_1 + x_2$

subject to the constraints

$2x_1 + 5x_2 \leq 16$

$6x_1 + 5x_2 \leq 30$

$x_1, x_2 \geq 0$ and Integer

6.8 Solve the problem using direct substitution method

Maximize $f = 3x_1^2 + 2x_2^2$

subject to the constraints

$x_1 + 3x_2 = 5$

6.9 Solve the problem using Lagrange multiplier method

Minimize $f = 3x_1^2 + 4x_2^2 + x_1 x_3 + x_2 x_3$

subject to

$x_1 + 3x_2 = 3$

$2x_1 + x_3 = 7$

6.10 Find the minimum value of the function

$f = x_1^2 + x_2^2 - x_1 x_2$

subject to the inequality constraint

$x_1 + 3x_2 \geq 3$

6.11 Why convex optimization problem is easier to solve compare to nonconvex problem?

6.12 We have 25 ft steel frame for manufacturing a window. The shape of window is shown in Figure. Estimate the optimum dimensions such that the area of the window will be maximum.

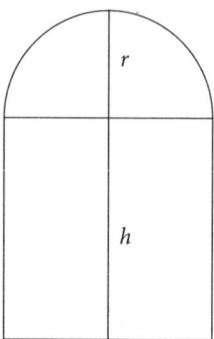

Fig. 6.9 Shape of the window

References

Bland, R. G., Goldfarb, D., Todd, M. J. 1981. *Feature Article-The Ellipsoid Method:* A Survey, Operations Research, 29(6): 1039–91.

Chong, E. K. P., Zak, S. H. 2001. *An Introduction to Optimization Second Edition,* John Wiley and Sons, Inc.

Dantzig, G. B. 1963. *Linear Programming and Extensions*, Princeton University Press.

Dantzig, G. B. 2002. *Linear Programming, Operations Research,* vol. 50, No. 1, January–February 2002, pp. 42–47.

Edgar, T. F., Himmelblau, D. M., Lasdon, L. S. 2001. *Optimization of Chemical Processes* McGraw–Hill.

Feroymson, M. A. and Ray, T. L. 1966. *A Branch-and-Bound Algorithm for Plant Location*, Opns Res 14, 361–68.

Fletcher, R. *Practical Methods of Optimization*, John Wiley and Sons.

Hancock, H., 1960. *Theory of Maxima and Minima*, New York: Dover.

Karmarkar, N. K. 1984. 'A New Polynomial Time Algorithm for Linear Programming', Combinatorica, 4, 373–95.

Khachiyan, L. G., 1979. *A Polynomial Algorithm in Linear Programming.* Doklady Akademiia Nauk SSSR, 244: 1093–96, English Translation: Soviet Mathematics Doklady, 20(1): 191–94,.

Land, A. H. and Doig, A. G. *An Automatic Method of Solving Discrete Programming Problems,* Econometrica, vol. 28, No. 3. (Jul., 1960), pp. 497–520.

Lemire, N. 1989. *An Introduction to Karmarkar's Method,* Windsor Mathematics Report WMR 89–10; July, 1989.

Mitten, L. G. 1970. *Branch-and-Bound Methods: General Formulation and Properties,* Operations Research, vol. 18, No. 1, pp. 24–34.

Rao, S. S. 2009. *Engineering Optimization: Theory and Practice (4e)*, Wiley and Sons, Inc.

Ravindran, A., Ragsdell, K. M., Reklaitis, G. V. 2006. *Engineering Optimization: Methods and Applications*, John Wiley and Sons, Inc.

Rebennack, S. 2008. *Ellipsoid Method, Encyclopedia of Optimization, Second Edition*, C. A. Floudas and P. M. Pardalos (Eds.), Springer, pp. 890–99.

Chapter 7

Optimization of Staged and Discrete Processes

7.1 Dynamic Programming

Multi-stage decision process arises during the study of different mathematical problems; the dynamic programming technique was created for taking care of these sorts of problems. Consider a physical system, at any time t, the state of this system is found out by a set of quantities that is called state variables or state parameters. At certain times, we are call upon to make a decision that will have an effect on the state of the system. This time may be prescribed in advance or the process itself determines the time. The transformations of the state variables are equivalent to these decisions, the selection of a decision being identical with the choice of a transformation. The future state is guided by the outcome of the preceding decision, which maximizes some function of the parameters that describe the final state.

There are many practical problems where sequential decisions are made at various points in time, at various points in space, and at various levels, say, for a system, for a subsystem, or even for a component level. The problems for which the decisions are required to make sequentially are termed as sequential decision problems. They are also called multistage decision problems because for these problems decisions are to be made at a number of stages. Dynamic programming is a mathematical technique, which is most suitable for the optimization of multistage decision problems. Richard Bellman in the early 1950s [Bellman, (1953)] developed this technique. During application of dynamic programming method, a multi-dimensional decision problem was decomposed into a series of single stage decision problems. In this way, an N-variable problem can be expressed as a series of N single-variable problems. These N number of single-variable problems can be solved sequentially. Most of the time, these single-variable subproblems are quite easier to solve compared to the original problem. This decomposition to N single-variable subproblems is accomplished in such a way that the optimal solution of the original N-dimensional problem can be attained from the optimal solutions of the N one-dimensional problems. It is essential to make a note that the methods used to solve these one-dimensional stages do not affect the overall result.

Any method can be used for this purpose; these methods may vary from an easy enumeration method to a relatively complicated technique like differential calculus or a nonlinear programming.

Classical optimization techniques can also be employed directly to solve the multistage decision problems. However there are some limitations, this is possible when small number of variables is there, the concerned functions should be continuous and continuously differentiable, and also the optimum points should not be positioned on the boundary line. Furthermore, the form of the problem needs to be quite simple such that the set of resultant equations can be solved by either analytical method or numerical method. To work out with somewhat more complex multistage decision problems we can use the nonlinear programming techniques. However, prior knowledge on the region of the global maxima or minima is required for their application, and the variables should also be continuous. For all these instances, the presence of stochastic variability changes the problem to a very complicated one and yields the problem unsolvable except by employing some sort of an approximation e.g., chance constrained programming. In contrast, the dynamic programming is able to solve the problems with discrete variables, and non-continuous, non-convex, and non-differentiable in nature. Usually it also considers the stochastic variability by a mere change of the deterministic procedure. There is a major drawback of the dynamic programming method that is recognized as the *curse of dimensionality*. Regardless of this weakness, it is quite appropriate for solving various types of complicated problems in diverse areas of decision-making.

Principle of optimality Irrespective of the initial state and initial decisions, the remaining decisions should formulate an optimal policy with regard to the state resulting from the first decisions [Bellman, (1954)].

7.1.1 Components of dynamic programming

A dynamic programming functional equation can be represented by the following equation

$$f(S) = \text{opt}_{d \in D(S)} \{R(S,d) \circ f(T(S,d))\} \quad (7.1)$$

where S represents a state within a state space $\mathbf{S}$, d signifies a decision selected from a decision space $D(S)$, $R(S, d)$ is called reward function (or decision cost), $T(S, d)$ is the transition or transformation function for the next-state, and "$\circ$" stands for a binary operator. In this chapter, only the discrete dynamic programming is discussed, where both the state space and decision space are discrete sets.

State The state S usually includes information on the series of decisions made until now. The state may be a complete sequence for some situations but, only partial information is satisfactory for other situations; for instance, when the set of all states can be divided into similar modules, all of them expressed by the last decision. For some uncomplicated problems, the extent of the sequence that is also called the stage at which the next decision is to be made, becomes sufficient. The state at the beginning that reflects the condition in which no decision has been made yet, is called the goal state and it is represented by S^*.

Decision Space The decision space $D(S)$ is defined by the set of possible or "eligible" choices for the next decision d. This decision space is a function of S (the state) within which the decision d is to be made. Constraints on probable transformations to the next-state from a state S can be implemented by properly restricting $D(S)$. The S is called a terminal state when there are no eligible decisions in state S i.e., $D(S) = 0$.

Objective Function The objective function f is a function of S. During the application of dynamic programming, we need to optimize this objective function. This is the optimal profit or cost resulting from creating a progression of decisions when within the state S, i.e., after creating the series of decisions correlated with S. The aim of a dynamic programming problem is to find $f(S)$ for the goal state S^*.

Reward Function The reward function R can be represented as a function of the state S and the decision d. Reward function is the cost or profit, which can be attributed to the subsequent decision d made in state S. This is also known as "Return Function". The reward $R(S, d)$ have to be separable from the costs or profits that are attributed to all other decisions. The value of $f(S^*)$, the objective function for the goal state, is the combination of the rewards for the entire optimal series of decisions starting from the goal state.

Transformation Function(s) The transition or transformation function (T) can be represented as a function of the state S and the decision d. This function identifies the next-state that results from making a decision d in state S. There may be more than one transformation function for a non-serial dynamic programming problem.

Operator The operator (°) is a binary operation, which enables us to connect the returns of various separate decisions. Usually this operator is either addition or multiplication. The operation should be associative whenever the returns of decisions are to be independent of the order in which they are made.

7.1.2 Theory of dynamic programming

As discussed above, the fundamental concept for the theory of dynamic programming is that of representing an optimal strategy as one finding out the required decision at every time in terms of the system's present state. The mathematical equation of dynamic programming problem has been developed based on this idea.

Mathematical formulation

In this section, we will formulate the mathematical equation of a deterministic discrete process as described by Bellman [Bellman, (1954)]. To demonstrate this kind of functional equation that appears from an application of the principle of optimality, let us start with a very simple example of a deterministic process in which the system is expressed at any time by a vector in M-dimensional space $p = (p_1, p_2, ..., p_M)$, constrained to be positioned within some region D. Let $T = \{T_k\}$, where k runs over a set that may be continuous, finite, or enumerable, be a set of transformations with the characteristic that $p \in D$ indicates that $T_k(p) \in D$ for all k.

Let us consider an N-stage process to be performed to maximize some scalar function, $R(p)$ of the final state. This function is called the N-stage return. The strategy is composed of a selection of N transformations, $P = (T_1, T_2, ..., T_N)$, yielding successively the states

$$p_1 = T_1(p)$$

$$p_2 = T_2(p_1) \tag{7.2}$$

....

$$p_N = T_N(p_{N-1})$$

When, the region D is finite, if every $T_k(p)$ is continuous in p, and if $R(p)$ is a continuous function of p for $p \in D$, then there must exists an optimal policy. The maximum value of the return $R(p_N)$, which is decided by an optimal policy, only be a function of the number of stages N and the initial vector p. Then we can determine

$$f_N(p) = \underset{p}{\text{Max}}\, R(p_N) \tag{7.3}$$

which is the return achieved after N-stage by utilizing an optimal policy that starts from the initial state p.

To develop a functional equation for $f_N(p)$, we use the principle mentioned above. Suppose that we have chosen some transformation T_k, which is the result of our first decision, in this manner we are obtaining a new state $T_k(p)$. Therefore, by definition, the maximum return from the following $(N-1)$ stages is $f_{N-1}(T_k(p))$. It follows that k must now be chosen so as to maximize this. The result is the basic functional equation

$$f_N(p) = \underset{k}{\text{Max}}\, f_{N-1}(T_k(p)) \quad N = 1, 2, \ldots \tag{7.4}$$

It is obvious that a knowledge of any particular optimal policy, necessarily not unique, will produce $f_N(p)$ that is unique. On the other hand, given the series $\{f_N(p)\}$, all optimal policies may be developed.

7.1.3 Description of a multistage decision process

A single-stage decision process is represented by the rectangular block in Fig 7.1. This single-stage is a part of some multistage problem. The whole decision process is characterized by the input data (or input parameters), S, decision variables (X), and the output parameters (T) that represent the outcome achieved as an effect of making the decision. The input parameters and the output parameters are defined as input state variables and output state variables respectively. Lastly, we get an objective function or return (R) that evaluates the effectiveness of the decisions made and the output as the consequence of these decisions.

The output of single-stage decision process as represented in Fig. 7.1 is associated with the input parameters through a stage transformation function represented by

$$T = t(X, S) \tag{7.5}$$

As the system's input state variable influences the decisions made, then we can express the return function as

$$R = r(X, S) \tag{7.6}$$

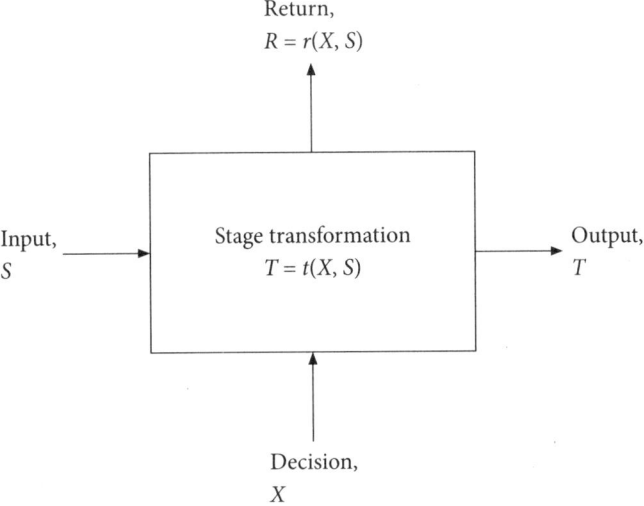

Fig. 7.1 Single-stage decision process

Schematic representation of a sequential multistage decision process is given in Fig. 7.2. Due to some convenience, the stages are labeled in decreasing order like $n, n-1, \ldots, i, \ldots, 2, 1$. Therefore, the input state vector for the ith stage is indicated by S_{i+1} accordingly the output state vector as S_i. For a serial system, it is quite obvious that the output from $i+1$ stage should be identical with the input to the next stage i. Thus, the state transformation and return functions for the ith stage can be expressed as

$$S_i = t_i\left(S_{i+1}, X_i\right) \tag{7.7}$$

$$R_i = r_i\left(S_{i+1}, X_i\right) \tag{7.8}$$

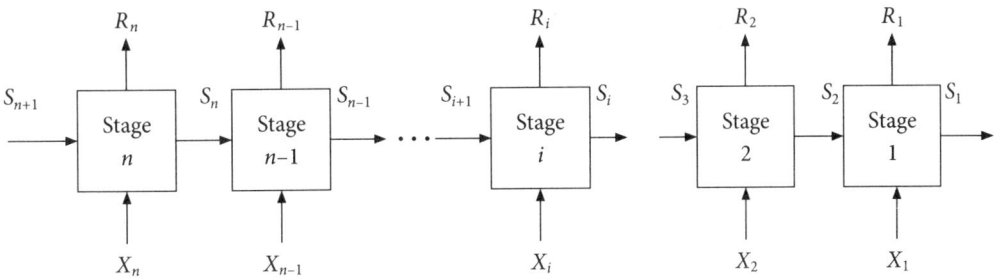

Fig. 7.2 Multi-stage decision process

where X_i represents the vector of decision variables for ith stage. The state transformation equations represented by Eq. (7.7) are also identified as design equations.

The purpose of a multistage decision problem is to determine a series $X_1, X_2, \ldots, X_n$ in order to optimize $f(R_1, R_2, \ldots, R_n)$ which is some function of the individual stage returns, and satisfy

Eqs (7.7) and (7.8). If a given multistage problem is solvable by using the dynamic programming are determined by the characteristic of the n-stage return function (f).

The monotonicity and separability of the objective function are required as this method runs as a decomposition technique. The objective function will be separable when it is possible to express the objective function as the composition of the individual stage returns. The additive objective functions satisfy this requirement:

$$f = \sum_{i=1}^{n} R_i = \sum_{i=1}^{n} R_i(X_i, S_{i+1}) \tag{7.9}$$

where X_i are real, and when the objective functions are multiplicative

$$f = \prod_{i=1}^{n} R_i = \prod_{i=1}^{n} R_i(X_i, S_{i+1}) \tag{7.10}$$

where X_i are nonnegative and real. Conversely, the following objective function given by Eq. (7.11) is not separable:

$$f = \left[R_1(X_1, S_2) + R_2(X_2, S_3) \right]\left[R_2(X_2, S_3) + R_3(X_3, S_4) \right]\left[R_3(X_3, S_4) + R_4(X_4, S_5) \right] \tag{7.11}$$

There are various practical problems in the field of chemical engineering that fulfill the condition of separability. For a monotonic objective function, values of all a and b that show

$$R_i(X_i = \mathbf{a}, S_{i+1}) \geq R_i(X_i = b, S_{i+1}) \tag{7.12}$$

the given inequality in Eq. (7.13) is satisfied:

$$f(X_n, X_{n-1}, \ldots, X_{i+1}, X_i = \mathbf{a}, X_{i-1}, \ldots, X_1, S_{n+1}) \geq f(X_n, X_{n-1}, \ldots, X_{i+1}, X_i = b, X_{i-1}, \ldots, X_1, S_{n+1})$$
$$i = 1, 2, \ldots, n \tag{7.13}$$

The application of DP method in chemical engineering can be illustrated with the following example.

Example 7.1

Optimal Replacement of equipment

The dynamic programming method can be discussed with this example of equipment replacement in a chemical industry. In this example, time (each year) is considered as a discrete stage.

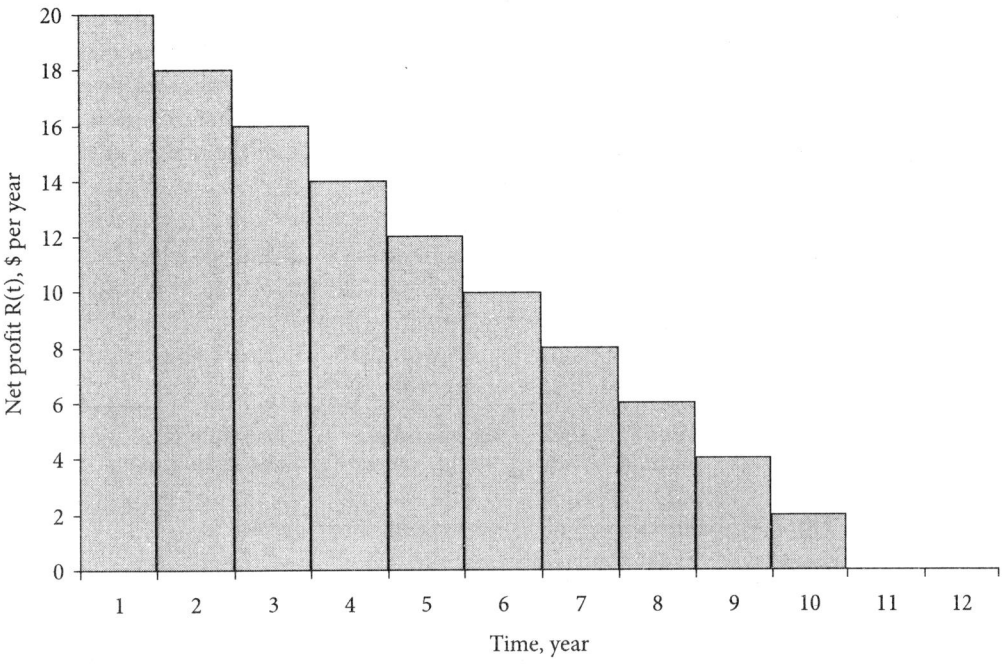

Fig. 7.3 The annual net profit vs. time plot

Figure 7.3 shows the data of the annual net profit for the operation of a continuous distillation unit over a time span of 12 year, where 10th year shows the break-even point for this distillation process. The replacement cost of this distillation unit with a new one employing up to date technology is considered to be the same as the net profit made during the 1st year of operation of the new unit, i.e., $20,000. An annual assessment is conducted at the starting of every year, and a decision is taken whether to continue with the old unit or replace it with a new contemporary model to get the maximum profit over a period of five years. At this moment, the distillation unit is four years old, and it is required to find out the best alternative now and for each of the following four years to maximize the profit, i.e., establish the optimal replacement strategy. This procedure starts with the consideration of the possible decisions to be made at the beginning of the fifth year (end of the fourth year), i.e., either continue with the process or replace it. Stage 1 is the time period from the beginning to the end of the fifth year. The algorithm for dynamic programming at stage one can be represented as:

$$f_1(S_1) = \operatorname*{Max}_{d_1}\left[R_1(S_1, d_1)\right] = \operatorname*{Max}_{\substack{K \\ or \\ R}}\begin{cases} R_1(t) \\ -20 + R(0) = 0 \end{cases} \qquad (7.14)$$

where d_1 represents the decision, keep (K) or replace (R) the distillation unit to maximize the profit, $R_1(S_1, d_1)$. Moreover the profit from the unit depends on the age of this process, the state variable S_1. Figure 7.4 shows the stages of the dynamic programming. The optimum decisions are shown in Table 7.1 for stage one as a function of the state variable, and they are to keep the process operating. At the beginning of the fifth year, the range of state variables goes from a process one year old to having a process ten years old. Similarly, for a ten year old process there is a tie between keeping the old distillation unit and replacing it with a new one, and the decision made here is the one that is easier, i.e., keep the process operating. The values shown in Table 7.1 were obtained from Fig. 7.3. The output state variable from stage one S_0 is the age of the process at the end of the year.

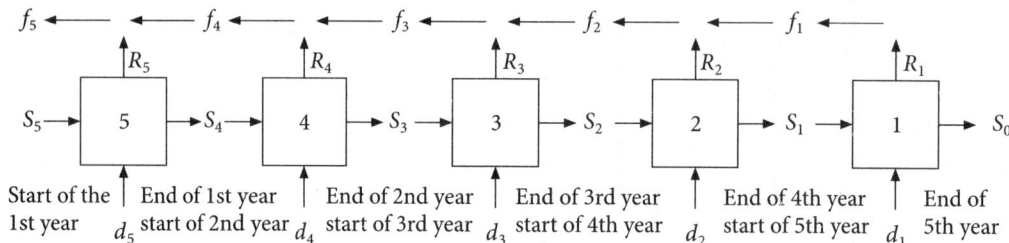

Fig. 7.4 Representation of 5 stage dynamic programming

Table 7.1 Values of the decision variables (as per Eq. (7.14))

S_0	2	3	4	5	6	7	8	9	10	11
d_1	K	K	K	K	K	K	K	K	K	K
f_1	18	16	14	12	10	8	6	4	2	0
S_1	1	2	3	4	5	6	7	8	9	10

At stage 2, we have taken the optimal decisions, which maximize the sum of the return at stage two and the optimal return from stage 1 for a range of the state variable values at stage 2. The algorithm of dynamic programming at stage 2 is:

$$f_2(S_2) = \underset{d_2}{\text{Max}}\left[R_2(S_2,d_2) + f_1(S_1)\right] = f_2(t) = \underset{\substack{K \\ or \\ R}}{\text{Max}} \begin{cases} R_2(t) + f_1(t+1) \\ -20 + R(0) + f_1(1) = 18 \end{cases} \quad (7.15)$$

If the decision is to keep (K) and continue with the same distillation unit, $f_2(t)$, the optimal return, is the sum of the return during the 4th year for a process t years old, $R_2(t)$, and $f_1(t + 1)$, the optimal return from stage 1 for a process whose age is in $t + 1$. If the decision is to replace (R) the distillation unit, the optimal return $f_2(t)$ is the sum of the cost of a new distillation unit, -20, the profit from operating a new unit for a year, $R(0) = 20$ and $f_1(1)$, the optimal return from stage 1 for a process one year old. The optimal decisions are given in Table 7.2 at stage two for a process whose age can be from 1 to 7 years, S_2. It is obvious from the figure that the optimal decisions are to continue to the operation of old distillation unit if its age is from 1 to 5 years old. However, if the unit is 6 years old or older, the profit will be higher for the 4th and 5th years if the old unit is replaced with a new one.

Table 7.2 Values of the decision variables (as per Eq. (7.15))

S_1	2	3	4	5	6	1	1
d_2	K	K	K	K	K	R	R
f_2	34	30	26	22	18	18	18
S_2	1	2	3	4	5	6	7

The similar method is employed at stage three to determine the maximum profit for the 3rd, 4th, and 5th years. The algorithm for dynamic programming is represented as

$$f_3(S_3) = f_3(t) = \underset{\substack{K \\ or \\ R}}{\text{Max}} \begin{cases} R_3(t) + f_2(t+1) \\ -20 + R(0) + f_2(1) = 34 \end{cases} \qquad (7.16)$$

The optimal decisions for stage three are given in Table 7.3. At this stage, the optimal decisions are to continue the operation of the distillation unit if its age is from 1 to 3 years old; and if it is older, the maximum profit over the 3 year period will be achieved by replacing the old unit with a new one.

Table 7.3 Values of the decision variables (as per Eq. (7.16))

S_2	2	3	4	1	1
d_3	K	K	K	R	R
f_3	48	42	36	34	34
S_3	1	2	3	4	5

Continuing to stage 4, we need to repeat the procedure to determine the maximum profit for the 4 year period. The algorithm of the dynamic programming to determine the optimal decisions for various age processes (the state variable) is represented by the following equation:

$$f_4(t) = \underset{\substack{K \\ or \\ R}}{\text{Max}} \begin{cases} R_4(t) + f_3(t+1) \\ -20 + R(0) + f_3(1) = 48 \end{cases} \qquad (7.17)$$

As shown in Table 7.4, the optimal decisions are to keep and run the distillation unit if it is from 1 to 3 years older and if it is older, replace the unit with a new one.

Table 7.4 Values of the decision variables (as per Eq. (7.17))

S_3	2	3	4	1	1
d_4	K	K	K	R	R
f_4	60	52	48	48	48
S_4	1	2	3	4	5

The results for the 5th or final stage are obtained by using the similar method as before. However, it is required to consider only one value of the state variable for the existing 4 years old distillation unit. The algorithm of the dynamic programming is:

$$f_5(t) = \underset{\substack{K \\ or \\ R}}{\text{Max}} \begin{cases} R_5(t) + f_4(t+1) \\ -20 + R(0) + f_4(1) = 60 \end{cases} \tag{7.18}$$

Table 7.5 Values of the decision variables (as per Eq. (7.18))

S_3	2	3	4	5	1
d_4	K	K	K	K	R
f_4	70	64	62	60	60
S_4	1	2	3	4	5

As described by Eq. (7.18) and Table 7.1–7.5 the maximum profit for the period of five years is $60,000 when the process is 4 year old. The optimal decisions are taken as continue the process for the 1st year and at the beginning of the second year replace it with a new one and to operate this new process for the remaining three years. Other cases are also described above example that were found by using the dynamic programming algorithm for processes that are 1, 2, 3 and 5 years old. For example, the maximum profit would be $70,000 for a 1-year-old process. The optimal decisions for this process would be to continue with the same process for the five year period. Whenever the process becomes 5 years old, the maximum profit would be $60,000; and the optimal decisions would be to replace the existing process with a new one and run this new process for the 5 year period. As a result, the dynamic programming algorithm produced other probably useful information without considerable additional computational load.

7.2 Integer and Mixed Integer Programming

As MILP has already been discussed in chapter 6 (6.2.4), we will focus on the mixed integer nonlinear programming in this section. In chemical engineering, there are many applications of nonlinear optimization problems that involve integer or discrete variables besides the continuous variables. These types of optimization problems are denoted as Mixed-Integer Nonlinear Programming (MINLP) problems. The integer variables can be utilized for modeling, for example, series of events, alternative candidates, existence or nonexistence of units (in their zero-one representation), number of trays for distillation column design etc. Discrete variables can also be used for model development, for example, different equipment sizes like standard pipe diameter, standard capacity of pump etc.

The coupling of the continuous domain and integer variables along with their associated nonlinearities make the category of MINLP problems quite challenging from the theoretical, algorithmic, and computational standpoint. Despite this challenging nature, there exists a broad spectrum of applications that can be formulated as mixed-integer nonlinear programming problems. MINLPs have been adopted in different fields, including the financial sector and the process industry, management science, engineering, and operations research sectors. Process engineers are using this method for process synthesis which include: (i) the synthesis of heat recovery networks (Floudas and Ciric, 1989); (ii) membrane system for multicomponent gas mixtures (Qi and Henson, 2000) (iii) the synthesis of complex reactor networks (Kokossis and Floudas, 1990); (iv) the synthesis of utility systems (Kalitventzeff and Marechal, 1988); and the synthesis of total process systems (Kocis and Grossmann, 1988). An excellent review of the mixed integer nonlinear optimization frameworks and applications in Process Synthesis are provided in Grossmann (1990). Algorithmic advances for logic and global optimization in Process Synthesis are reviewed in Floudas and Grossmann (1994). Key applications of MINLP approaches have also emerged in the area of Design, Scheduling, and Planning of Batch Processes in chemical engineering and include: (i) the design of multiproduct plants (Grossmann and Sargent, 1979; Birewar and Grossmann, 1990). Excellent recent reviews of the advances in the design, scheduling, and planning of batch plants can be found in Reklaitis (1991), and Grossmann et al. (1992).

7.2.1 Formulation of MINLP

The primary objective of this section is to present the general formulation of MINLP problems, discuss the difficulties, and give an overview of the algorithmic approaches developed for these problems.

Mathematical description

The general form of a MINLP is

$$\min_{x,y} f(x, y) \tag{7.19a}$$

$$\text{subject to } h(x, y) = 0 \tag{7.19b}$$

$$g(x, y) \leq 0 \tag{7.19c}$$

$$x \in X \subseteq \Re^n \tag{7.19d}$$

$$y \in Y \quad \text{integer} \tag{7.19e}$$

Here, $f(x, y)$ is a nonlinear objective function (e.g., annualized total cost, profit, energy consumption), x represents a vector of n continuous variables (e.g., flow, pressure, composition, temperature), and y is a vector of integer variables (e.g., alternative solvents or materials, number of units in HEN, and reactor network, standard pipe diameter); $h(x, y) = 0$ denote the m equality constraints (e.g., mass, and energy balances, equilibrium relationships); $g(x, y) \leq 0$ are the p inequality constraints (e.g., specifications on purity of distillation products, environmental regulations such as maximum limit of pollutant discharge, feasibility constraints in heat recovery systems, logical constraints).

Remark 1

The integer variables y with given lower and upper bounds,

$$y^L \leq y \leq y^U \qquad (7.20)$$

This variable y can be expressed through 0–1 variables (i.e., binary) denoted as z, by the following formula:

$$y = y^L + z_1 + 2z_2 + 4z_3 + \ldots + 2^{N-1} z_N \qquad (7.21)$$

where N is the minimum number of 0–1 variables needed. This minimum number is given by

$$N = 1 + INT \left\{ \frac{\log(y^U - y^L)}{\log 2} \right\} \qquad (7.22)$$

where INT function truncates its real argument to an integer value. This approach however may not be practical when the bounds are large.

Then, formulation (7.19a–7.19e) can be written in terms of 0–1 variables:

$$\min_{x,y} f(x,y) \qquad (7.23a)$$

$$\text{subject to } h(x,y) = 0 \qquad (7.23b)$$

$$g(x,y) \leq 0 \qquad (7.23c)$$

$$x \in X \subseteq \Re^n \qquad (7.23d)$$

$$y \in Y = \{0,1\}^q \qquad (7.23e)$$

where y now is a vector of q, 0 – 1 variables (e.g., existence of a process unit ($y_i = 1$) or non-existence ($y_i = 0$)). We will focus on (7.23a–7.23e) in the majority of the subsequent developments.

Solution Approaches

A collection of MINLP algorithms has given below. These algorithms are developed for solving MINLP models of the form (7.23a–7.23e) or restricted classes of (7.23a–7.23e) includes the following in chronological order of development:

1. Generalized Benders Decomposition, GBD (Geoffrion, 1972);
2. Branch and Bound, BB (Beale, 1977);
3. Outer Approximation, OA (Duran and Grossmann, 1986);

4. Outer Approximation with Equality Relaxation, OA/ER (Kocis and Grossmann, 1987);
5. Outer Approximation with Equality Relaxation (Viswanathan and Grossmann, 1990)
6. Generalized Outer Approximation, GOA (Fletcher and Leyffer, 1994)
7. Generalized Cross Decomposition, GCD (Holmberg, 1990);

The following section elucidates the Generalized Benders Decomposition method with problem formulation.

7.2.2 Generalized Benders Decomposition

The decomposition method was first developed by Benders (Benders, 1962) to solve the mixed-variable programming problems. After that, the Generalized Benders Decomposition (GBD) was formulated for MINLP. However, some limitations were identified regarding the convexity and other properties of the functions involved. In a pioneering work of Geoffrion (1972) on the Generalized Benders Decomposition (GBD) two sequences of updated upper (nonincreasing) and lower (non-decreasing) bounds are created that converge within ε, in a finite number of iterations. The upper bounds correspond to solving subproblems in the x variables by fixing the y variables, whereas the lower bounds are based on duality theory. A review paper by Floudas et al. (Floudas et al. 1989) enlightened the potential applications of this technique for chemical process design. It also suggests a computational implementation for identifying global optimum point in nonconvex NLP and MINLP problems.

GBD requires the successive solution of an associated MIP problem. The algorithm decomposes the MINLP into a NLP subproblem with the discrete variables fixed and a linear MIP master problem. The major difference between OA and GBD is in the definition of the MIP master problem. OA relies on linearizations or tangential planes that reduce each subproblem effectively to a smaller feasible set, while the master MIP problem for a GBD is represented by a dual representation of the continuous space.

7.2.2.1 Formulation

Geoffrion (1972) generalized the approach proposed by Benders (1962), for exploiting the structure of mathematical programming problems (7.23a–7.23e), to the class of optimization problems stated as

$$\min_{x,y} f(x,y) \tag{7.23a}$$

$$\text{subject to } h(x,y) = 0 \tag{7.23b}$$

$$g(x,y) \leq 0 \tag{7.23c}$$

$$x \in X \subseteq \Re^n \tag{7.23d}$$

$$y \in Y = \{0,1\}^q \tag{7.23e}$$

under the following conditions:

C1 X is a nonempty, convex set and the functions

$$f : \Re^n \times \Re^q \to \Re \qquad (7.24a)$$

$$g : \Re^n \times \Re^q \to \Re^p \qquad (7.24b)$$

are convex for each fixed $y \in Y = \{0,1\}^q$, while the function $h : \Re^n \times \Re^q \to \Re^m$ are linear for each fixed $y \in Y = \{0,1\}^q$.

C2 The set

$$Z_y = \{z \in \Re^p : h(x, y) = 0, g(x, y) \leq z \text{ for some } x \in X\} \qquad (7.25)$$

is closed for each fixed $y \in Y$

C3 For every fixed $y \in Y \cap V$, where

$$V = \{y : h(x, y) = 0, g(x, y) \leq 0 \text{ for some } x \in X\} \qquad (7.26)$$

one of the following two conditions holds:

i. the resulting problem (7.23a–7.23e) possesses a finite solution and has an optimal multiplier vector for the equalities and inequalities.
ii. the resulting problem (7.23a–7.23e) is unbounded, that is, its objective function value goes to $-\infty$.

The basic idea in Generalized Benders Decomposition (GBD) is the generation of an upper bound and a lower bound at each iteration for the required solution of the MINLP model. The primal problem gives us the upper bound, whereas the lower bound results from the master problem. The primal problem is similar to the problem (7.23–7.23e) with fixed y-variables (i.e., it is in the x space only), and the solution gives information about the upper bound and the Lagrange multipliers associated with the equality and inequality constraints. The master problem is obtained by the use of nonlinear duality theory, makes use of the Lagrange multipliers obtained in the primal problem. The solution of this master problem gives information about the lower bound, and the next set of fixed y-variables to be used subsequently in the primal problem. As the iterations continue, it is noticed that the sequence of updated upper bounds is non-increasing, the sequence of lower bounds is non-decreasing, and that the sequences converge in a finite number of iterations.

7.2.2.2 Theoretical development

This section presents the theoretical development of the Generalized Benders Decomposition (GBD). The primal problem is analyzed first for the feasible and infeasible cases. Subsequently, the theoretical analysis for the derivation of the master problem is presented.

The primal problem

The primal problem is obtained by fixing the y variables to a particular 0–1 combination, which can be denoted as y^k where k stands for the iteration counter. The formulation of the primal problem $P(y^k)$, at iteration k is

$$\left. \begin{aligned} &\min_{x} f\left(x, y^k\right) \\ &\text{s.t.} \quad h\left(x, y^k\right) = 0 \\ &\quad\quad g\left(x, y^k\right) \leq 0 \\ &\quad\quad x \in X \subseteq \Re^n \end{aligned} \right] \quad P\left(y^k\right) \tag{7.27}$$

Remark 1 Note that due to conditions C1 and C3(i), the solution of the primal problem $P(y^k)$ is its global solution.

We will distinguish the two cases of (i) feasible primal, and (ii) infeasible primal, and describe the method of analysis for each case separately.

Case (i): Feasible Primal

When the primal problem at iteration k is feasible, then its solution gives information on x_k, $f(x^k, y^k)$, which is the upper bound, and the optimal multiplier vectors λ^k, μ^k for the equality and inequality constraints. Subsequently, using this information we can formulate the Lagrange function as

$$L\left(x, y, \lambda^k, \mu^k\right) = f\left(x, y\right) + \lambda^{k^T} h\left(x, y\right) + \mu^{k^T} g\left(x, y\right) \tag{7.28}$$

Case (ii): Infeasible Primal

If the primal is detected by the NLP solver to be infeasible, then we consider its constraints

$$h\left(x, y^k\right) = 0 \tag{7.29a}$$

$$g\left(x, y^k\right) \leq 0 \tag{7.29b}$$

$$x \in X \subseteq \Re^n \tag{7.29c}$$

where the set X, for instance, consists of lower and upper bounds on the x variables. To identify a feasible point we can minimize an l_1 or l_∞ sum of constraint violations. An l_1-minimization problem can be formulated as

$$\min_{x \in X} \sum_{i=1}^{p} \alpha_i \tag{7.30a}$$

$$\text{subject to } h\left(x, y^k\right) = 0 \tag{7.30b}$$

$$g_i(x, y^k) \le \alpha_i, \quad i = 1, 2, \ldots, p \qquad (7.30c)$$

$$\alpha_i \ge 0, \quad i = 1, 2, \ldots, p \qquad (7.30d)$$

Note that if $\sum_{i=1}^{p} \alpha_i = 0$, then a feasible point has been determined.

Also note that by defining as

$$\alpha^+ = \max(0, \alpha) \qquad (7.31)$$

and

$$g_i^+(x, y^k) = \max\left[0, g_i(x, y^k)\right] \qquad (7.32)$$

the l_1-minimization problem is stated as

$$\min_{x \in X} \sum_{i=1}^{P} g_i^+ \qquad (7.33a)$$

$$\text{subject to } h(x, y^k) = 0 \qquad (7.33b)$$

An l_∞-minimization problem can be stated similarly as:

$$\min_{x \in X} \max_{1, 2, \ldots P} g_i^+(x, y^k) \qquad (7.34a)$$

$$\text{subject to } h(x, y^k) = 0 \qquad (7.34b)$$

Alternative feasibility minimization approaches aim at keeping feasibility in any constraint residual once it has been established. An l_1-minimization in these approaches takes the form:

$$\min_{x \in X} \sum_{i \in I'} g_i^+(x, y^k) \qquad (7.35a)$$

$$\text{subject to } h(x, y^k) = 0 \qquad (7.35b)$$

$$g_i(x, y^k) \le 0, \quad i \in I \qquad (7.35c)$$

where I is the set of feasible constraints; and I' is the set of infeasible constraints. Other methods seek feasibility of the constraints one at a time whilst maintaining feasibility for inequalities indexed by $i \in I$. This feasibility problem is formulated as

$$\min_{x \in X} \sum_{i \in I'} w_i g_i^+ \left(x, y^k\right) \qquad (7.36a)$$

$$\text{subject to } h\left(x, y^k\right) = 0 \qquad (7.36b)$$

$$g_i\left(x, y^k\right) \leq 0, \quad i \in I \qquad (7.36c)$$

and it is solved at any one time.

To include all mentioned possibilities Fletcher and Leyffer (1994) formulated a general feasibility problem (FP) defined as

$$\min_{x \in X} \sum_{i \in I'} w_i g_i^+ \left(x, y^k\right) \qquad (7.37a)$$

$$\text{subject to } h\left(x, y^k\right) = 0 \qquad (7.37b)$$

$$g_i\left(x, y^k\right) \leq 0, \quad i \in I \qquad (7.37c)$$

The weights w_i are non-negative and not all are zero. Note that with $w_i = 1$ $i \in I'$ we obtain the l_1-minimization. Also in the l_∞-minimization, nonnegative weights are existing at the solution such that

$$\sum_{i \in I'} w_i = 1 \qquad (7.38)$$

and $w_i = 0$ if $g_i\left(x, y^k\right)$ does not reach the maximum value.

Note that infeasibility in the primal problem is detected when a solution of (FP) is obtained for which its objective value is greater than zero.

The solution of the feasibility problem (FP) gives information on the Lagrange multipliers for the equality and inequality constraints that are denote as $\bar{\lambda}^k, \bar{\mu}^k$ respectively. Then, the Lagrange function resulting from on infeasible primal problem at iteration k can be defined as

$$\bar{L}^k \left(x, y, \bar{\lambda}^k, \bar{\mu}^k\right) = \bar{\lambda}^{k^T} h(x, y) + \bar{\mu}^{k^T} g(x, y) \qquad (7.39)$$

Remark 2

It should be noted that two different types of Lagrange functions are defined depending on whether the primal problem is feasible or infeasible. In addition, the upper bound is obtained only from the feasible primal problem.

The master problem

The derivation of the master problem in the GBD utilizes nonlinear duality theory and is characterized by the following three key ideas:

i. Projection of problem (7.23a–7.23e) onto the y-space;
ii. Dual representation of v; and
iii. Dual representation of the projection of problem (7.23a–7.23e) on the y-space.

In the sequel, the theoretical analysis involved in these three key ideas is presented.

i. Projection of (7.23a–7.23e) onto the y-space

Problem (7.23a–7.23e) can be written as

$$\min_{y} \inf_{x} f(x,y) \tag{7.40a}$$

$$\text{subject to } h(x,y) = 0 \tag{7.40b}$$

$$g(x,y) \leq 0 \tag{7.40c}$$

$$x \in X \subseteq \Re^n \tag{7.40d}$$

$$y \in Y = \{0,1\}^q \tag{7.40e}$$

where the min operator has been written separately for y and x. Note that it is infimum with respect to x since for given y the inner problem may be unbounded. Let us define $v(y)$

$$v(y) = \inf_{x} f(x,y) \tag{7.41a}$$

$$h(x,y) = 0 \tag{7.41b}$$

$$g(x,y) \leq 0 \tag{7.41c}$$

$$x \in X$$

Note that $v(y)$ is parametric in the y variables and therefore, from its definition corresponds to the optimal value of problem (7.23a–7.23e) for fixed y (i.e., the primal problem $P(y^k)$ for $y = y^k$).

Figure 7.5 shows the flowchart for solving GBD method.

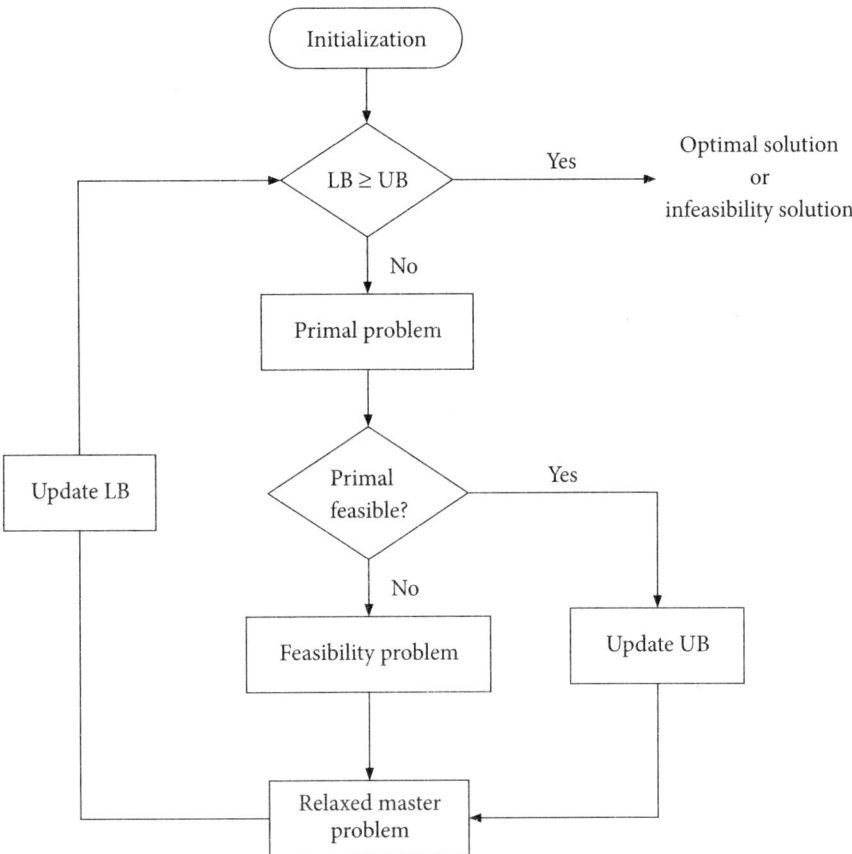

Fig. 7.5 Generalized Benders Decomposition method

7.2.2.3 Convergence and termination criteria

The convergence criterion of Generalized Benders Algorithm's as proposed by Geoffrion (1972) is:

The GBD algorithm terminates with an upper bound representing the optimal solution when the optimal value of the master problem approaches from below the upper bound given by the primal, within a convergence error ε.

The solutions of the master problem may go beyond the values provided by the primal problem during the implementation of the Generalized Benders Decomposition. This occurs as the Property (P) and/or L-Dual-Adequacy do not hold. Therefore, to tackle these situations a termination criterion is added. The new termination criterion is given as follows: [http://dx.doi.org/10.1016/0098-1354%2891%2985015-M]

The algorithm terminates with the upper bound representing the optimal solution, when the optimal value of the master problem, $\hat{y}^0$, go above the current upper bound (Floudas et al. 1989).

Whenever the criterion for termination is not activated and the relaxed master sequence of the NP-GBA converges, the converged value of the master problem is equal to the upper bound achieved from the solution of the primal. This is also true for non-convex problems which may show a gap between $v(y)$ and its dual. To be sure, after convergence of the relaxed master sequence, the Kuhn-Tucker conditions of the corresponding primal imply that $u^{(k_1)T}G\left(x^{(k_1)}, y\right) = 0$. In turn, this shows that one of the master's constraints becomes $y_0 \geq F(x^{(k_1)}, y)$. The criterion for termination is not activated and thus, $y_0 \geq F(x^{(k_1)}, y)$ will render $y_0 = F(x^{(k_1)}, y)$.

Application

Mixed-integer nonlinear programming (MINLP) has a wide range of applications for process design problems that allow simultaneous optimization of the process configuration as well as operating conditions (Floudas, 1995). A mixed-integer nonlinear programming (MINLP) is proposed for designing the optimal configuration of membrane networks to separate multi-component gas mixtures based on an approximate permeator model. The MINLP design model is developed for minimizing the total annual cost of the process by simultaneous optimization of the permeator configuration and operating conditions [Qi and Henson, (2000)].

The nonlinear nature of these mixed-integer optimization problems may arise from (i) nonlinear relations in the integer domain exclusively (e.g., products of binary variables in the quadratic assignment model), (ii) nonlinear relations in the continuous domain only (e.g., complex nonlinear input-output model in a distillation column or reactor unit), (iii) nonlinear relations in the joint integer-continuous domain (e.g., products of continuous and binary variables in the scheduling/planning of batch processes, and retrofit of heat recovery systems). Synthesis of reactor network and heat exchanger network, design of optimal distillation tower are common application of MINLP in chemical industry.

Summary

- Discrete variables like standard pipe diameter, number of HE in a HEN, number of plates in the distillation column, etc., are used to develop optimization problem. Optimization methods with discrete variables are discussed in this chapter. An optimization problem was divided into N stages and solved using dynamic programming algorithm. A problem of equipment replacement in chemical industry is represented as a discrete process; where each year is considered as a stage. Generalized Benders Decomposition method was used to solve an MINLP problem. The flowchart, algorithm, and termination criteria have been considered in this chapter.

Exercise

7.1 Find the value of x_1 and x_2, which minimizes

$$F = x_1^2 + x_2^2 - x_1 x_2 - 3x_2$$

subject to

$$-x_1 - x_2 \geq -2$$

$x \geq 0$, x_2 integer

7.2 The example proposed by Yuan *et al.* (1988). It involves three continuous variables and four binary variables. The formulation is

$$\min_{x,y} (y_1-1)^2 + (y_2-2)^2 + (y_3-1)^2 - \ln(y_4-1)$$

$$(x_1-1)^2 + (x_2-2)^2 + (x_3-3)^2$$

subject to

$$y_1 + y_2 + y_3 + x_1 + x_2 + x_3 \leq 5$$

$$y_1^2 + x_1^2 + x_2^2 + x_3^2 \leq 5.5$$

$$y_1 + x_1 \leq 1.2$$

$$y_2 + x_2 \leq 1.8$$

$$y_3 + x_3 \leq 2.5$$

$$y_4 + x_1 \leq 1.2$$

7.3 A linear programming problem with n decision variables and m constraints can be considered as an n-stage dynamic programming problem with m state variables-Justify this statement with proper example.

7.4 Minimize the function (using dynamic programming)

$$F = (50x_1 - 0.2x_1^2) + (50x_2 - 0.2x_2^2) + 8(x_1 - 80)$$

subject to the constraints

$$x_1 \geq 75$$

$$x_1 + x_2 = 220$$

$$x_1, x_2 \geq 0$$

7.5 What is the difference between an initial value problem and a final value problem?

7.6 A refinery needs to supply 50 barrels of petrol at the end of the first month, 80 barrels at the end of second month, and 100 barrels at the end of third month. The production cost of x barrels of petrol in any month is given by $\$(450x + 20x^2)$. It can produce more petrol in any month and supply it in the next month. However, there is an inventory carrying cost of $10 per barrel per month. Find the optimal level of production in each of the three periods and the total cost involved by solving it as an initial value problem.

7.7 Find the maximum value of the function

$$F = x_1 x_2$$

subject to the constraints

$$x_1^2 + x_2^2 \leq 4$$

$x_1, x_2 \geq 0$ and integer

7.8 Discuss the termination criteria of GBD method.

References

Beale, E. M. L. 'Discrete Optimization II' *Branch and Bound Methods for Mathematical Programming Systems Annals of Discrete Mathematics* 5(1979): 201–19.

Bellman, R. 1953. *An Introduction to the Theory of Dynamic Programming,* The RAND Corporation, Report R–245.

Bellman, R. 1954. *The Theory of Dynamic Programming,* Bull. Am. Math. Soc., 60: 503–16.

Benders, J. F., *Partitioning Procedures for Solving Mixed-variables Programming Problems,* Numerische Mathematik, 4(1962): 238–52.

Birewar, D. B. and Grossmann, I. E. 1990. *Simultaneous Production Planning and Scheduling in Multiproduct Batch Plants,* Ind. Eng. Chem. Res., 29: 570–80.

Duran, M. A. and Grossmann, I. E. *An Outer Approximation Algorithm for a Class of Mixed Integer Nonlinear Programs,* Mathematical Programming, 36(1986): 307–39.

Fletcher, R. and Leyffer S. 1994. *Solving Mixed Integer Nonlinear Programs by Outer Approximation.* Mathametical Programming, 66: 327–49.

Floudas, C. A. and Ciric, A. R. 1989. *Strategies for Overcoming Uncertainties in Heat Exchanger Network Synthesis,* Computers and Chemical Engineering, 13(10): 1133–52.

Floudas, C. A. and Grossmann, I. E. 1994. *Algorithmic Approaches to Process Synthesis: Logic and Global Optimization.* In Proceedings of Foundations of Computer-Aided Design, Colorado: FOCAPD '94, Snowmass.

Geoffrion, A. M., *Generalized Benders Decomposition, Journal of Optimization Theory and Applications* 10(4): 1972.

Grossmann, I. E. and Sargent, R. W. H. 1979. *Optimum Design of Multipurpose Chemical Plants,* Ind. Eng. Chem. Process Des. Dev., 18(2): 343.

Grossmann, I. E. 1990. *Mixed-integer Nonlinear Programming Techniques for the Synthesis of Engineering Systems.* Res. Eng. Des., 1: 205.

Grossmann, I. E. Quesada, I. Ramon, R. and Voudouris, V. T. 1992. *Mixed-integer Optimization Techniques for the Design and Scheduling of Batch Processes.* 'In Proc. of NATO Advanced Study Institute on Batch Process Systems Engineering '.

Holmberg, K. 1990. *On the Convergence of Cross Decomposition,* 'Mathematical Programming', 47: 269–96.

Kalitventzeff, B. and Maréchal, F. 1988. *The Management of a Utility Network,* Process System Engineering.

Kalitventzeff, B. 1991. *Mixed Integer Non-linear Programming and its Application to the Management of Utility Networks,* 'Engineering Optimization' vol. 18(1–3): 183–207.

Kocis, G. R. and Grossmann, I. E. 1869. *Relaxation Strategy for the Structural Optimization of Process Flow Sheets,* Ind. Eng. Chem. Res., 26, 1987.

Kocis, G. R. and Grossmann, I. E. *Global Optimization of Non-convex Mixed-integer Nonlinear Programming (MINLP) Problems in Process Synthesis.* 'Industrial Engineering Chemistry Research', 27(1988): 1407–21.

Kokossis, C. Floudas, C. A. 1990. *Optimization of Complex Reactor Networks—I.* ' Isothermal operation Antonis Chemical Engineering Science', 45(3): 595–614.

Qi, R. Henson, M. A. *Membrane System Design for Multi-component Gas Mixtures via Mixed-integer Nonlinear Programming,* 'Computers and Chemical Engineering', 24(2000): 2719–37.

Reklaitis, G. V. 1991. *Perspectives on Scheduling and Planning of Process Operations,* 'Presented at the Fourth international symposium on process systems engineering', Canada: Montebello.

Viswanathan, J. and Grossmann, I. E. *A Combined Penalty Function and Outer–Approximation Method for MINLP Optimization,* Computers and Chemical Engineering, 14(1990): 769–82.

Chapter

8

Some Advanced Topics on Optimization

8.1 Stochastic Optimization

Global optimization algorithms are categorized into two major classes: Deterministic and stochastic (Dixon *et al.* 1975). In some chemical engineering optimization problems, the model cannot be entirely specified because, it depends on quantities that are random or probabilistic at the time of formulation. For instance, uncertainty governs the prices of fuels, the availability of electricity, and the demand for chemicals in the market [N. V. Sahinidis, (2004)]. These ever-changing market conditions require a high flexibility for chemical process industries under various product specifications and various feedstocks. Furthermore, the properties of processes change themselves during process operation, e.g., tray efficiencies and fouling of the equipment (heat exchanger, catalytic reactor), which leads to a decrease of product quality with the same operating conditions [Henrion *et al.* (2001)]. For deterministic optimization methods, the expected values of uncertain variables are generally used. In real life application, the values of uncertain variables deviate from their expected values. When we are introducing a stochastic element within deterministic algorithms, the deterministic assurance that the global optimum can be determined is relaxed into a confidence measure. Stochastic methods can be utilized to evaluate the probability of obtaining the global minimum. Stochastic ideas are generally used to development the stopping criteria, or for the approximation of the regions of attraction as utilized by some methods [Arora *et al.* 1995].

8.1.1 Uncertainties in process industries

Uncertainty is a part of real systems; in reality, every phenomenon has a certain degree of uncertainty. During their operation, chemical processes are frequently faced with uncertain conditions. From the perspective of chemical process operation, there exist two types of uncertainties. These uncertainties can occur due to variation either in external parameters, such as fluctuation of inlet flow, temperature, pressure, composition of the feedstock, supply of utilities and other environment parameters, or in internal process parameters for example heat transfer coefficients, tray efficiency and reaction rate constants, etc. The optimization problems become much more complex due to the existence of uncertain parameters and have considerable implications on process feasibility and other quality measures for instance controllability, safety, and environmental compliance [Hou *et al.* (2000)]. Uncertainty also arises in the planning problem. Uncertainty in planning problems can be categorized into two groups: endogenous uncertainty and exogenous uncertainty. Problems where stochastic processes are affected by decisions are said to possess endogenous uncertainty while problems where stochastic processes are independent of decisions are said to have exogenous uncertainty (e.g., yields, demands) [Jonsbraten, (1998)]. A typical example of stochastic process is discussed in the following section.

Fluidization Fluidization is a very common application in chemical process plant. In the chemical process industry, fluidized bed reactors are among those that have been utilized extensively for various applications. During fluidization, movement of the particles are highly stochastic; different particle has different residence time. Dehling *et al.* [Dehling *et al.* (1999)] investigated residence time distributions in continuous fluidized beds employing stochastic modeling. They proposed a stochastic model for fluidization operation.

To elucidate the problem, a batch fluidized bed has been considered. There is no inflow and outflow of the particles when the fluidized bed is in operation. The discretized form of a fluidized bed has shown in the Fig. 8.1(b). The movement of a single particle is considered, and the transport processes are transformed to transition probabilities between cells as shown in Fig. 8.1(b). The behavior of a particle has been reflected by the probability distribution for the particle's position as a function of time. The model is developed based on Markov chains so that the transition probability distribution of a single particle does not depend on the past history of the system.

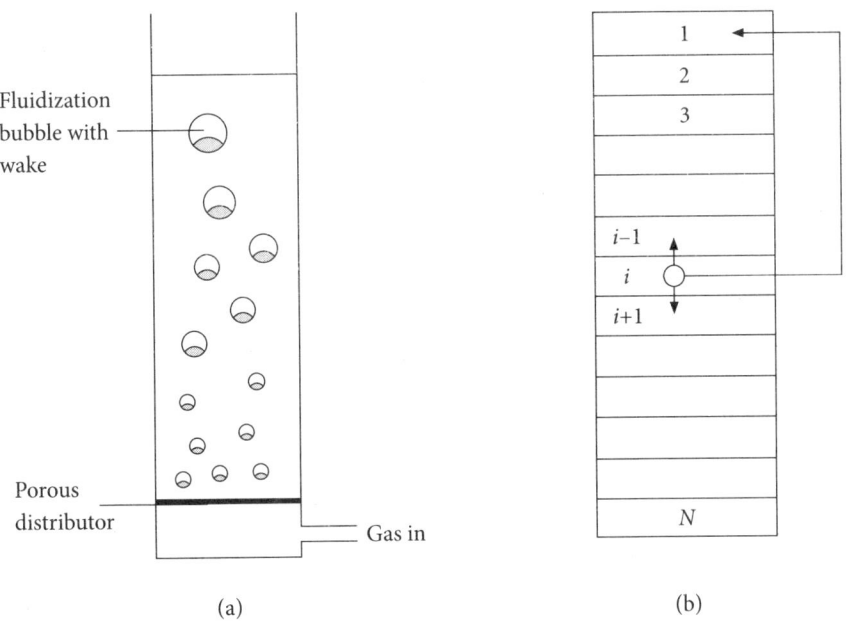

Fig. 8.1 Fluidization column

In this discrete Markov model, the fluidized bed reactor is partitioned into N horizontal cells, and the position of the particle was modeled at discrete times only. The cell numbers are given from the top toward the bottom as shown in Fig. 8.1. This model determines the probability distribution of the particle's axial position as a function of time. The probable transitions are as follows:

i. staying in the same cell (i)
ii. traveling to the next (above) cell ($i-1$)
iii. moving back to the previous (below) cell ($i+1$)
iv. deposited at the top of the bed (cell 1) after getting caught up in a bubble wake

8.1.2 Basic concept of probability theory

Probability and statistics are concerned with events that occur by chance. Examples consist of an occurrence of events such as opening of an off-on valve, errors of measurements, concentration of contaminants in wastewater, demand of a particular petroleum product. In each event, we may have some information on the likelihood of different probable outcomes, but, we are not able to predict with any certainty the outcome of any particular trial. Probability and statistics are applied all over engineering applications. Signals and noise are analyzed using probability theory in electrical engineering. Mechanical, civil, and industrial engineers use probability and statistics to test and account for variations in materials and goods. To control and improve the chemical processes, chemical engineers use probability and statistics to evaluate experimental data. In recent times, it is an essential requirement for all engineers [DeCoursey (2003)].

Preliminaries

Probability Every incident in reality possesses a certain element of uncertainty.

A basic requirement in probability theory is random experiment: an experiment whose outcome cannot be decided in advance. The set of all potential outcomes of an experiment is called the sample space of that experiment, and represented by S.

An event is a subset of a sample space, and said to occur if the outcome of the experiment is an element of that subset. We can define the probability in terms of the likelihood of a particular event. If an event is denoted by E, the probability of occurrence of the event E is generally represented by $P(E)$. The probability of occurrence depends on the number of trials or observations. It can be written as

$$P(E) = \lim_{n \to \infty} \frac{m}{n} \tag{8.1}$$

where m indicates the number of successful occurrence of the event E and the total number of trials is denoted by n. From Eq. (8.1) we can say that probability is a non-negative number and

$$0 \leq P(E) \leq 1 \tag{8.2}$$

$$P(S) = 1 \tag{8.3}$$

For any series of events $E_1, E_2,\ldots$ that are mutually exclusive, that is, events for which $E_i E_j = \phi$ when $i \ne j$ (where ϕ is the null set)

$$P\left(\bigcup_{i=1}^{\infty} E_i\right) = \sum_{i=1}^{\infty} P(E_i) \tag{8.4}$$

the probability of the event E is represented as $P(E)$.

Random variable

Consider a random experiment having sample space S. A random variable X is a function that assigns a real value to each outcome in S. An event can be defined as a probable outcome of an experiment. Let us consider that a random event is the measurement of a quantity X that has different values within the range $-\infty$ to ∞. This quantity (X) is called a random variable. Random variable is denoted by a capital letter and the particular value taken by it is represented by a lowercase letter.

For any real number x, the distribution function F of the random variable X can be defined by

$$F(x) = P\{X \le x\} = P\{X \in (-\infty, x)\} \tag{8.5}$$

we shall denote $1 - F(x)$ by $\overline{F}(x)$, and so

$$\overline{F}(x) = P\{X > x\} \tag{8.6}$$

There are two types of random variables: (i) discrete and (ii) continuous. When the random variable is permitted to use only discrete values $x_1, x_2, \ldots, x_n$, it is called a discrete random variable. A random variable X is called discrete when its set of possible values is countable. For discrete random variables,

$$F(x) = \sum_{y \le x} P\{X = y\} \tag{8.7}$$

In contrast, when the random variable is allowed to use any real value within a specified range, it is called a continuous random variable. A random variable is called continuous if there exist a function $f(x)$, called the probability density function, such that

$$P\{X \text{ is in } B\} = \int_B f(x) dx \tag{8.8}$$

for every set B. Since, $F(x) = \int_{-\infty}^{x} f(x) dx$, it follows that

$$f(x) = \frac{d}{dx} F(x) \tag{8.9}$$

For example, the number of worker present in any day, number of tray in a distillation column are discrete random variables, while the inlet composition of a distillation column can be considered as a continuous random variable.

Expected value

The expected value or mean of the random variable X is represented by $E(X)$

$$E(X) = \int_{-\infty}^{\infty} x dF(x) \qquad (8.10)$$

$$= \begin{cases} \int_{-\infty}^{\infty} xf(x)dx & \text{if } X \text{ is continuous} \\ \sum_{x} xP\{X = x\} & \text{if } X \text{ is discrete} \end{cases} \qquad (8.11)$$

provide the above integral exists.

Equation (8.11) also defines the expected value of any function of X, say $h(X)$. Since, $h(X)$ itself is a random variable, it follows from Eq. (8.11) that

$$E[h(X)] = \int_{-\infty}^{\infty} x dF_h(x) \qquad (8.12)$$

where F_h is the distribution function of $h(X)$. Though, we can show that this is identical to $\int_{-\infty}^{\infty} h(x) dF(x)$. That is

$$E[h(X)] = \int_{-\infty}^{\infty} h(x) dF(x) \qquad (8.13)$$

The variance of the random variable X is defined by

$$\text{Var } X = E\left[(X - E[X])^2\right]$$

$$= E[X^2] - E^2[X] \qquad (8.14)$$

Two jointly distributed random variables X and Y are said to be uncorrelated if their covariance, define by

$$\text{Cov}(X,Y) = E[(X - EX)(Y - EY)]$$

$$= E[XY] - E[X]E[Y] \qquad (8.15)$$

is zero. It follows that independent random variables are uncorrelated. However, the converse need not be true.

An important property of expectations is that the expectation of a sum of random variables is equal to the sum of the expectations.

$$E\left[\sum_{i=1}^{n} X_i\right] = \sum_{i=1}^{n} E[X_i] \tag{8.16}$$

The corresponding property for variances is that

$$\text{Var}\left[\sum_{i=1}^{n} X_i\right] = \sum_{i=1}^{n} \text{Var}(X_i) + 2\sum\sum_{i<j} \text{Cov}(X_i X_j) \tag{8.17}$$

Standard Deviation

The expected value or mean is a measurement of the central tendency, which indicates the location of the distribution on some coordinate axis. The variability of the random variable is measured by a quantity that is called the standard deviation.

The variance or mean-square deviation of a random variable X can be written as

$$\sigma_X^2 = \text{Var}(X) = E\left[(X - \mu_X)^2\right]$$

$$= E\left[X^2 - 2X\mu_X + \mu_X^2\right]$$

$$= E(X^2) - 2\mu_X E(X) + E(\mu_X^2)$$

$$= E(X^2) - \mu_X^2 \tag{8.18}$$

the standard deviation is defined as

$$\sigma_X = +\sqrt{\text{Var}(X)} = \sqrt{E(X^2 - \mu_X^2)} \tag{8.19}$$

Example 8.1

Number of worker present in a chemical industry and their probability is given in Table 8.1

Table 8.1 Number of worker present with probability

x_i	0	1	2	3	4	5	6	7
$P_x(x_i)$	0.01	0.12	0.22	0.25	0.15	0.13	0.07	0.05

Find the mean and standard deviation of x

Solution

Using the data from the above table,

$$\bar{x} = \sum_{i=0}^{7} x_i P_x(x_i) \tag{8.20}$$

$$= 0(0.01) + 1(0.12) + 2(0.22) + 3(0.25) + 4(0.15) + 5(0.13) + 6(0.07) + 7(0.05)$$

$$= 0 + 0.12 + 0.22 + 0.75 + 0.60 + 0.65 + 0.42 + 0.35$$

$$= 3.33$$

and

$$\sum_{i=0}^{7} x_i^2 P_x(x_i) = 0(0.01) + 1(0.12) + 4(0.22) + 9(0.25) + 16(0.15) \tag{8.21}$$
$$+ 25(0.13) + 36(0.07) + 49(0.05)$$

$$= 0 + 0.12 + 0.88 + 2.25 + 2.40 + 3.25 + 2.52 + 2.45$$

$$= 13.87$$

Standard deviation $= 13.87 - (3.33)^2$

$$= 2.7811$$

Stochastic Process A stochastic process is a collection of random variables that can be represented as $\underline{X} = \{X(t), t \in T\}$. Where $X(t)$ is a random variable for each t in the index set T. Most often, t is interpreted as time and describe $X(t)$ the state of the process at time t. If the index set T is a countable set, we call $\underline{X}$ a discrete-time stochastic process, and if T is a continuum, we call this process a continuous-time process.

8.1.3 Stochastic linear programming

A simple stochastic linear programming problem is represented as

$$\text{Minimize } f(X) = C^T X = \sum_{j=1}^{n} c_j x_j \tag{8.22a}$$

subject to

$$A_i^T X = \sum_{j=1}^{n} a_{ij} x_j \leq b_i \quad i = 1,, 2.......m \qquad (8.22b)$$

$$x_j \geq 0, \quad j = 1,, 2.......n \qquad (8.22c)$$

where c_j, a_{ij}, and b_i are random variables (for simplicity we are assuming that the decision variables x_j are deterministic) with known probability distributions. Numerous methods are available for solving the problem given by Eqs (8.22a)–(8.22c). Chance-constrained programming technique has been considered in this section.

Charnes and Cooper [Charnes and Cooper, (1959)] originally developed the method chance-constrained programming technique. It is clear from the name that the method can be employed to solve problems concerning chance constraints that is, constraints may be violated with finite probability. The stochastic programming problem can be represented as follows:

$$\text{Minimize } f(X) = \sum_{j=1}^{n} c_j x_j \qquad (8.23a)$$

subject to

$$P\left[\sum_{j=1}^{n} a_{ij} x_j \leq b_i\right] \geq p_i, \quad i = 1,, 2.......m \qquad (8.23b)$$

$$x_j \geq 0, \quad j = 1,, 2.......n \qquad (8.23c)$$

where p_i are specified probabilities and c_j, a_{ij}, b_i are random variables. Notice that Eq. (8.23b) indicate that $\sum_{j=1}^{n} a_{ij} x_j \leq b_i$ (the ith constraint) has to be satisfied with a probability of at least p_i where $0 \leq p_i \leq 1$. To make it simple, we consider that the design variables x_j are deterministic and c_j, a_{ij}, and b_i are random variables. We also consider that all the random variables are normally distributed with known mean and standard deviations.

Since, c_j are normally distributed random variables, the objective function $f(X)$ will also be a normally distributed random variable. The value of mean and variance of the function f are as follows

$$\overline{f} = \sum_{j=1}^{n} \overline{c}_j x_j \qquad (8.24)$$

$$\text{Var}(f) = X^T V X \qquad (8.25)$$

where $\bar{c}_j$ denotes mean value of c_j and the covariance matrix of c_j is defined by the matrix V (Eq. (8.26)).

$$V = \begin{bmatrix} \text{Var}(c_1) & \text{Cov}(c_1,c_2) & \cdots & \text{Cov}(c_1,c_n) \\ \text{Cov}(c_2,c_1) & \text{Var}(c_2) & \cdots & \text{Cov}(c_2,c_n) \\ \vdots & \vdots & \ddots & \vdots \\ \text{Cov}(c_n,c_1) & \text{Cov}(c_n,c_2) & \cdots & \text{Var}(c_n) \end{bmatrix} \qquad (8.26)$$

where Var(c_j) representing the variance of c_j and covariance between c_i and c_j are represented by Cov(c_i, c_j). The formulated new deterministic objective function for minimization is given by

$$F(X) = k_1 \bar{f} + k_2 \sqrt{\text{Var}(f)} \qquad (8.27)$$

where the constants k_1 and k_2 are nonnegative. The values of these constants signify the relative importance of the mean of the function, $\bar{f}$ and standard deviation of f during minimization. When we minimize the expected value of f without considering the standard deviation of f, the value of $k_2 = 0$. In contrast, $k_1 = 0$ signifies that we are paying attention to minimize the variability of f about its mean value without caring for the mean value of f. When equal importance is given to both the minimization of the mean and the standard deviation of f, we have to consider $k_1 = k_2 = 1$. Equation (8.27) shows that the new objective function is a nonlinear function of X.

The constraints of Eq. (8.23b) can be represented as

$$P[h_i \leq 0] \geq p_i, \quad i = 1,2\ldots\ldots m \qquad (8.28)$$

where the new random variable h_i is given as

$$h_i = \sum_{j=1}^{n} a_{ij} x_j - b_j = \sum_{k=1}^{n+1} q_{ik} y_k \qquad (8.29)$$

where

$$q_{ik} = a_{ik}, \quad k = 1,2,\ldots\ldots,n \qquad q_{i,n+1} = b_i \qquad (8.30)$$

$$y_k = x_k, \quad k = 1,2,\ldots\ldots,n \qquad y_{n+1} = -1 \qquad (8.31)$$

Here, for convenience we introduce the constant y_{n+1}. The h_i will follow normal distribution as it is given by a linear combination of the normally distributed random variables q_{ik}. The mean and the variance of h_i are given below

$$\bar{h}_i = \sum_{k=1}^{n+1} \bar{q}_{ik} y_{ik} = \sum_{j=1}^{n} \bar{a}_{ij} x_j - \bar{b}_i \qquad (8.32)$$

$$\mathrm{Var}(h_i) = Y^T V_i Y \qquad (8.33)$$

where

$$Y = \begin{Bmatrix} y_1 \\ y_2 \\ \vdots \\ y_{n+1} \end{Bmatrix} \qquad (8.34)$$

$$V_i = \begin{bmatrix} \mathrm{Var}(q_{i1}) & \mathrm{Cov}(q_{i1}, q_{i2}) & \cdots & \mathrm{Cov}(q_{i1}, q_{i,n+1}) \\ \mathrm{Cov}(q_{i2}, q_{i1}) & \mathrm{Var}(q_{i2}) & \cdots & \mathrm{Cov}(q_{i2}, q_{i,n+1}) \\ \vdots & \vdots & \ddots & \vdots \\ \mathrm{Cov}(q_{i,n+1}, q_{i1}) & \mathrm{Cov}(q_{i,n+1}, q_{i2}) & \cdots & \mathrm{Var}(q_{i,n+1}) \end{bmatrix} \qquad (8.35)$$

More explicitly, we can write this as

$$\mathrm{Var}(h_i) = \sum_{k=1}^{n+1} \left[y_k^2 \, \mathrm{Var}(q_{ik}) + 2 \sum_{l=k+1}^{n+1} y_k y_l \, \mathrm{Cov}(q_{ik}, q_{il}) \right]$$

$$= \sum_{k=1}^{n} \left[y_k^2 \, \mathrm{Var}(q_{ik}) + 2 \sum_{l=k+1}^{n} y_k y_l \, \mathrm{Cov}(q_{ik}, q_{il}) \right]$$

$$+ y_{n+1}^2 \mathrm{Var}(q_{i,n+1}) + 2 y_{n+1}^2 \mathrm{Cov}(q_{i,n+1}, q_{i,n+1})$$

$$+ \sum_{k=1}^{n} \left[2 y_k y_{n+1} \mathrm{Cov}(q_{ik}, q_{i,n+1}) \right]$$

$$= \sum_{k=1}^{n} \left[x_k^2 \, \mathrm{Var}(a_{ik}) + 2 \sum_{l=k+1}^{n} x_k x_l \, \mathrm{Cov}(a_{ik}, a_{il}) \right] \qquad (8.36)$$

$$+ \mathrm{Var}(b_i) - 2 \sum_{k=1}^{n} x_k \mathrm{Cov}(a_{ik}, b_i)$$

Therefore, the constraints in Eq. (8.28) can be rewritten as

$$P\left[\frac{h_i - \overline{h}_i}{\sqrt{\text{Var}(h_i)}} \leq \frac{-\overline{h}_i}{\sqrt{\text{Var}(h_i)}}\right] \geq p_i, \quad i = 1,,2\ldots\ldots m \tag{8.37}$$

where $\left[(h_i - \overline{h}_i)\right]/\sqrt{\text{Var}(h_i)}$ signifies a standard normal variable which have a mean value of zero and a variance of one. Therefore, if s_i represents the value of the standard normal variable at which $f(s_i) = p_i$ the constraints of Eq. (8.37) can be written as

$$\phi\left(\frac{-\overline{h}_i}{\sqrt{\text{Var}(h_i)}}\right) \geq \phi(s_i), \quad i = 1,,2\ldots\ldots m \tag{8.38}$$

The following deterministic nonlinear inequalities should be satisfied to satisfy the inequalities given in Eq. (8.38):

$$\frac{-\overline{h}_i}{\sqrt{\text{Var}(h_i)}} \geq s_i \quad i = 1,2\ldots\ldots m \tag{8.39a}$$

or

$$\overline{h}_i + s_i\sqrt{\text{Var}(h_i)} \leq 0, \quad i = 1,,2\ldots\ldots m \tag{8.39b}$$

So, the stochastic linear programming problem of Eqs (8.23a) to (8.23c) can be written as an equivalent deterministic nonlinear programming problem as

$$\text{Minimize } F(X) = k_1 \sum_{j=1}^{n} \overline{c}_j x_j + k_2 \sqrt{X^T V X}, \quad k_1 \geq 0, \quad k_2 \geq 0 \tag{8.40a}$$

subject to

$$\overline{h}_i + s_i\sqrt{\text{Var}(h_i)} \leq 0, \quad i = 1,,2\ldots\ldots m \tag{8.40b}$$

$$x_j \geq 0, \quad j = 1,,2\ldots\ldots n \tag{8.40c}$$

A stochastic linear programming problem has been developed by Ai-Othman et al. [Al-Othman et al. (2008)] for supply chain optimization of petroleum organization under market demands and prices uncertainty.

8.1.4 Stochastic nonlinear programming

A general optimization problem need to be developed as a stochastic nonlinear programming problem, when some of the parameters within the objective function and constraints vary about their mean values [S. S. Rao, (2009)]. In this present discussion, we consider that all the random variables are independent and follow normal distribution. The standard form of a stochastic nonlinear programming problem is given as

$$\text{Find values of } X \text{ that will minimize } f(Y) \tag{8.41a}$$

subject to

$$P\left[g_j(Y) \geq 0\right] \geq p_j, \quad j = 1, 2 \ldots m \tag{8.41b}$$

where Y represents the vector of N random variables $y_1, y_2, \ldots, y_N$ and it consist of the decision variables $x_1, x_2, \ldots, x_n$. A special case can be obtained during the present formulation, when X is deterministic. The Eq. (8.41b) indicate that the probability of realizing $g_j(Y)$ greater than or equal to zero should be greater than or equal to the specified probability p_j. By using the change constrained programming technique, the problem given in Eqs (8.41a) and (8.41b) can be transformed into an equivalent deterministic nonlinear programming problem as follows.

Objective Function

The objective function $f(Y)$ is expanded with the mean values of y_i, $\bar{y}_i$ as

$$f(Y) = f(\bar{Y}) + \sum_{i=1}^{N} \left(\frac{\partial f}{\partial y_i}\bigg|_{\bar{Y}}\right)(y_i - \bar{y}_i) + \text{higher-order derivative terms} \tag{8.42}$$

We can approximate the objective function $f(Y)$ by the first two terms of Eq. (8.42), if the standard deviations of y_i $\left(\sigma_{y_i}\right)$ are small

$$f(Y) \simeq \bar{Y} - \sum_{i=1}^{N} \left(\frac{\partial f}{\partial y_i}\bigg|_{\bar{Y}}\right)\bar{y}_i + \sum_{i=1}^{N} \left(\frac{\partial f}{\partial y_i}\bigg|_{\bar{Y}}\right)y_i = \psi(Y) \tag{8.43}$$

The $\psi(Y)$, a linear function of Y follows the normal distribution when all $y_i (i = 1, 2, \ldots, N)$ follow the normal distribution. The mean and the variance of ψ are written as

$$\bar{\psi} = \psi(\bar{Y}) \tag{8.44}$$

$$\text{Var}(\psi) = \sigma_\psi^2 = \sum_{i=1}^{N} \left(\frac{\partial f}{\partial y_i}\bigg|_{\bar{Y}}\right)^2 \sigma_{y_i}^2 \tag{8.45}$$

as all y_i are independent. For the optimization purpose, a new objective function $F(Y)$ can be formulated as

$$F(Y) = k_1 \bar{\psi} + k_2 \sigma_\psi \qquad (8.46)$$

where $k_1 \geq 0$ and $k_2 \geq 0$, and the numerical values of k_1 and k_2 indicate the relative importance of $\bar{\psi}$ and σ_ψ during minimization. The standard deviation of ψ can be dealt in another way by minimizing $\bar{\psi}$ subject to the constraints $\sigma_\psi \leq k_3 \bar{\psi}$, where k_3 is a constant, along with the other constraints.

Constraints

When some of the parameters are random, then the constraints will also be probabilistic and one may wish to have the probability that a given constraint is satisfied to be larger than a certain value. This is exactly what is mentioned in Eq. (8.41b). The constraint inequality (8.41b) can be represented as

$$\int_0^\infty f_{g_j}(g_j) dg_j \geq p_j \qquad (8.47)$$

where $f_{g_j}(g_j)$ denotes the probability density function of the random variable g_j (as the function of some random variables is also a random variable) with a range considered as $-\infty$ to ∞. The constraint function $g_j(Y)$ has been expanded around $\bar{Y}$, the vector of mean values of the random variables by the following equation

$$g_j(Y) \simeq g_j(\bar{Y}) + \sum_{i=1}^{N} \left(\frac{\partial g_j}{\partial y_i} \bigg|_{\bar{Y}} \right)(y_i - \bar{y}_i) \qquad (8.48)$$

From the Eq. (8.48), we can get $\bar{g}_j$, the mean value and σ_{g_j}, the standard deviation of g_j as given below

$$\bar{g}_j = g_j(\bar{Y}) \qquad (8.49)$$

$$\sigma_{g_j} = \left\{ \sum_{i=1}^{N} \left(\frac{\partial g_j}{\partial y_i} \bigg|_{\bar{Y}} \right)^2 \sigma_{y_i}^2 \right\}^{1/2} \qquad (8.50)$$

a new variable is introduced as

$$\theta = \frac{g_j - \bar{g}_j}{\sigma_{g_j}} \qquad (8.51)$$

and noting that

$$\int_{-\infty}^{\infty} \frac{1}{\sqrt{2\pi}} e^{-t^2/2} dt = 1 \tag{8.52}$$

equation (11.90) can be expressed as

$$\int_{-(\bar{g}_j/\sigma_{g_j})}^{\infty} \frac{1}{\sqrt{2\pi}} e^{-\theta^2/2} d\theta \geq \int_{-\phi_j(p_j)}^{\infty} \frac{1}{\sqrt{2\pi}} e^{-t^2/2} dt \tag{8.53}$$

where $\phi_j(p_j)$ is the value of the standard normal variate corresponding to the probability p_j. Thus,

$$-\frac{\bar{g}_j}{\sigma_{g_j}} \leq -\phi_j(p_j)$$

or

$$-\bar{g}_j + \sigma_{g_j} \phi_j(p_j) \leq 0 \tag{8.54}$$

Equation (8.54) can be written as

$$\bar{g}_j - \phi_j(p_j) \left[\sum_{i=1}^{N} \left(\frac{\partial g_j}{\partial y_i} \bigg|_{\bar{Y}} \right)^2 \right]^{1/2} \geq 0, \; j=1,2,\ldots,m \tag{8.55}$$

Therefore, the optimization problem given in Eqs (8.41a) and (8.41b) can be given in its corresponding deterministic form as: Minimize $F(Y)$ as given in Eq. (8.46) subject to the m number of constraints given by Eq. (8.55).

8.2 Multi-Objective Optimization

Multi-objective optimization (MOO) involves the simultaneous optimization of more than one objective function. This is quite common in the field of Chemical Engineering and Biochemical Engineering. In most of the cases, the objective functions are defined in incomparable units, and they present some degree of conflict among them. Fettaka *et al.* [Fettaka *et al.* (2013)] have described the design of heat exchanger with optimized heat transfer area and pumping power. Mass of the pipeline (kg/m) and pumping power (HP) were used as objective variables. The units of these objective functions are different. The multi-objective optimization becomes more difficult as improving one objective will worsen another. Improvement of one objective function is not possible without deteriorating at least another one.

The general multi-objective optimization problems can be represented by Eqs (8.56a)–(8.56e). Here, both equality and inequality constrains have been considered.

$$\text{Find } X = \begin{Bmatrix} x_1 \\ x_2 \\ \vdots \\ x_n \end{Bmatrix} \quad \quad (8.56a)$$

which minimizes $F(X) = \left[F_1(X), F_2(X), \cdots, F_k(X) \right]^T$ (8.56b)

subject to: $g_i(X) = 0; \ i = 1, 2, \cdots, m$ (8.56c)

$h_j(X) \leq 0; \ j = 1, 2, \cdots, q$ (8.56d)

$X_i^L \leq X_i \leq X_i^U$ (8.56e)

where k is the number of objective functions, m is the number of equality constraints and q is the number of inequality constraints. $X \in E^n$ is a vector of design variables, and $F(X) \in E^n$ is a vector of objective functions $F_i(X): E^n \to E^1$. X_i^L and X_i^U represent the lower and upper boundary of the variable X_i.

Before discussing the theories of MOO, we should have ample knowledge on Pareto optimal point and utopia point (ideal point). The multi-objective optimization problem is also called the vector minimization problem. Unlike single objective optimization where there is one best solution, MOO generates a set of optimal solutions. These solutions are called Pareto optimal solutions or non-dominated solutions. The Pareto domain refers to the set of non-dominated solutions that present trade-offs among the different objectives. A solution is said to be non-dominated if it is not worse than another solution in all its objectives and it is better with respect to at least one objective (Deb, 2001).

Pareto set:

In the 1970s, Stadler [Stadler, W., (1979)] applied the concept of Pareto optimality to the fields of science and engineering. A feasible solution X^* is called Pareto optimal while there exists no other feasible solution Y such that $f_i(Y) \leq f_i(X^*)$ for $i = 1, 2, \ldots, k$ with $f_j(Y) < f_j(X^*)$ for at least one j. An objective vector $f(X^*)$ is called Pareto optimal if the corresponding decision vector X is Pareto optimal. This can be explained easily by the following example. Consider two functions $f_1 = 3 + (x-4)^2$ and $f_2 = 16 + (x-10)^2$. The minimum values of these functions are $f_{1\min} = 3$ at $x_{1\min} = 4$ at and $f_{2\min} = 16$ at $x_{2\min} = 10$ as shown in Fig. 8.2. Therefore, all the values of x between 4 and 10 are called Pareto optimal solutions. If we plot the graph f_1 vs. f_2 (Fig. 8.3), the segment a, c, d, b represents the Pareto domain.

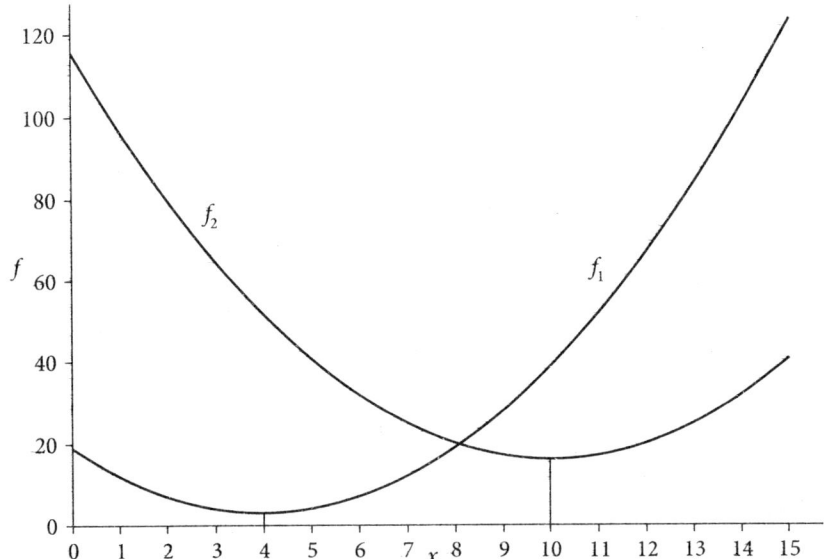

Fig. 8.2 Multi-objective optimization

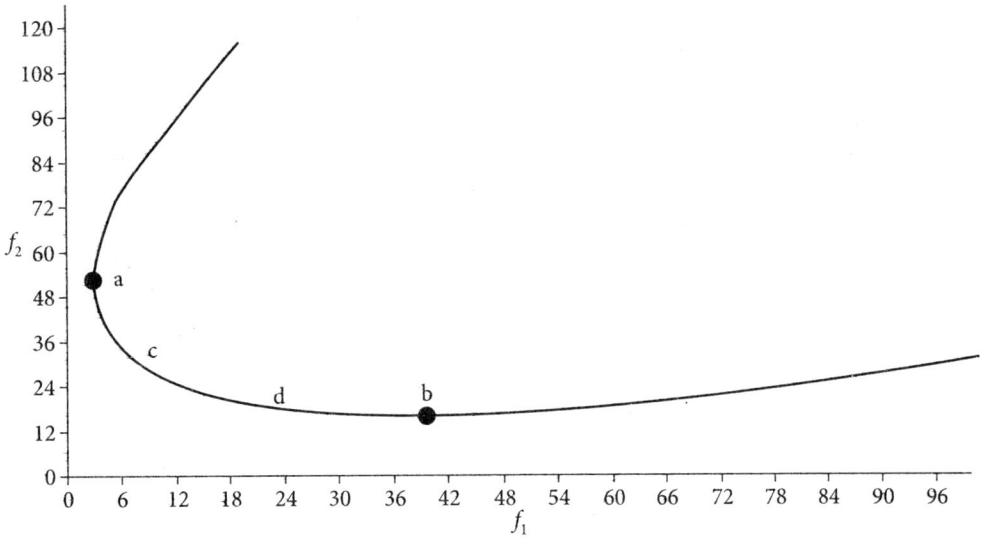

Fig. 8.3 Pareto optimal set

A very common example of MOO is simultaneous optimization of yield and selectivity of the intermediate desired product B for an isothermal batch reactor with series reaction $A \xrightarrow{k_1} B \xrightarrow{k_2} C$. We are interested to maximize both the yield and the selectivity of the desired product, B, simultaneously. The optimization problem for this system can be formulated as

$$\text{Max } Y_B = \frac{C_B}{C_A} \tag{8.57}$$

$$\text{Max } S_B = \frac{C_B}{C_B + C_C} \tag{8.58}$$

Maximization of the yield is required since it leads to higher amounts of B. Whereas the maximization of the selectivity is desired to reduce the downstream separation costs. For any feed concentration, the yield and selectivity can be formulated as a function of time (t) and temperature, by integrating the mass balance equations for this system (use similar method of section 2.4.1 for batch reactor). Figure 8.4 shows the graph between yield and selectivity.

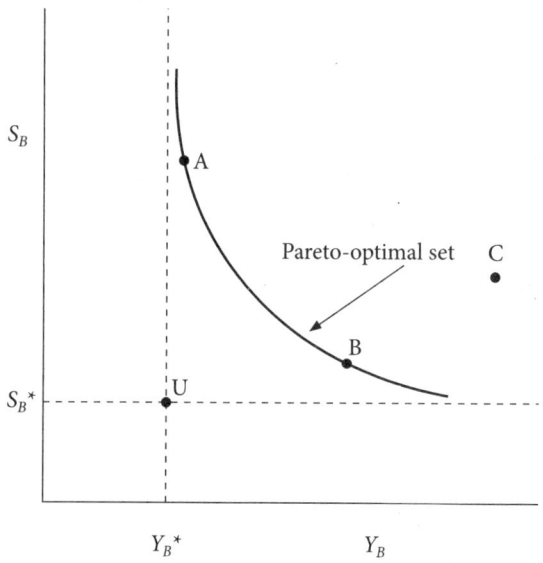

Fig. 8.4 Utopia point

Utopia point

Utopia points are points that optimize (minimize or maximize) all the objective functions of the multi-objective problem at once. This is given by the intersection of the minima of all independent objectives; this is also known as ideal point. The utopian point F is defined by $F^* = \{F_1^*, F_2^*, \ldots, F_k^*\}$. In Fig. 8.4, point U represents the utopia point for a bi-objective optimization problem. This point is the intersection of lines $Y_B = Y_B^*$ and $S_B = S_B^*$.

The utopia point is unachievable because the objectives are contradictory. Though, it can be utilized as a reference point for further calculation. For instance, we can find the closest point of utopia point on the Pareto front (compromise solution). Mathematically compromise solution can be represented as

$$\min_{Y_B, S_B} \|\Phi(Y_B, S_B) - U\| \tag{8.59}$$

8.2.1 Basic theory of multi-objective optimization

Although, there are multiple Pareto optimal solutions, only one solution has to be selected for practical implementation. Therefore, the multi-objective optimization process can be distinguished into two tasks: i) find a set of Pareto optimal solutions, and ii) select the most preferred solution from this set. The latter task requires a "Decision Maker" to choose the best solution in a particular instance of MOO problem as all Pareto optimal solutions are mathematically equivalent. A decision maker is one who can express preference information based on the requirement about the mathematically equivalent Pareto optimal solutions.

There are two main approaches for solving multi-objective optimization problems. The first one is Multi-Criteria Decision Making (MCDM) approach that can be characterized by the use of mathematical programming techniques and a decision making method in an intertwined manner. The decision maker plays a vital role in providing information to build a preference model. Evolutionary Optimization is a useful approach to solve MOO problems. Since evolutionary algorithms are population-based approach, they usually find an approximation of the Pareto front in a single run.

In this section, we will discuss various methods like Lexicographic Method, Linear Weighted Sum Method, Evolutionary Multi-objective Optimization, and Utopia-Tracking approach.

8.2.1.1 Lexicographic method

In this method, the objectives are ranked in order of importance (from best to worst) by the decision maker. The optimal solution X is found by minimizing the objective functions successively. The optimal value F_i^* $(i = 1, 2, \ldots, k)$ is found by minimizing the objective functions sequentially, starting with the most important objective function and proceeding according to the order of importance. Additionally, the optimal value (F_i^*) found of objective $F_i(X)$ is added as a constraint for optimization of $F_{i+1}(X)$. In this way, we move towards the optimal value of the most important objectives.

The subscripts of the objectives signify the objective function number as well as the priority of the objective. Therefore, $F_1(X)$ and $F_k(X)$ represent the most and least important objective functions, respectively. Therefore, the first problem is developed as

$$\text{Minimize } F_1(X) \tag{8.60a}$$

$$\text{subject to } g_j(X) \leq 0, \quad j = 1, 2, \ldots, m \tag{8.60b}$$

The solution of this problem is X_1^* and $F_1^* = F_1(X_1^*)$. After that the second problem is constructed as

$$\text{Minimize } F_2(X) \tag{8.61a}$$

$$\text{subject to } g_j(X) \leq 0, \quad j = 1, 2, \ldots, m \tag{8.61b}$$

$$F_1(X) = F_1^* \tag{8.61c}$$

the solution of this problem is found as X_2^* and $F_2^* = F_2(X_2^*)$.

This process repeats k times, until all the k objectives have been considered. The ith problem can be written as

Minimize $F_i(X)$ (8.62a)

subject to $g_j(X) \leq 0$, $j = 1, 2, \ldots, m$ (8.62b)

$F_l(X) = F_l^*$, $l = 1, 2, \ldots, i-1$ (8.62c)

and the solution of this ith problem is X_i^* and $F_i^* = F_i(X_i^*)$. Finally, the solution obtained at the end (X_k^*) is considered as the desired solution X^* of the original multi-objective optimization problem.

It is proved that the optimal solution obtained by the lexicographic problem is Pareto optimal. For this reason, the lexicographic method is usually adopted as an additional optimization approach in methods that can only guarantee weak optimality by themselves [Ehrgott, (2005); MIETTIEN, (1999)].

8.2.1.2 Linear weighted sum method

Multi-objective optimization can be modeled and solved by transforming it into a single objective problem using different methods, namely restriction method, ideal point method, and linear weighted sum method. These methods largely depend on the values assigned to the weighted factors or the penalties used, which are done quite arbitrarily. Another disadvantage of the above methods is that these algorithms obtain only one optimal solution at a time and may miss some useful information [Li et al. (2003)]. The weighted sum method is used extensively for solving MOO problems. This method converts multiple objective functions into an aggregated objective. This is done by multiplying each objective function with a weighting factor and summing up all weighted objective functions. Problem stated in Eq. (8.56b) can be written as [Kim, (2004)]

$$F_{WS} = \sum_{i=1}^{k} w_i F_i = w_1 F_1 + w_2 F_2 + \cdots + w_k F_k \qquad (8.63)$$

where $w_i = (i = 1, 2, \ldots k)$ represents the weighting factor for the ith objective function (potentially also dividing each objective by a scaling factor $w_i = \alpha_i / sf_i$). The weighted sum is said to be a convex combination of objectives when $\sum_{i=1}^{k} w_i = 1$ and $0 \leq w_i \leq 1$. Each single objective optimization provides one particular optimal solution point on the Pareto front. The weighted sum method then modifies weights systemically, and each different single objective optimization determines a different optimal solution. The solutions obtained estimate the Pareto front [Kim, (2004)].

The major drawbacks of the weighted sum method are (1) often the distribution of the optimal solution is not uniform, and (2) this method is not suitable for non-convex regions (Figs 8.5(a), 8.5(b)). Adaptive weighted sum method has been developed to optimize non-convex objective space.

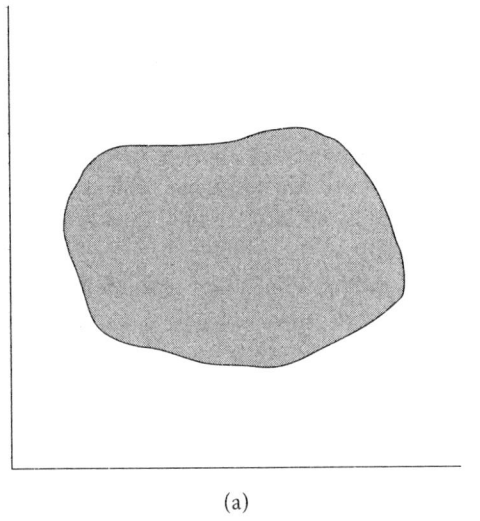

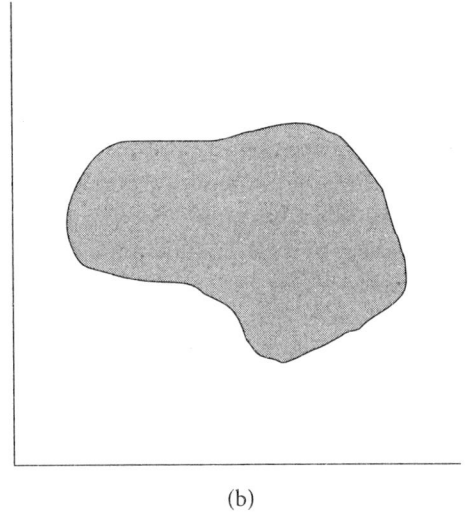

Fig. 8.5(a) Convex objective space   **Fig. 8.5(b)** Non-convex objective space

Finding the proper weights for Eq. (8.63) is a crucial job for the decision makers. Setting proper weights is just one way of expressing preferences, and this approach is applicable to many different methods. Misinterpretation of the practical and theoretical significance of the weights can make the method of intuitively choosing non-arbitrary weights an ineffective task [Marler and Arora, (2004)]. Therefore, understanding how the weights are affecting the solution to the weighted sum method has implications about other approaches that involve similar method parameters [Marler and Arora, (2010)]. Many researchers have developed methodical approaches for deciding weights.

In ranking methods, the objective functions are arranged according to their importance. The least important objective be given a weight of 1 and integer weights with subsequent increments are given to objective functions that are more important. The similar way is followed in categorization methods, where various objectives are segregated in broad categories for instance highly important, moderately important, and less important. The decision maker allocates independent values of relative importance to each objective function in the ranking methods [Marler and Arora, (2004)].

A study shows that values of weights can be decided using a preference function. A preference function is an abstract function, which is made by the points in the criterion space. This function perfectly incorporates the preferences, which is in the mind of the decision-maker. The majority of the MOO methods that minimizes a single combined objective function, which try to approximate the preference function with some mathematical formulation, called a utility function. The gradients of $P[F(X)]$, the preference function and $U = \sum_{i=1}^{k} w_i F_i(X)$, the utility function are given in Eqs (8.64) and (8.65) respectively:

$$\nabla_X P[F(X)] = \sum_{i=1}^{k} \frac{\partial P}{\partial F_i} \nabla_X F_i(X) \qquad (8.64)$$

$$\nabla_X U = \sum_{i=1}^{k} w_i \nabla_X F_i(X) \qquad (8.65)$$

Each part of the gradient $\nabla_x P$ qualitatively shows how the satisfaction of decision-makers changes with the change in design point consequently change in function values. Comparing Eqs (8.64) and (8.65) shows that when the weights are chosen correctly, the utility function and preference function can have gradients that are parallel to each other. This is very significant as the purpose of any utility function is approximating the preference function, a more accurate representation of one's preferences can result by imposing similarities between these two functions such as parallel gradients. When the utility function and the preference function are ideally same, then definitely the gradients of these two functions should be the same as well.

Equation (8.64) and (8.65) show that w_i corresponds to $\partial P/\partial F_i$ value. The value of $\partial P/\partial F_i$ is the approximate change in the preference function value that results from a change in the objective function value for F_i. Thus, $\partial P/\partial F_i$ gives us a mathematical definition for the importance of F_i. It will be better if we consider the significance of an objective function or change in preference function value in relative terms. Consequently, the value of a weight is significant relative to the values of other weights; the independent absolute magnitude of a weight is not relevant in terms of preferences [Marler and Arora, (2010)].

8.2.1.3 Evolutionary multi-objective optimization

In some cases, computational cost for generating the Pareto set is high and often it is not feasible; the difficulty of the underlying application makes the exact methods inapplicable. Therefore, a number of stochastic optimization techniques like simulated annealing, tabu search, ant colony optimization, etc. have been used to generate the Pareto set. The term evolutionary algorithm (EA) is used for a class of stochastic optimization methods that replicate the process of natural evolution. These methods have been used successfully for solving MOO problem [Deb (2001)]. The EAs are able to find a complete set of Pareto optimal solutions in a single run as these algorithms simultaneously deal with a set of possible solutions (population). Whereas the conventional mathematical programming techniques, need a series of separate runs. Even though the fundamental mechanisms are simple, the EAs have established themselves as a general, robust and powerful search mechanism [Bäck et al. (1997)] Particularly they have some characteristics that are required for problems involving i) objectives with multiple conflicts, and ii) obstinately large and very complex search spaces. The main advantage of EAs are, they are useful when we have inadequate knowledge about the problem being solved, easy to implement, robust, could be employed in a parallel environment, and are less susceptible to the continuity or shape of the Pareto front.

The EA is initiated with a population of solution candidates. Then to generate new solutions, reproduction process is used. The reproduction process enables to combine the existing solutions. Finally, natural selection decides which individuals of the current population join in the new population. The flowchart (Fig. 8.6) and algorithm can be written as follows:

Step 1 Initialize the random population, and additional parameters
Step 2 Calculate the fitness of each individual in the population
Step 3 Choose best-ranking individuals for reproducing
Step 4 Generate a new offspring through crossover operator
Step 5 Generate a new offspring through mutation operator
Step 6 Calculate the fitness of the individual offspring

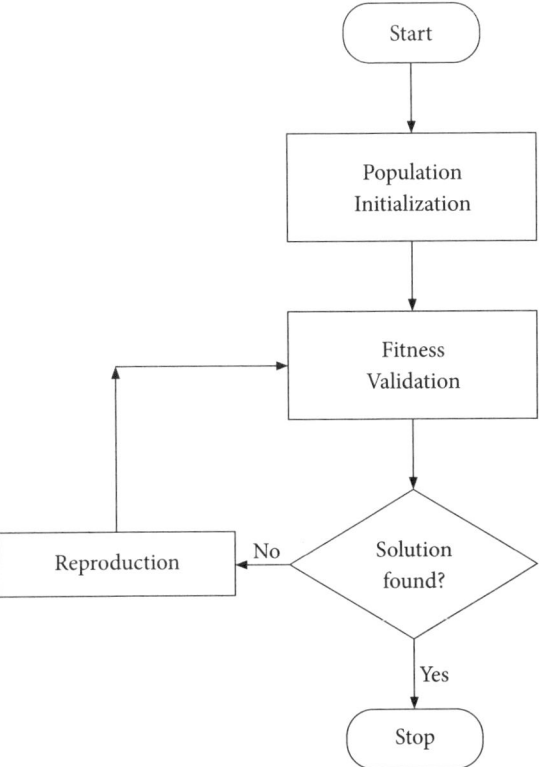

Fig. 8.6 Flowchart for evolutionary algorithm

Step 7 Replace worst ranked part of population with offspring
Step 8 Repeat, until terminating criteria matched

Details of different algorithm like genetic algorithm, simulated annealing, differential evolution have been elucidated in chapter 9.

8.2.1.4 Utopia-tracking approach

The main technical challenge in tackling the multiple objectives is that the construction of Pareto front is computationally expensive, mainly with multiple dimensions. Moreover, expert knowledge is still required to achieve a preferred solution even when such a front is built. Conventional approaches such as weighting and expert systems are limited as system conditions and priorities change under various operating modes [Zavala and Flores-Tlacuahuac, (2012)].

The key idea of the Utopia-Tracking approach is to minimize the distance between the cost function and the utopia point. This problem can be formulated as

$$\min\|F-U\|_p = \left(\sum_{i=1,2}\left|f_1 - f_1^L\right|^p + \left|f_2 - f_2^L\right|^p\right)^{1/p} \quad (8.66)$$

Here, $\|\bullet\|_p$ is the p-norm. This method is illustrated in Fig. 8.7. Where F (compromise solution) is any point on the Pareto front, U is the utopia point.

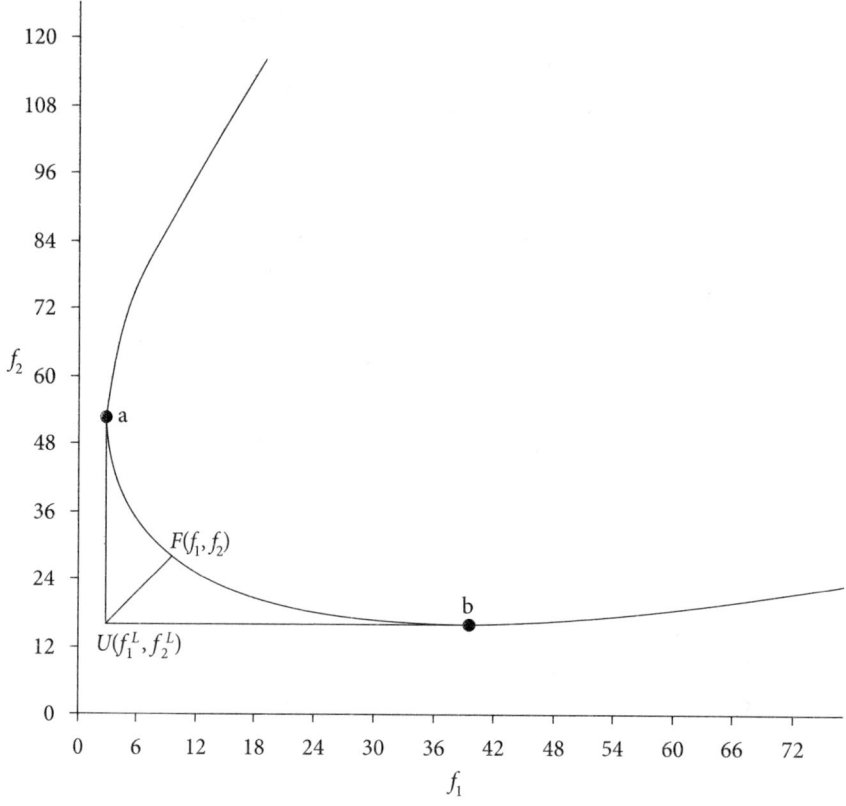

Fig. 8.7 Utopia-tracking approach

Traditional weighting method minimizes the function $w_1 f_1 + w_2 f_2$, where the weights are usually fixed. The weights have to be adapted to stay near to the utopia point as conditions change. The utopia-tracking is a method in which the weights of the compromise solution is determined automatically which is the closest point to the utopia point at each point in time [V.M. Zavala, (2013)]. This method has been applied successfully for optimizing control scheme like model predictive control (MPC) [Zavala and Flores–Tlacuahuac (2012)].

8.2.2 Multi-objective optimization applications in chemical engineering

Multi-objective optimization is very useful during the design of process equipments. In this chapter, multi-objective optimization of the fed batch reactor, PID controller tuning, and optimization of catalytic reforming process have been discussed.

8.2.2.1 Fed-batch bioreactor

To improve real-time decision making efficient approaches are required to determine the Pareto set in a rapid and precise way.

We consider the problem of finding the optimal feeding rate and batch duration for a fed-batch lysine fermentation process as described by Logist *et al.* [Logist *et al.* (2009)]. The state variables and the control variable of the process are given as follows.

x_1 : biomass [g],

x_2 : substrate [g],

x_3 : product, lysine [g],

x_4 : fermenter volume [l],

u : volume rate of the feed stream [l/h],

Two competing objective functionals are to be maximized, in search for Pareto-optimal feeding strategy, $u(t)$, and batch duration, $t_f [h]$.

i. The ratio of the product formed and the process duration, i.e., the productivity:

$$-\varphi_1\left(x(t_f), t_f\right) = \frac{x_3(t_f)}{t_f} \quad (8.67)$$

ii. The ratio of the product formed and the mass of the substrate, i.e., the yield:

$$-\varphi_2\left(x(t_f), t_f\right) = \frac{x_3(t_f)}{2.8\left(x_4(t_f) - 5\right)} \quad (8.68)$$

Note that clearly t_f is free, but, it is subject to the constraint

$$20 \leq t_f \leq 40 \quad (8.69)$$

Constraints are also imposed on the fermenter volume, the feed rate and the amount of substrate to be added, respectively, as follows.

$$5 \leq x_4(t) \leq 20 \quad (8.70a)$$

$$0 \leq u(t) \leq 2 \quad (8.70b)$$

$$20 \leq 2.8\left(x_4(t_f) - 5\right) \leq 42 \quad (8.70c)$$

The objective functions, constraints, and the process equations as given in Logist et al. [Logist et al. (2009)] yield the following bi-objective optimal control problem.

$$\min \left(-\frac{x_3(t_f)}{t_f}, \frac{x_3(t_f)}{2.8(x_4(t_f)-5)} \right) \quad (8.71a)$$

subject to

$$\dot{x}_1 = \left(0.125 \frac{x_2}{x_4} \right) x_1, \quad x_1(0) = 0.1 \quad (8.71b)$$

$$\dot{x}_2 = -\left(\frac{0.125\, x_2}{0.135\, x_4} \right) x_1 + 2.8u, \quad x_2(0) = 14 \quad (8.71c)$$

$$\dot{x}_3 = \left[-384 \left(0.125 \frac{x_2}{x_4} \right)^2 + 134 \left(0.125 \frac{x_2}{x_4} \right) \right] x_1, \quad x_3(0) = 0 \quad (8.71d)$$

$$\dot{x}_4 = u, \quad x_4(0) = 5 \quad (8.71e)$$

$$0 \le u(t) \le 2,\ 5 \le x_4(t) \le 20,\ 20 \le t_f \le 40,\ 20 \le 2.8(x_4(t_f)-5) \le 42 \quad (8.71f)$$

8.2.2.2 MOO for PID controller tuning

PID controller tuning can be formulated as a multi-objective optimization problem. During controller tuning we need to find the parameters (e.g., k_P, τ_I, τ_D) of any PID controller. For this purpose, we need to minimize the error functions like ISE, IAE, and ITAE. These error functions are represented as

Integral Square Error (ISE): $\text{ISE} = \int_0^\infty \varepsilon^2(t)\,dt$ \quad (8.72)

Integral Absolute Error (IAE): $\text{IAE} = \int_0^\infty |\varepsilon(t)|\,dt$ \quad (8.73)

Integral of the Time-weighted Absolute Error (ITAE): $\text{ITAE} = \int_0^\infty t|\varepsilon(t)|\,dt$ \quad (8.74)

The optimization scheme has been shown in Fig. 8.8

Fig. 8.8 PID controller as MOO problem

Mathematically, it can be written as

$$\text{Min. ISE} = \int_0^\infty \varepsilon^2(t)\,dt \tag{8.75a}$$

$$\text{Min. IAE} = \int_0^\infty |\varepsilon(t)|\,dt \tag{8.75b}$$

$$\text{Min. ITAE} = \int_0^\infty t|\varepsilon(t)|\,dt \tag{8.75c}$$

8.2.2.3 Industrial naphtha catalytic reforming process

The variables that affect the catalytic reforming process are the volume flow of naphtha charge to the volume of the catalyst (liquid hourly space velocity, LHSV), the latent aromatics content of naphtha charge (LA), the inlet temperatures of four reactor (T_1, T_2, T_3, T_4), the reaction pressure (p_r), the mole flow of hydrogen in the recycle gas to the mole flow of naphtha charge (hydrogen to oil molar ratio, n_{H_2}/n_{HC}), the product separator temperature (T_s), etc. Among these nine process variables chosen using mechanism analysis, the sensitivity analysis of each variable is performed using the process. It is shown that the appropriate set point value of one variable for maximizing the aromatics yield may not be suitable for minimizing the yield of heavy aromatics. So, the suitable trade-off solutions for the two optimal objectives should be considered.

For the continuous catalytic reforming process in this study, the unit is in full load operation and the value of LHSV cannot be further increased. Similarly, the quality of naphtha feedstock (e.g., LA) cannot be changed artificially for most domestic petroleum-refining enterprises. The product separator temperature T_s is not independent of other variables. Moreover, for further lowering of the temperature, coolers need to be included in the system, which in turn increase the operation costs. Hence, the remaining process variables are chosen as the decision variables for optimization of this process. These are the four reactor inlet temperatures (T_1, T_2, T_3, T_4), the reaction pressure (p_r), and the hydrogen-to-oil molar ratio (n_{H_2}/n_{HC}).

Thus, the two independent objectives, namely, the maximization of the aromatics yield (AY) and the minimization of the yield of heavy aromatics (HAY) are formulated mathematically as follows:

$$\text{maximize AY}(T_1, T_2, T_3, T_4, p_r, n_{H_2}/n_{HC}) \tag{8.76a}$$

$$\text{minimize HAY}\left(T_1, T_2, T_3, T_4, p_r, n_{H_2}/n_{HC}\right) \tag{8.76b}$$

$$\text{subject to } 520 \leq T_1, T_2, T_3, T_4 \leq 530 \tag{8.76c}$$

$$0.8 \leq p_r \leq 0.9 \tag{8.76d}$$

$$3.0 \leq n_{H_2}/n_{HC} \leq 4.0 \tag{8.76e}$$

$$65 \leq AY \leq 68 \tag{8.76f}$$

$$18 \leq HAY \leq 23 \tag{8.76g}$$

Weifeng *et al.* [Weifeng *et al.* (2007)] used neighborhood and archived genetic algorithm (NAGA) to this MOO problem.

8.3 Optimization in Control Engineering

Application of optimization techniques in the field of control engineering is very popular in recent times. Tuning of PID controller, application of MPC, and designing of optimal control systems need sufficient knowledge of optimization techniques. Some applications are discussed in the following section.

8.3.1 Real time optimization

The online calculation of optimal set points is known as Real Time Optimization (RTO). RTO deals with techniques of maximizing an economic objective related to the operation of a continuous process while satisfying other operating constraints [De Souza *et al.* (2010)]. Most RTO systems require non-linear steady state models of the process system combined with data reconciliation. This data reconciliation system updates the major parameters such as feed compositions and efficiencies using some parameter estimation techniques [Marlin and Hrymak (1996)]. The RTO system optimizes the process operating conditions (i.e., temperature, pH) and updates the set points to local MPCs that are developed based on linear dynamic models. If the set of active constraints with significant economic importance changes frequently, then a steady state RTO may not be sufficient. For these processes, it is more appropriate to use dynamic optimization with a nonlinear model developed by using dynamic RTO (DRTO) or nonlinear MPC with economic objective [Tosukhowong *et al.* (2004). Even though for practical application, steady state RTO is more appropriate than the DRTO.

As RTO relies on a static model, the plant should reach at sufficiently steady state to update the plant model properly. Typical steady-state detection methods include statistical parameters like mean, variance, or slope for selected signals over a moving window [Shrikant and Saraf (2004)].

Implementation of RTO system in process industries is a big challenge. Despite of many advantages, some companies feel that RTO is not viable to them. However, some companies are convinced of the profit from RTO systems and continue to invest in research and practical applications [Darby *et al.* (2011)].

Figure 8.9 shows the Plant decision hierarchy and approximate time level for different steps.

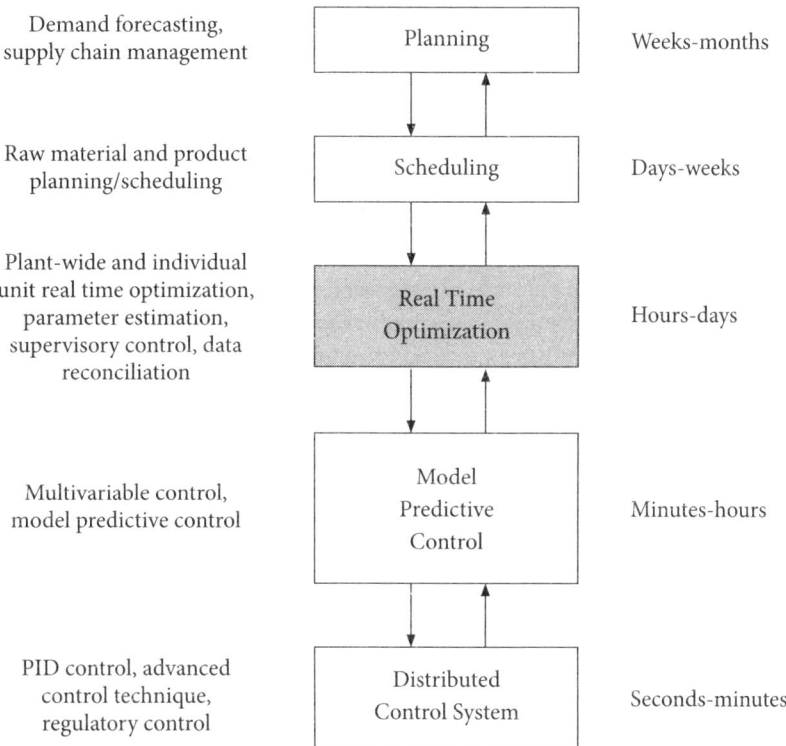

Fig. 8.9 Plant decision hierarchy

However, the execution rate of RTO depends on two factors; the frequency of unmeasured disturbances and the time necessary for an MPC to move the process to a new steady state. There is no clear-cut distinction of time scales, i.e., open-loop response times are similar because both consider the same variables (e.g., compositions). The MPC first discard unmeasured disturbances in those variables before a new steady state can be attained [Darby *et al.* (2011)].

Modern industrial practice validates input data primarily by using bound checks, rather than by statistical data reconciliation/gross error detection procedures. When this validity check fails, alternate measurements or estimates may be replaced automatically if it is accessible. For significant variables, if appropriate alternative value is not available for replacement, the implementation of the optimizer is aborted until suitable data is available. To minimize the deviation between measured variables and model variables, the RTO model is updated (calibrated) at each cycle. Model parameters such as heat transfer coefficients, distillation efficiencies, and unmeasured flows are usually adjusted to achieve this objective. When the model is updated properly, then optimization is executed. Usually, the objective function is developed based on profit. After successful convergence of the optimization algorithm, the optimal decisions are passed to MPC. Most of the industrial MPC applications work based on linear models, which are build empirically from a dedicated plant test [Darby et al. 2009]. Linear models (or linearized transform) have shown to be satisfactory for numerous control problems encountered in the chemical process industries [Gattu et al. 2003].

There are different techniques to accommodate RTO set points for controlled and manipulated variables in MPC. To penalize deviations from RTO set points, an additional term is included in the MPC objective function. Another formulation is done by using a separate target selection layer to determine the best, feasible set points for the MPC [Muske and Rawlings, (1993); Ying and Joseph (1999)]. We are considering the target selection is a linear programming (LP) optimizer, which is executing at a same frequency of the MPC that is common in the industrial application. The main advantage of this two-stage implementation over the single stage formulation is that it can better respond to disturbances and maintain feasibility between RTO executions [Ying and Joseph (1999)]. The following form can represent the typical LP formulation:

$$\min_{y^{TG}, u^{TG}} c_y^T y^{TG} + c_u^T u^{TG} \tag{8.77a}$$

subject to

$$y^{TG} = G_{ss} u^{TG} + b \tag{8.77b}$$

$$y^{\min} \leq y^{TG} \leq y^{\max} \tag{8.77c}$$

$$u^{\min} \leq u^{TG} \leq u^{\max} \tag{8.77d}$$

where c_y and c_u represent the costs associated with the outputs and inputs (controller manipulated variables), respectively; y^{TG} and u^{TG} are the optimal targets for the outputs and inputs, respectively; superscripts min and max give the limits of constraint for the inputs and outputs; G_{ss} re[resents the steady-state gain matrix obtained from the dynamic model of MPC; and b is the steady-state model bias term that is updated based on present output measurements and predictions from the dynamic model.

8.3.2 Optimal control of a batch reactor

In this section we will discuss the optimal control of a batch reactor used for bio biodiesel production. This optimization scheme was developed by Benavides and Diwekar [Benavides and Diwekar, 2012].

Batch reactor model

Biodiesel was produced by the transesterification of triglycerides and methanol in the presence of an alkaline catalyst for example sodium hydroxide. This reversible reaction composed of 3 steps, where triglycerides (TG) are converted to diglycerides (DG), diglycerides to monoglycerides

(MG), and lastly monoglycerides to glycerol (GL). An ester (R_iCOOCH_3) is produced in each step; resulting in 3 molecules of ester from one molecule of triglycerides. Steps of the reaction are given by Eqs (8.78a)–(8.78c) where k_1 to k_6 are the rate constants; and the overall reaction is represented by Eq. (8.79). All reactions are taken place at atmospheric pressure [Noureddini and Zhu (1997)].

$$TG + CH_3OH \xrightleftharpoons[]{k_1, k_2} DG + R_1COOCH_3 \tag{8.78a}$$

$$DG + CH_3OH \xrightleftharpoons[]{k_3, k_4} MG + R_2COOCH_3 \tag{8.78b}$$

$$MG + CH_3OH \xrightleftharpoons[]{k_5, k_6} GL + R_3COOCH_3 \tag{8.78c}$$

the overall reaction is given by the following equation

$$TG + 3CH_3OH \longleftrightarrow 3RCOOCH_3 + GL \tag{8.79}$$

The mathematical model of this batch reactor is represented by the following Ordinary Differential Equations (ODEs) (8.80)–(8.85) obtained from the mass balance of the batch reactor [Noureddini and Zhu (1997)].

$$F_1 = \frac{dC_{TG}}{dt} = k_1 C_{TG} C_A + k_2 C_{DG} C_E \tag{8.80}$$

$$F_2 = \frac{dC_{DG}}{dt} = k_1 C_{TG} C_A - k_2 C_{DG} C_E - k_3 C_{DG} C_A + k_4 C_{MG} C_E \tag{8.81}$$

$$F_3 = \frac{dC_{MG}}{dt} = k_3 C_{DG} C_A - k_4 C_{MG} C_E - k_5 C_{MG} C_A + k_6 C_{GL} C_E \tag{8.82}$$

$$F_4 = \frac{dC_E}{dt} = k_1 C_{TG} C_A - k_2 C_{DG} C_E + k_3 C_{DG} C_A - k_4 C_{MG} C_E + k_5 C_{MG} C_A - k_6 C_{GL} C_E \tag{8.83}$$

$$F_5 = \frac{dC_A}{dt} = -\frac{dC_E}{dt} \tag{8.84}$$

$$F_6 = \frac{dC_{GL}}{dt} = k_5 C_{MG} C_A - k_6 C_{GL} C_E \tag{8.85}$$

where C_{TG}, C_{DG}, C_{MG}, C_E, C_A, and C_{GL} are concentrations of triglycerides, diglycerides, monoglycerides, methyl ester, methanol, and glycerol respectively.

Optimal control problem

Optimal control provides us useful information for designing and controlling the reaction process. It has numerous applications in the academic and industrial field. Usually, a time dependent profiles of the control variable are required to optimize a particular performance index [Rico-Ramirez and Diwekar (2004)]. For example, one of the most common cases of an optimal control problem is that the best temperature profile is found so that the performance index, conversion or yield, is optimized at a particular final time. Some other methods can be employed to solve these problems are the dynamic programming, maximum principle, and calculus of variations [Diwekar (2003)].

These methods depend on suitable mathematical representations, which cannot be solved directly by simple mathematical models. Methods with second order differential equations or partial differential equations are difficult to solve so the maximum principle that uses ODEs, can be employed in most of cases [Diwekar (1995)]. At each time step, the maximum principle method adds adjoint variables and corresponding adjoint differential equations, and a Hamiltonian that requires to be optimized to get the control variable. In this section, an optimal control problem is formulated for the production of biodiesel in a batch reactor. Generally two different types of optimization problems are considered: the maximization of concentration and minimization of time. Though, only the maximum concentration is discussed in this section. We need to maximize the biodiesel concentration. For this optimal control problem, the best temperature profile in a given reaction time (100 min) is found to maximize the concentration of methyl ester (biodiesel). The maximum principle formulation is used for solving this problem. The objective function is reconstructed in this method. It is represented as a linear function in terms of final values of state variables (C_i) and the constant values (A_i), therefore, the objective function for this problem is given below:

$$\text{Maximize } J = \sum_{i=1}^{n} A_i C_i \left(t_f\right) = \overline{A}_i^T . \overline{C}_i \left(t_f\right) = C_E \left(t_f\right) \tag{8.86a}$$

subject to state Eqs (8.80) to (8.85) given in the generalized form below

$$\frac{dC_i}{dt} = f(C_i, T) \tag{8.86b}$$

for

$$C_i = \left[C_{TG}(t_0), C_{DG}(t_0), C_{MG}(t_0), C_E(t_0), C_A(t_0), C_{GL}(t_0) \right] \tag{8.86c}$$

where C_i denotes the state variable that corresponds to the concentration of each component: C_{TG}, C_{DG}, C_{MG}, C_E, C_A, and C_{GL}. Temperature T is a control variable. A_i represents the constants values for the linear representation of the maximum principle. Initial conditions:

Initial time (t_0) = 0 min.

$$C_i(t_0) = [0.3226; 0; 0; 0; 1.9356; 0] \text{ mol/l}$$

$A = [0;0;0;1;0;0]$

The maximum principle employs the addition of n adjoint variables (one adjoint variable per state variable), n adjoint equations, and a Hamiltonian that satisfies the following relations

$$H\left(\overline{z}_t, \overline{C}_t, T\right) = \overline{z}_t^T, F\left(\overline{C}_t, T_t\right) = \sum_{i=1}^{n} z_i F_i \left(\overline{C}_t, T_t\right) \tag{8.87}$$

$$\frac{dz_i}{dt} = \sum_{j=1}^{n} z_j \left(\frac{\partial F_j}{\partial C_i}\right) \tag{8.88}$$

where n is the number of components (six) and F_i is the right-hand side of the differential Eqs (8.90a) to (8.90f). For this problem Hamiltonian can be written as

$$H = z_1 F_1 + z_2 F_2 + z_3 F_3 + z_4 F_4 + z_5 F_5 + z_6 F_6 \tag{8.89}$$

The adjoint equations can be compute as

$$\frac{dz_1}{dt} = -z_1\left(-k_1 C_A\right) - z_2\left(k_1 C_A\right) - z_4\left(k_1 C_A\right) - z_5\left(-k_1 C_A\right) \tag{8.90a}$$

$$\frac{dz_2}{dt} = -z_1\left(k_2 C_A\right) - z_2\left(-k_2 C_E - k_3 C_A\right) - z_3\left(k_3 C_A\right) \\ - z_4\left(-k_2 C_E + k_3 C_A\right) - z_5\left(k_2 C_E - k_3 C_A\right) \tag{8.90b}$$

$$\frac{dz_3}{dt} = -z_2\left(k_4 C_E\right) - z_3\left(-k_4 C_E - k_5 C_A\right) - z_4\left(-k_4 C_E + k_5 C_A\right) \\ - z_5\left(k_4 C_E - k_5 C_A\right) - z_6\left(k_5 C_A\right) \tag{8.90c}$$

$$\frac{dz_4}{dt} = -z_1\left(k_2 C_{DG}\right) - z_2\left(-k_2 C_{DG} + k_4 C_{MG}\right) - z_3\left(-k_4 C_{MG} + k_6 C_{GL}\right) \\ - z_4\left(-k_2 C_{DG} - k_4 C_{MG} - k_6 C_{GL}\right) - z_5\left(k_2 C_{DG} + k_4 C_{MG} + k_6 C_{GL}\right) + z_6\left(k_6 C_{GL}\right) \tag{8.90d}$$

$$\frac{dz_5}{dt} = z_1\left(k_1 C_{TG}\right) - z_2\left(k_1 C_{TG} - k_3 C_{DG}\right) - z_3\left(k_3 C_{DG} - k_5 C_{MG}\right) \\ - z_4\left(k_1 C_{TG} + k_3 C_{DG} + k_5 C_{MG}\right) - z_5\left(-k_1 C_{TG} - k_3 C_{DG} - k_5 C_{MG}\right) - z_6\left(k_5 C_{MG}\right) \tag{8.90e}$$

$$\frac{dz_6}{dt} = -z_3\left(k_6 C_E\right) - z_4\left(k_6 C_E\right) - z_5\left(k_6 C_E\right) + z_6\left(k_6 C_E\right) \tag{8.90f}$$

For the adjoint variable, the boundary conditions are $z_i(t_f) = [0;0;01;0;0]$, which stands for the constant values of vector A. These equations can be solved easily using backward integration along with Runge–Kutta–Fehlberg (RKF) method. At last, the optimal decision vector $T(t)$ can be achieved by finding the extremum of the Hamiltonian at each time step, in other words, applying the optimality condition:

$$\left.\frac{dH}{dT}\right|_t = 0 \tag{8.91}$$

As it was stated earlier, the optimal control problem can also be described as the minimization of time problem. The objective is to minimize the batch time for a given final concentration. The objective function can be described as given in Eq. (8.92a) and the differential equations (Eqs (8.80)–(8.85)) can be converted by multiplying with dt/dC_E, as given away in Eq. (8.92b):

$$\text{Minimize } J = \int_{C_E(t_0)}^{C_E(t_f)} \frac{dt}{dC_E} = t_f \tag{8.92a}$$

$$\text{subject to } \frac{dC_i}{dC_E} = \frac{dC_i}{dt} \bigg/ \frac{dC_E}{dt} \tag{8.92b}$$

These equations can be solved using *Runge–Kutta–Fehlberg method* (RK 45).

8.3.3 Optimal regulatory control system

Consider a process described by a kth order linear differential equation,

$$y_n = a_1 y_{n-1} + a_2 y_{n-2} + \cdots + a_k y_{n-k} + b_1 m_{n-1} + \cdots b_k m_{n-k} \tag{8.93}$$

where $a_1, a_2, \ldots, a_k; b_1, \ldots, b_k$ are constant parameters with known values. Suppose that the purpose of the regulatory control system is to keep the closed-loop output as close as possible to a prescribed set point value y_{SP} in the presence of disturbance changes. The deviation can be specified by one of the following measures:

$$P_1 = \{y_n - y_{SP}\}^2 \tag{8.94}$$

$$P_2 = \frac{1}{N} \sum_{n=1}^{N} \{y_n - y_{SP}\}^2 \tag{8.95}$$

The first measure, P_1, is referred to as one-stage control; the second as N-stage control. Thus, the controller design problem can be formulated as follows:

Find a controller that minimizes P_1 or P_2 in the presence of load changes.

The control action that minimizes P_1 attempts to keep the output close to the set point by making individual control decisions at each stage. Minimum of P_2, on the other hand, relaxes the restriction above and plans the control action over a longer time horizon. Let us now solve the two design problems above.

Suppose that we are at the nth sampling instant and that we want to compute the control action m_n in such a way that y_{n+1} will be as close as possible to the desired y_{SP}. Using criterion P_1, we take

$$P_1 = \{y_n - y_{SP}\}^2 = \{a_1 y_n + a_2 y_{n-1} + \cdots a_k y_{n-k+1} + b_1 m_n + b_2 m_{n-1} + \cdots b_k m_{n-k+1} - y_{SP}\}^2 \quad (8.96)$$

The minimum of P_1 is found when $\partial P_1 / \partial m_n = 0$. Then, we have

$$2\{a_1 y_n + a_2 y_{n-1} + \cdots a_k y_{n-k+1} + b_1 m_n + b_2 m_{n-1} + \cdots b_k m_{n-k+1} - y_{SP}\}(b_1) = 0 \quad (8.97)$$

and the optimum regulatory control action at the nth instant is given by

$$m_n = \frac{1}{b_1}\{y_{SP} - a_1 y_n - a_2 y_{n-1} - \cdots - a_k y_{n-k+1} - b_2 m_{n-1} - \cdots - b_k m_{n-k+1}\} \quad (8.98)$$

The controller defined by Eq. (8.98) is physically realizable because it uses only current or past information on the manipulated variable and the controlled output.

Now let us turn our attention to the second design criterion, P_2. Consider the situation at the $(N - 1)$ sampling instant. The outputs $y_{N-1}, y_{N-2}, \ldots, y_1$ have been measured and the control problem is to determine the value of the manipulated variable m_{N-1}. Since m_{N-1} influence only the last term of P_2, we have

$$\min P_2 = \{y_N - y_{SP}\}^2 \quad (8.99)$$

Then the optimal value of m_{N-1} is given by Eq. (8.98) with $n = N - 1$.

Consider, the situation at $n = N - 2$. The output has been measured for $n = N - 2, N - 3, \ldots, 1$ and the problem is to determine the optimal value of m_{N-2}. Since m_{N-2} influences the last two terms of P_2, we have

$$\min P_2 = \{y_N - y_{SP}\}^2 + \{y_{N-1} - y_{SP}\}^2 \quad (8.100)$$

If the optimum value of m_{N-1} has been used for the last stage, the minimization problem Eq. (8.100) yields

$$\min P_2 = \{y_{N-1} - y_{SP}\}^2 \quad (8.101)$$

because $y_N - y_{SP} = 0$ for the optimum value of m_{N-1}. However, the optimum value of m_{N-2} solving the last problem is given again by Eq. (8.98) for $n = N - 2$. Therefore, we reach the following conclusion:

The optimal regulatory control action for a system described by a kth order difference model with constant and known parameters is given by Eq. (8.98) independently of which criterion, P_1 or P_2 is used.

8.3.4 Dynamic matrix control

Dynamic Matrix Control (DMC) was developed by Shell Oil Company in the 1970s. Afterward, DMC has been successfully used in industry for more than a decade. Several authors have reported improved control performance by using the DMC as compared to "traditional" control algorithms [Cutler and Ramaker, (1980)]. DMC belongs to the family of model predictive control (MPC) algorithms. This algorithm can be separated into two parts, a predictor and an optimizer [Lundstrom *et al.* (1995)].

The key features of any MPC are given below:

i. linear controller
ii. uses a process or plant step response model
iii. predicts the process dynamics over a time window
iv. minimizes an objective function
v. compensates for a relatively high degree of process non-linearity (good robustness)
vi. easily adapted to multivariable plants
vii. easy tuning

Dynamic matrix control is a popular technique for the control of dynamic systems, which has slow response. The most available predictive control algorithms are implemented by optimization method that minimizes a performance index. This result in the complexity to analyze the algorithms of stability of the predictive control consequently makes it difficult to implement the stability design of predictive controllers. The calculation of inverse matrix is associated in dynamic matrix control (DMC) and it holds down on-line application. The main advantage of DMC is its capability to handle control problem with constraints.

The objective of the DMC controller is to drive the output to track the set point in the least squares sense including a penalty term on the manipulated variable moves. This results in smaller computed input moves and a less aggressive output response [Qin *et al.* (2003)]. The method is strictly developed for linear systems (like any other conventional controller) and as a result, any analysis on DMC and its features must be made in a framework of linear system theory [Garcia and Morshedi (1986)].

Linear Input-Output Model

Without loss of generality, we can consider a dynamic linear system with an input I and one output O. In computer applications, only the system behavior at the sampling intervals is of interest. A discrete representation of the dynamics is utilized for this purpose. Such a representation is given in Eq. (8.102).

$$O(k+1) = \sum_{i=1}^{M} a_i \Delta I(k-i+1) + O_0 + d(k+1) \qquad (8.102)$$

where k represents discrete time; O_0 is initial condition of the output; $\Delta I(k)$ denotes the change in input (or manipulated variable) at different time intervals k; $O(k)$ represents the value of the controlled variable at time k; $d(k)$ accounts for unmodelled factors that influence $O(k)$; a_i denote the coefficients of unit step response of the system; and M is the number of time intervals necessary for the system to attain steady-state.

Therefore, $a_i = a_M$, for $i \geq M$.

The term $d(k + 1)$ has been added to the input-output description to take into consideration the unmodelled effects on the measured output. This is consisting of unmeasured disturbances and/or modelling errors. Addition of this factor is vital during the formulation of DMC as shown below.

Controller Design

The aim of any controller is to find the change in the manipulated variables, $\Delta I(k)$, which give the output $O(k)$ best match a target value O_s in the face of disturbances. Considering the present time interval to be $\bar{k}$, in DMC a projection of the output $O(k)$ over P future time intervals $\left(\bar{k}+1 \text{ to } \bar{k}+P\right)$ is matched to the set-point O_s by recommending a series of future moves.

Formulation of the DMC equations Considering Eq. (8.102), the projected output for any future time $\bar{k}+l$, $l > 0$ is

$$O\left(\bar{k}+l\right) = \sum_{i=1}^{l} a_i \Delta I\left(\bar{k}+l-i\right)$$

$$+ O_0 + \sum_{i=l+1}^{M} a_i \Delta I\left(\bar{k}+l-i\right) \tag{8.103}$$

$$+ d\left(\bar{k}+l\right)$$

For simplicity, let us define

$$O^*\left(\bar{k}+l\right) = O_0 + \sum_{i=l+1}^{M} a_i \Delta I\left(\bar{k}+l-i\right) \tag{8.104}$$

to be the contribution to $O\left(\bar{k}+l\right)$ owing to past input moves up to the present time $\bar{k}$. This term can always be calculated from the past history of moves.

This definition can be used to write Eq. (8.103) for times $\bar{k}+1$ up to $\bar{k}+P$ to generate a set of P equations for the output projections as given below:

$$\begin{bmatrix} O(\bar{k}+1) \\ \vdots \\ O(\bar{k}+P) \end{bmatrix} = \begin{bmatrix} O^*(\bar{k}+1) \\ \vdots \\ O^*(\bar{k}+P) \end{bmatrix} + A \begin{bmatrix} \Delta I(\bar{k}) \\ \vdots \\ \Delta I(\bar{k}+N-1) \end{bmatrix} + \begin{bmatrix} d(\bar{k}+1) \\ \vdots \\ d(\bar{k}+P) \end{bmatrix} \tag{8.105}$$

where

$$A = \begin{bmatrix} a_1 & 0 & 0 & 0 \\ a_2 & a_1 & 0 & 0 \\ \vdots & & \ddots & 0 \\ a_N & a_{N-1} & & a_1 \\ \vdots & & & \\ a_M & a_{M-1} & & a_{M-N+1} \\ \vdots & \vdots & & \vdots \\ a_M & a_M & \cdots & a_M \end{bmatrix}$$

This matrix is termed as the "dynamic matrix" of the system. It is clear that during the formulation of DMC, only N moves are calculated, i.e.,

$$\Delta I(k) = 0 \quad \text{for} \quad k > \bar{k} + N \tag{8.106}$$

Setting the values of these moves equal to zero imparts significant stability properties to the resulting controller. Particularly, we can tell from our experience that selecting $P = N + M$ yields a stable controller in most of the cases.

Estimation of unmodelled effects $d(k)$ in DMC

An estimation of the unmodelled effects $d(k)$ is required to solve the set of equations in (8.105). As future values of the "disturbance" $d(k)$ are unavailable, the best we can do is to use an estimate. From Eq. (8.102) for $k = \bar{k} - 1$, and Eq. (8.104) for $l = 0$, we get

$$O(\bar{k}) = O^*(\bar{k}) + d(\bar{k}) \tag{8.107}$$

Therefore, $d(k)$ can be estimated using the current feedback measurement $O_m(\bar{k})$ of O in conjunction with the moves information of past input. When we do not have any additional knowledge of $d(k)$ over future intervals (as is true in most cases), it is assumed that the predicted disturbance and the present, "measured" $d(\bar{k})$ are equal.

$$d(\bar{k} + l) = d(\bar{k})$$

$$= O_m(\bar{k}) - O^*(\bar{k}); \quad l = 1, \ldots, P \tag{8.108}$$

Solution of the DMC equations

For a known set of equations the DMC control problem is described as finding the N future input moves $\Delta I(\bar{k}) \cdots \Delta I(\bar{k} + N - 1)$ such that the sum of squared deviations between the projections $O(\bar{k} + l)$ and the target O_s are minimized. This is same as the least-squares solution of the DMC equations:

$$\begin{bmatrix} O_s - O^*(\bar{k}+1) - d(\bar{k}) \\ \vdots \\ O_s - O^*(\bar{k}+P) - d(\bar{k}) \end{bmatrix} = e(\bar{k}+1) = AX(\bar{k}) \qquad (8.109)$$

where $e(\bar{k}+1)$ represents a P-dimensional vector of projected deviations from the target and

$$X(\bar{k}) = \left[\Delta I(\bar{k}) \cdots \Delta I(\bar{k}+N-1) \right]^T \qquad (8.110)$$

is the vector of future moves. Such least-squares solution can be written as

$$X(\bar{k}) = \left(A^T A \right)^{-1} A^T e(\bar{k}+1) \qquad (8.111)$$

In Dynamic Matrix Control, only the move estimated for the current time interval $\bar{k}$ is employed. At every sampling time k, the computation is repeated. We obtain a new feedback measurement that is used to update $e(\bar{k}+1)$. The disturbance handling characteristics of the algorithm could be impaired, if it fails to estimate a move at each sampling time.

Formulation for multivariable systems

It should be mentioned that for a multivariable system the DMC equations could be derived in the same way as for the single-input single-output (SISO) system. For a s-input and r-output system, a linear dynamic system can be represented by

$$O(k+1) = \sum_{i=1}^{M} a_i \Delta I(k-i+1) + O_0 + d(k+1) \qquad (8.112)$$

where $O(k)$ denotes an r-dimensional vector of outputs, a_i is an $r \times s$ matrix of unit step response coefficients for the ith time interval, $\Delta I(k)$ represents the s-dimensioned vector of moves for all manipulated variables at a given time interval, O_0 is the vector of initial condition, and $d(k)$ is a vector of unmodelled factors. For $r = s = 1$ this equation reduces to model Eq. (8.102).

We can describe a multivariable system dynamic matrix A, which is composed of blocks of dimension $P \times N$ of step response coefficient matrices as in Eq. (8.105) relating the ith output to the jth input as follows:

$$A = \begin{bmatrix} A_{11} & A_{12} & \cdots & A_{1s} \\ A_{21} & A_{22} & \cdots & A_{2s} \\ \vdots & \vdots & \ddots & \vdots \\ A_{r1} & A_{r2} & \cdots & A_{rs} \end{bmatrix} \qquad (8.113)$$

where elements from matrices a_l have been regrouped accordingly. The matrix A_{ij} contains all the ij coefficients in matrices $a_l = l = 1$ to M arranged as in Eq. (8.105).

The corresponding vector of moves is

$$X(\bar{k}) = \left[X_1(\bar{k})^T \ X_2(\bar{k})^T \cdots X_s(\bar{k})^T \right]^T \tag{8.114}$$

and the output projection vector becomes:

$$e(\bar{k}+1) = \left[e_1(\bar{k}+1)^T \ e_2(\bar{k}+1)^T \cdots e_r(\bar{k}+1)^T \right]^T \tag{8.115}$$

Therefore, Eq. (8.110) is equally valid for multivariable systems.

Summary

- Uncertainty is a part of many chemical engineering processes. Fluidized bed, catalytic reactor, composition of available raw material, etc. are very common examples. We have discussed the formulation method of the optimization problem for stochastic method. Stochastic linear and stochastic nonlinear methods are discussed here. Basic knowledge of probability is required to understand the stochastic optimization. Besides this, in the chemical industry, we need to optimize more than one objective functions. Lexicographic method, Linear weighted sum method, evolutionary algorithm and utopia tracking approach can be used to solve multi-objective optimization problem. Controlling different processes is a major task in the chemical process plant. Some advanced control system has been developed based on optimization theories. The error between the measured value and estimated value is minimized.

Review Questions

8.1 Give some examples of uncertainties in chemical industry.

8.2 What is meant by expected value of any random variable?

8.3 When a chemical process is called stochastic process? Give some examples.

8.4 The effluent stream from a wastewater process is monitored to make sure that two process variables, the biological oxidation demand (BOD) and the solid content, meet specifications.

Sample no	1	2	3	4	5	6	7	8	9	10	11	12	13	14	15	16	17	18	19	20
BOD (mg/l)	17.7	23.6	13.2	25.2	13.1	27.8	29.8	14.3	26.0	23.2	22.8	20.4	17.5	18.4	16.8	13.8	19.4	24.7	16.8	27.6
Solids (mg/l)	1380	1458	1322	1448	1334	1485	1503	1341	1448	1426	1417	1384	1380	1396	1345	1349	1398	1426	1361	1476

Find the average BOD and solid content of the effluent.

8.5 Develop the optimization problem of a fluid flow system with two objectives, minimization of installation cost as well as minimization of pumping cost.

8.6 What do you mean by utopia point? What is the significance of that point?

8.7 Write down the algorithm for solving a multi-objective optimization problem using Lexicographic method.

8.8 Why Evolutionary Multio-bjective Optimization is superior to conventional multi-objective optimization?

8.9 Develop optimization problem for controlling a batch reactor with a reaction

$$A + B \xrightarrow{k_1} C + D$$

where C is the desired product.

8.10 What are the advantages of MPC over PID control? Explain your answer considering both economic and technical point.

8.11 Write down the algorithm of a regulatory control system.

References

Al-Othman, W. B. E., Lababidi, H. M. S., Alatiqi, I. M., Al-Shayji, K. *Supply Chain Optimization of Petroleum Organization Under Uncertainty in Market Demands and Prices,* European Journal of Operational Research, 189(2008): 822–40.

Bäck, T., Hammel, U. and Schwefel, H. P. 1997. *Evolutionary Computation: Comments on the History and Current State.* 'IEEE Transactions on Evolutionary Computation', 1(1): 3–17.

Benavides, P. T. and Diwekar U. *Optimal Control of Biodiesel Production in a Batch Reactor Part I: Deterministic Control',* Fuel, 94(2012): 211–17.

Charnes, A. and Cooper, W. W. 1959. *Chance Constrained Programming, Management Science,* vol. 6, pp. 73–79.

Cutler, C. R. and B. L. Ramaker. 1980. *Dynamic Matrix Control-a Computer Control Algorithm.* Proc. Joint Automatic Control Conf., San Francisco, CA, Paper WP5–B.

Darby, M. L., Nikolaou, M., Jones J., Nicholson D., RTO: *An Overview and Assessment of Current Practice,* 'Journal of Process Control', 21(2011): 874–84.

Darby, M. L., Harmse, M., Nikolaou, M. 2009. *MPC: Current Practice and Challenges,* Istanbul: ADCHEM.

Deb, K. 2001. *Multi-Objective Optimization using Evolutionary Algorithms,* UK: John Wiley and Sons.

DeCoursey W. J. 2003. *Statistics and Probability for Engineering Applications;* Newnes.

Dehling, H. G., Hoffmann, A. C. and Stuut, H. W. 1999. *Stochastic Models for Transport in a Fluidized Bed.* SIAM J. Appl. Math. 60, pp. 337–58.

De Souza, G., Odloak D., Zanin A. C. *Real Time Optimization (RTO) with Model Predictive Control (MPC);* 'Computers and Chemical Engineering', 34(2010): 1999–2006.

Diwekar, U. 2003. *Introduction to Applied Optimization. 2nd ed.* Boston, MA: Kluwer Academic Publishers.

Diwekar U. 1995. *Batch Distillation: Simulation, Optimal Design and Control. 2nd ed.* Washington D. C.: Taylor and Francis.

Dixon, L. C. W. and Szegö, G. P. 1975. *Towards Global Optimisation,* 1–2, North-Holland, Amsterdam, The Netherlands.

Ehrgott, M. 2005. *Multicriteria Optimization,* Second edn., Springer, Berlin.

Fettaka S., Thibault J. and Gupta Y. *Design of Shell and Tube Heat Exchangers using Multi-objective Optimization,* 'International Journal of Heat and Mass Transfer', 60(2013): 343–54.

Garcia, C. E. and Morshedi, A. M. 1986. *Quadratic Programming Solution of Dynamic Matrix Control (QDMC),* 'Chemical Engineering Communications', 46: 1–3: 73–87.

Gattu, G., Palavajjhala, S. and Robertson, D. B. 2003. *Are Oil Refineries Ready for Non-linear Control and Optimization?* in: International Symposium on Process Systems Engineering and Control, January, Mumbai.

Henrion, R., LI, P. Möller, A. *et al. Stochastic Optimization for Operating Chemical Processes Under Uncertainty,* ZIB-Report 01–04 (May 2001).

Hou, K., Li, Y., Shen, J. and Hu, S. *Delayed Sampling Approach to Stochastic Programming in Chemical Processes,* 'Computers and Chemical Engineering' 24(2000): 619–24.

Jonsbraten, T. W. 1998. *Optimization Models for Petroleum Field Exploitation.* Ph.D. thesis. Norwegian School of Economics and Business Administration, Norway: Stavanger College.

Kim, I. Y. 2004. *Adaptive Weighted Sum Method for Multi-objective Optimization* 10th AIAA/ISSMO Multidisciplinary Analysis and Optimization Conference.

Li, S. J. Wang, H. and Qian F. 2003. *Multi-objective Genetic Algorithm and its Applications in Chemical Engineering,* Comput. Appl. Chem., 20(6): 755–60.

Logist, F. Van Erdeghem, P. Van Impe, J. 2009. *Efficient Deterministic Multiple Objective Optimal Control of (bio)Chemical Processes.* Chem. Eng. Sci., 64: 2527–38.

Lundstrom, P., Lee, J. H. Morari, M. and Skogestad S. 1995. *Limitations of Dynamic Matrix Control,* Computers Chemical Engineering, 19(4): 409–21.

Marler, R. T. Arora, J. S. *The Weighted Sum Method for Multi-objective Optimization: New Insights,* Struct Multidisc Optim., (2010)41: 853–62.

Marler, R. T., Arora, J. S. *Survey of Multi-objective Optimization Methods for Engineering.* Struct Multidiscipl Optim., 26(2004): 369–95.

Marlin, T. E. and Hrymak, A. N. 1997. *Real-time Operations Optimization of Continuous Processes.* In Proceedings of CPC V, AIChE Symposium Series, vol. 93, pp. 156–64.

Miettien, K. 1999. *Nonlinear Multi-objective Optimization,* Kluver Academic Publishers, Boston, London, Dordrecht.

Muske, K. R., Rawlings, J. B. 1993. *Model Predictive Control with Linear Models,* AIChE Journal, 39(2): 262–87.

Noureddini, H., Zhu D. 1997. *Kinetic of Transesterification of Soybean Oil.* J Am Oil Chem Soc, 74: 1457.

Qin, S. J., Badgwell T. A. 2003. *A Survey of Industrial Model Predictive Control Technology.* Control Engineering Practice, vol. 11, pp. 733–64.

Rao, S. S. 2009. *Engineering Optimization: Theory and Practice,* John Wiley and Sons, Inc.

Rico-Ramirez, V., Diwekar, U. 2004. *Stochastic Maximum Principal for Optimal Control Under Uncertainty.* 'Computers Chemical Engineering', 28: 2845.

Sahinidis, N. V. *Optimization Under Uncertainty: State-of-the-art and Opportunities;* 'Computers and Chemical Engineering', 28(2004): 971–83.

Shrikant, A. B., Saraf D. N. *Steady State Identification, Gross Error Detection, and Data Reconciliation for Industrial Process Units,* 'Industrial and Engineering Chemistry Research', 43(2004): 4323–36.

Stadler, W. 1979. *A Survey of Multicriteria Optimization, or the Vector Maximum Problem, Journal of Optimization Theory and Applications,* vol. 29, pp. 1–52.

Tosukhowong, T. Lee, J. Lee, J. and Lu J. *An Introduction to a Dynamic Plantwide Optimization Strategy for an Integrated Plant.* 'Computers and Chemical Engineering', 29(2004): 199–208.

Weifeng, H., Hongye, S., Shengjing, M. and Jian C. 2007. *Multi-objective Optimization of the Industrial Naphtha Catalytic Re-forming Process,* Chin. J. Chem. Eng., 15(1): 75–80.

Ying, C. M., Joseph, B. 1999. *Performance and Stability Analysis of LP-MPC and QP-MPC Cascade Control Systems,* AIChE Journal, 45(7): 1521–34.

Zavala, V. M. 2013. *Real-Time Optimization Strategies for Building Systems, Industrial and Engineering Chemistry Research,* Ind. Eng. Chem. Res., 52: 3137-50.

Zavala, V. M., Flores-Tlacuahuac, A. *Stability of Multi-objective Predictive Control: A Utopia-tracking Approach,* Automatica, 48(2012): 2627–32.

Chapter 9

Nontraditional Optimization

Traditional methods have a tendency to be stuck in the local optimum point; mostly, they find the local or relative optimum point. Nontraditional methods are able to find the global optimization. These optimization algorithms are very useful to handle complicated problems in chemical engineering. In this chapter, we will discuss four methods namely Genetic Algorithm (GA), Particle Swarm Optimization (PWO), Simulated Annealing (SA), and Differential Evolution (DE) that are able to find the global optimizer. These stochastic algorithms are susceptible to premature termination. "Premature optimization is the root of all evil." – Donald Ervin Knuth (Art of Computer Programming, Volume 1: Fundamental Algorithms). Therefore, we should be very careful about the termination criteria of these stochastic optimization algorithms; otherwise, we will get wrong information from the optimization study.

9.1 Genetic Algorithm

Many useful optimum design problems in chemical engineering are described by mixed discrete–continuous variables, discontinuous and non-convex design spaces. When the traditional nonlinear programming methods are applied for these problems, they will be ineffective and computationally expensive. Mostly, they find a local (relative) optimum point that is nearby to the starting point. Genetic algorithms (GAs) are suitable for solving of this kind of problems, and in most instances, they are able to locate the global optimum solution with high accuracy. GAs are stochastic techniques whose search procedures are modeled similar to the natural evolution. Philosophically, Genetic algorithms work based on the theory of Darwin "Survival of the fittest" in which the fittest species will persist and reproduce while the less fortunate tend to disappear. To preserve the critical information, GAs encode the optimization problem to a chromosome-like simple data structure and employ recombination operators to these structures.

The GA was first proposed by Holland in 1975 [Holland, (1975)]. This approach works based on the similarity of improving a population of solutions by transforming their gene pool.

Two types of genetic modification, crossover, and mutation are utilized and the elements of the optimization vector, X, are expressed as binary strings. Crossover operation (Figs 9.1–9.2) deals with random swapping of vector elements (among parents that have highest objective function value or other ranking populations) or any linear combination of two parents. Mutation operation (Fig. 9.3) involved with the incorporation of a random variable to an element of the vector. The GAs have been used efficiently in process engineering, and numerous codes are readily available. For example, Edgar, Himmelblau, and Lasdon (2002) have been described a related GA algorithm that is available in Excel. Some case studies in process engineering consist of scheduling of batch process [Jung et al. (1998), Löeh et al. (1998)], and synthesis of mass transfer network [Garrard and Fraga, (1998)].

Unlike conventional optimization processes, GA is a robust, global and is not bound to domain-specific heuristics. GA is very useful for solving complex multi-objective optimization problems because

i. the objective function need not to be continuous and/or differentiable
ii. extensive problem formulation is not required
iii. they are less sensitive to the starting point (X^0)
iv. they generally do not get trapped into suboptimal local optima and
v. they are capable of finding out a function's true global optimum

These advantages of GAs increase their application to various fields [Babu and Angira (2006)].

9.1.1 Working principle of GAs

GAs are well suited for both maximization and minimization problems. The fitness function essentially measures the quality of each candidate solution and its magnitude is proportional the fitness of objective function.

GA for a constrained optimization problem can be represented as

$$\text{Maximize } f(X) \tag{9.1}$$

$$\text{subject to } h_j(X) = 0, \qquad j = 1, 2, \ldots q \tag{9.2a}$$

$$g_j(X) \le 0, \qquad j = q+1, \ldots m \tag{9.2b}$$

$$x_i^l \le x_i \le x_i^u, \qquad i = 1, 2, \ldots n \tag{9.2c}$$

GAs work based on the theories of natural genetics and natural selection. The essential elements of a natural genetics are employed in the genetic search process are reproduction, crossover, and mutation. The GA algorithm starts with a population of random strings representing the design variable. A fitness function is found to evaluate each string. The three main GA operators – reproduction, crossover, and mutation are employed to the random population to generate a new population. The new population is evaluated and tested until the termination criterion is met, iteratively altered by the GA operators. The term "generation" in GA represents the cycle of operation by the genetic operators and the evaluation of the fitness function. The steps of Genetic Algorithms are discussed below.

9.1.1.1 Initialization

Genetic Algorithms start with the initialization of a population of suitable size of chromosome length. All the strings are evaluated for their fitness values using specified fitness function. The objective function is interpreted as the problem of minimization and maximization and becomes the fitness function. They initiate with parent chromosomes selected randomly within the search space to generate a population. The population "evolves" towards the superior chromosomes by employing operators that are imitating the genetic processes taking place in the nature: selection or reproduction, recombination or crossover, and mutation.

Coding of GA

The representation of chromosomes (design variables) or coding of GA can be classified into two main categories: real and binary coding. In real coding, each variable is represented by a floating point number whereas binary coding transforms the variables within the chromosomes into a binary representation of zero and one before carrying out crossover and mutation operations. Both methods have some advantages and had been proven effective for certain problems [Herrera *et al.* (1998)]. The benefit of the binary representation is that conventional crossover and mutation operations become very simple. The drawback is that the binary numbers need to be transformed to real numbers when the design variable is to be estimated.

Example 9.1

Convert these decimal numbers into binary code

13, 19, 27, 30, and 46

Solution

The converted values are as follows

Table 9.1 Decimal to binary conversion

Sl. No	Decimal value	Binary code
1	13	001101
2	19	010011
3	27	011011
4	30	011110
5	46	101110

9.1.1.2 Reproduction

Selection process compares the chromosomes within the population intending to pick out those that will take part in the reproduction process. Reproduction selects good strings in a population and forms a mating pool. The reproduction operator will pick the above-average strings from the current population. The selection takes place with a specified probability on the basis of fitness functions. This fitness function plays a significant role to differentiate between bad and good solutions. The string that has a higher fitness value will represent a larger range in the cumulative probability values and consequently has a higher probability of being copied into the mating pool. Therefore, individuals that have higher fitness should have more chances to reproduce. Conversely, a string with a smaller fitness value signifies a smaller range in the cumulative probability values and has less probability of being copied into the mating pool.

The selection of members for the next generation could be done by the Roulette Wheel selection technique [Goldberg (1989)] in which the members with high fitness value have a higher probability of being replicated in the next generation.

Fitness evaluation

After defining the genetic representation, the next step is to find out the fitness functions related to the solutions. They are computed on the basis of the chromosomes phenotype. A fitness function evaluates the difference between bad and good solutions by mapping the solution within a non-negative interval. Two mappings corresponding to the maximization and minimization are as follows [Handbook of Evolutionary Computation (1997)]

i. During the minimization of problems, for some solution i, the fitness function is estimated according to the equation:

$$\text{fitness}(\vec{x}_i(t)) = \frac{1}{1 + f(\vec{x}_i(t)) - f_{\min}(t)}, \quad \vec{x}_i(t) \in X_x \tag{9.3}$$

where X_x represents a design space of the vector of control variables $\vec{x}_i$; and $f_{\min}$ is the minimum value obtained for the objective function up to generation t.

ii. For maximization problems, it estimates the fitness function by the following relation:

$$\text{fitness}(\vec{x}_i(t)) = \frac{1}{1 + f_{\max}(t) - f(\vec{x}_i(t))}, \quad \vec{x}_i(t) \in X_x \tag{9.4}$$

where $f_{\max}$ denotes the maximum value attained for the objective function up to generation t.

Roulette-wheel selection

Roulette-wheel is the easiest method for proportionate selection. In this method, the individuals of each population are considered as slots of the Roulette-wheel. The width of each slot is proportional to the probability for selecting the corresponded chromosome. The scaled fitness function is applied to determine corresponding selection probabilities:

$$\Pr(i) = \frac{\text{fitness}(\vec{x}_i)}{\sum_{i=1}^{\mu} \text{fitness}(\vec{x}_i)} \tag{9.5}$$

where μ represents the population size.

In Roulette-wheel selection, emphasis is given to the better individuals within the population; therefore, a large pressure has been exerted on the search procedure. The probable number of copies in the sampling pool that some solution could obtain varies proportionally to its selection probability. This might cause loosing of genetic diversity, which leads to premature convergence to a suboptimal solution. Picking the samplings for reproduction is done by utilizing the so-called Roulette-wheel sampling algorithm. The process is repeated until the sampling pool is completed.

Elitism

Elitism is used to copy best parents to the next generation to replace worst offspring. In elitist selection, the best member in the population can be forwarded to the next generation without any modifications. When genetic operators are applied on the population, there is a chance of losing the best chromosome. Therefore, the best-fit chromosome is preserved at each generation. The fitness of offspring is compared with their parent chromosomes.

9.1.1.3 Crossover

The recombination (crossover) is carried out after completion of the selection process. In the crossover operation, new strings are created by exchanging information between strings of the mating pool. Crossover is the operation that ensure the genetic diversity of the population. Two strings involved in the crossover operation are recognized as parent strings whereas the resulting strings are identified as child strings. The child strings produced may be good or not, which depends on the performance of the crossover site. The outcome of crossover may be beneficial or detrimental. In order to conserve some good strings that are already existing in the mating pool, all strings in the mating pool are not used in the crossover.

The crossover operation can take place at a single point or at more than one point (multipoint) [Davis (1991)].

a	b	c	d	e	f	g	h	i	j
A	B	C	D	E	F	G	H	I	J

Fig. 9.1 Parent chromosomes

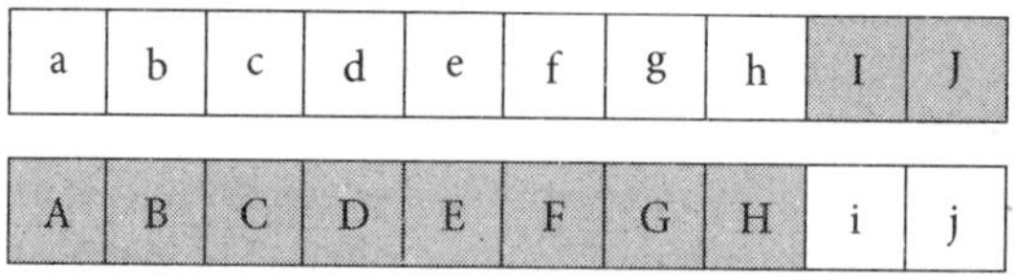

a	b	c	d	e	f	g	h	I	J
A	B	C	D	E	F	G	H	i	j

Fig. 9.2 Single point crossover

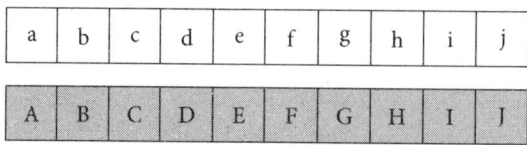

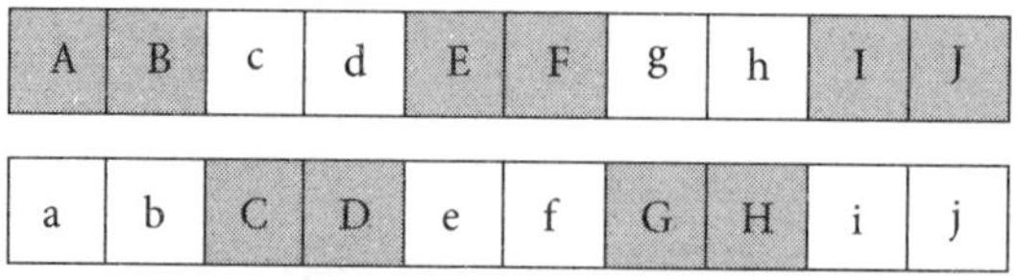

Fig. 9.3 Multipoint crossover

Example 9.2

Perform crossover operation between the 1–2 and 3–4 binary codes of Example 9.1.

Solution

The binary codes of Example 9.1 are

Table 9.2 Single and multiple point crossover

Sl. No	Binary code
	Single point crossover
1	001101
2	010011
	Multiple point crossover
3	011011
4	011110

after crossover, the chromosomes becomes as in Table 9.3

Table 9.3 Chromosomes after crossover

Sl. No	Binary code	Decimal value
1	010101	21
2	001011	11
3	011010	26
4	011111	31

9.1.2.4 Mutation

After recombination operation, offspring undergoes to mutation process (Fig. 9.4). The mutation operation is the creation of a new chromosome from one and only one individual with a predefined probability. Mutation is required to generate a point in the neighborhood of the current point, thus, achieving local search around the current solution. This mutation operation is also used to preserve diversity in the population.

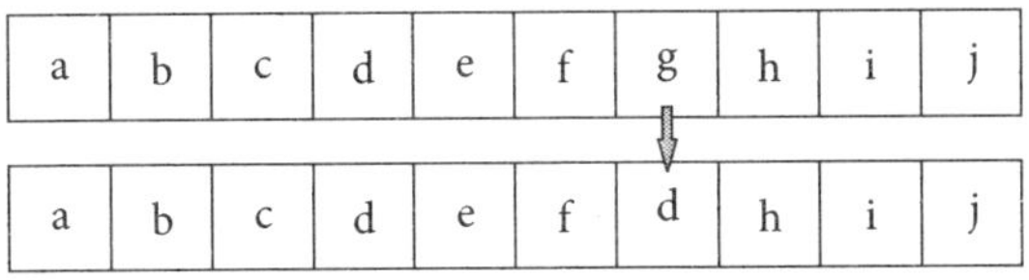

a	b	c	d	e	f	g	h	i	j
a	b	c	d	e	f	d	h	i	j

Fig. 9.4 Mutation operation

After completion of these three operations, the offspring is inserted into the population that produces a new generation by replacing the parent chromosomes from which they were derived (see Fig. 9.5). This cycle is carried out until the optimization criterion is met [Shopova and Vaklieva-Bancheva (2006)]. Theoretically, there is no clear explanation for the distinction between generations. This restriction is merely an implementation model that makes the computation straightforward. During the search, to facilitate preservation of a good solution, sometimes we employ an elitism procedure, i.e., the best individuals from the old population are directly copied into the new one. This procedure ensures that the overall solution of the GA will not become worse.

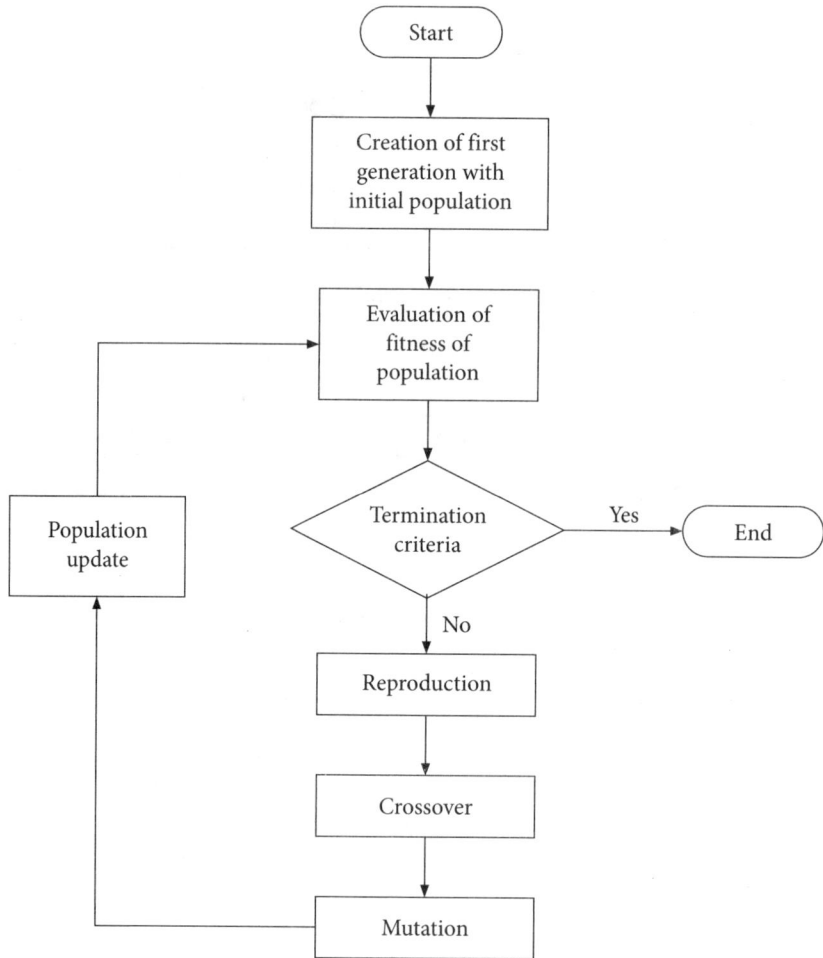

Fig. 9.5 Flowchart of genetic algorithm

9.1.2 Termination

Unlike simple neighborhood search techniques that terminate when they reach a local optimum, GAs are stochastic search methods that could in principle run forever. However, a termination criterion is required for practical application. The standard practices are setting a limit on the number of fitness evaluations or the clock time of computer, or tracking the diversity of the population and stop when this diversity falls below a predetermined threshold.

The significance of diversity in the latter case is not always clear, and it could relate to either the genotype or the phenotype or even, possibly, to the fitnesses. However, the most common way, of measuring this, is by genotype statistics. For instance, we could make a decision to terminate a run when at every locus the proportion of one particular allele raised above 90 per cent.

9.2 Particle Swarm Optimization

Particle Swarm Optimization (PSO) is a population-based stochastic search method which is proposed by Kennedy and Eberhart in 1995 [Kennedy and Eberhart (1995)]. This is an evolutionary technique similar to genetic algorithm. Population based searches may be a useful alternative when the search space is very large and we need exhaustive search. However, population based search techniques do not guarantee you the best (optimal) solution. The PSO method was originally devised based on the social behaviour of a bird flocks, school of fish etc. In PSO, each member of the population is called a particle and the population is called a swarm. A swarm of particles moves around in the search-space influenced by the improvements identified by the other particles. This method is very helpful when calculation of the gradient is too laborious or even impossible to derive. This technique can be readily employed for a wide range of optimization problems because it does not utilize the gradient of the objective function to be optimized. The original PSO algorithm has some drawbacks. This algorithm does not work well continuously and it may require tuning of its behavioural parameters such that it will perform well on the problem at hand. Numerous research works have been carried out to improve the performance of the PSO method. A conventional and easy way of improving the PSO algorithm is by manually adjusting its behavioural parameters. Some researchers [Clerc and Kennedy (2002); I.C. Trelea (2003)] have done mathematical analysis to show how the contraction of the particles to a single point is influenced by its behavioural parameters.

Both Particle Swarm Optimizers and Genetic Algorithms have some common feature; both initialize a population in a similar fashion, both utilize a fitness function to determine potential solution and both are generational, that is both repeat the same set of processes for a predetermined amount of time. However, PSO is somewhat better than genetic algorithm (GA) for solving constrained optimization problems, because GA, which has been generally employed for solving such problems has disadvantage of slow convergence due to mutation operator leading to demolition of good genes consequently poor convergence. Particle Swarm Optimization is a computational intelligence-based method that is not considerably affected by the size and nonlinearity of the optimization problem. When most analytical methods fail to converge, PSO can converge to the optimal solution for some problems. As compared with other optimization methods, it is faster, cheaper and more efficient. It requires less memory space and lesser speed of CPU. In addition, adjustment of few parameters is required in PSO. For this reason, PSO is a useful optimization method for practical application. This method is well suited to solve the problems such as non-linear, non-convex, continuous, discrete, integer variable etc.

9.2.1 Working principle

In PSO method, particles move through the search space using an equation (Eq. (9.6)) which is a combination of the individual best solution and the best solution that has found by any particle in their neighborhood. The neighborhood of an individual particle is defined as the subset of particles with which it can communicate. For a particle communication, the first PSO model used a Euclidian neighborhood where the actual distance between particles are measured to determine which were closer to be in communication. This was performed by imitating the behavior of birds flocks, like the biological models where individual birds are only able to communicate with other birds in the immediate vicinity [Reynolds (1987); Heppner and Grenander (1990)].

PSO is an evolutionary computation technique, which was modeled similar to the social behavior of fish schooling, and bird flocking. PSO algorithm is motivated from the artificial life and social psychology, as well as engineering and computer science. It uses a "population" of particles that travel through the hyperspace of the problem with specified velocities. The velocities of the individual particles are adjusted stochastically at each iteration. The historical best position of the particle itself and the neighborhood best position are used for this purpose. Both the particle best and the neighborhood best are determined by considering a user defined fitness function. The movement of each particle usually results in an optimal or near-optimal solution. The word "swarm" originates from the random movements of the particles in the problem space, now it is more comparable to a swarm of mosquitoes rather than a school of fish or a flock of birds. The PSO can be explained by Fig. 9.6. There are three velocity components cognitive velocity V^t_{pbest}, social velocity V^t_{gbest} and inertia velocity V^t_i. Particle moves from initial position X^t_i to the new position X^{t+1}_i. The new velocity can be calculated using Eq. (9.7).

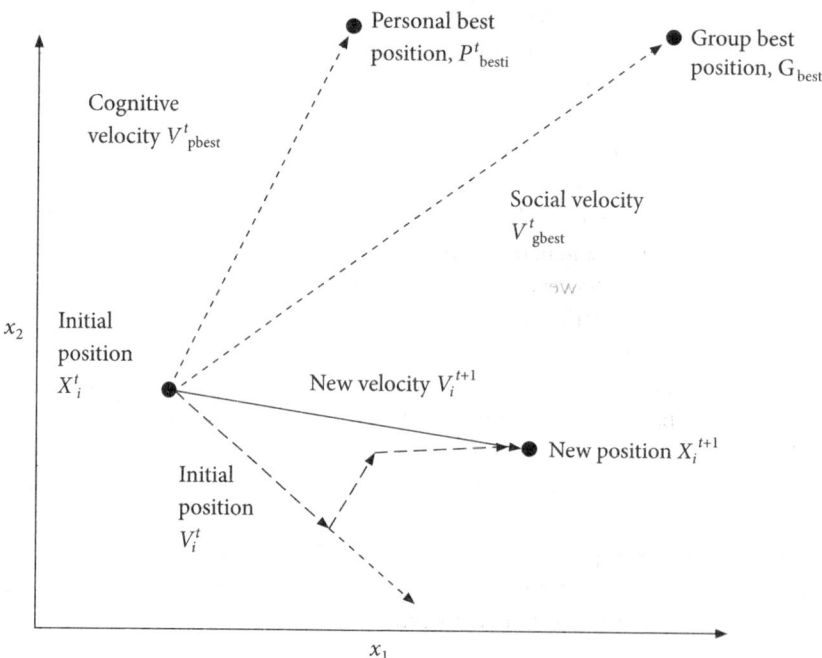

Fig. 9.6 Movement of particles in PSO

The PSO algorithm includes some tuning parameters that significantly influence the performance of the algorithm, often stated as the exploration–exploitation tradeoff. The capability of a search algorithm to examine the diverse region of the search space in order to establish a good optimum is called exploration. In contrast, exploitation is the capability to focus the search around a promising region to facilitate refinement of a candidate solution [Ghalia (2008)]. Using their exploration and exploitation, the particles of the swarm fly through hyperspace and have two essential reasoning capabilities: their memory of their own best position – personal best (P_{best}) and knowledge of the global or their neighborhood's best – global best (G_{best}). Position of the particle is determined by its velocity. Let X_i^t represent the position of particle i within the search space at time step t; unless otherwise stated, t indicates discrete time steps. The updated position of the particle is determined by adding a velocity V_i^t to the current position.

$$X_i^{t+1} = X_i^t + V_i^{t+1} \tag{9.6}$$

where

$$V_i^t = V_i^{t+1} + c_1 r_1 \left(P_{best}^t - X_i^{t-1} \right) + c_2 r_2 \left(G_{best}^t - X_i^{t-1} \right) \tag{9.7}$$

The PSO algorithm is similar to GA as this system is also initialized with a population of random solutions. The main difference with GA is that; in PSO, a randomized velocity is allotted to each potential solution, and the probable solutions, called particles, are then traveled through the problem space.

9.2.2 Algorithm

The steps for implementing the PSO are as follows:

Step 1 Initialize a population of particles (array) with random position and velocities on D-dimensional problem space.

Step 2 For each particle, estimate the desired optimization fitness function in D variables.

Step 3 Compare the fitness of particles at current position with particle's P_{best}. If current value is better than P_{best}, then replace current value with P_{best} value, then the P_{best} location turns into the current location in D-dimensional space.

Step 4 Compare values of fitness function with the overall previous best of the population. If current value is better than G_{best}, then reset G_{best} to the current particle's array index and value.

Step 5 Modify the position and velocity of the particle as expressed by Eqs (9.6) and (9.7) respectively.

Step 6 Go to Step 2 until a criterion is met. Usually a satisfactorily good fitness or a maximum number of iterations (generations) are considered as termination criteria.

9.2.3 Initialization

In PSO algorithm, initialization of the swarm is very important because proper initialization may control the exploration and exploitation tradeoff in the search space more efficiently and find the better result. Initially the particles are placed at arbitrary locations within the search-space and then moving in randomly defined directions. Usually, a uniform distribution over the search space is used for initialization of the swarm. The initial diversity of the swarm is important for PSO's performance, it denotes that how well particles are distributed and how much of the search space is covered. Therefore, when the initial swarm does not cover the entire search space, the PSO algorithm will have difficultly to find the optimum if the optimum is located outside the covered area. Then, the PSO will only discover the optimum if a particle's momentum takes the particle into the uncovered area. Therefore, the optimal initial distribution to be located within the domain defined by X_{min} and X_{max} which represent the minimum and maximum ranges of X for all particles i in dimension j respectively. Then the initialization method for the position of each particle is given by

$$X^0 = X_{min,j} + r_j \left(X_{max,j} - X_{min,j} \right) \tag{9.8}$$

where $r_j \sim U(0,1)$.

The selected size of the population depends on the problem considered. The most common population size is 20–50. It is found that smaller populations are common for evolutionary algorithms like evolutionary programming and genetic algorithms. It was optional for PSO in terms of minimizing the total number of evaluations (size of population times the number of generations) required to achieve a satisfactory solution.

The velocities of the particles can be initialized to zero, i.e., $V_i^0 = 0$, since randomly initialized particle's positions already ensure random positions and moving directions. Then, the direction of a particle is changed gradually hence, it will start moving in the direction of the best previous positions of itself and its peers, searching in their surrounding area and hopefully finding even better positions with reference to some fitness measure $f : \mathbb{R}^n \to \mathbb{R}$. The trajectory of every particle is determined by a simple rule combining the current velocity and exploration histories of the particle and its neighbors. The particles may also be initialized with nonzero velocities, however, it must be done with care and such velocities must not be very large. In general, large velocity has large momentum and accordingly large position update. Therefore, such large initial position updates can cause particles to move away from boundaries of the feasible region, and the algorithm needs to consider more iteration before settling the best solution.

9.2.4 Variants of PSO

There are many variants of PSO, which are developed to overcome the limitations of the original PSO algorithm. Originally, the inertia weight coefficient was not an element of the PSO algorithm, however, later it was modified that became commonly accepted. Some PSO variants do not comprise a single global best characteristic of the algorithm; rather they utilize multiple global best that are provided by separate subpopulations of the particles. In addition, PSO is also influenced by velocity clamping, and velocity constriction and these parameters are described in the following sections.

9.2.4.1 PSO with inertia weight

In 1998, Shi and Eberhart came up with a variant of PSO called PSO with inertia. A nonzero inertia weight introduces a choice for the particle for continuing their movement in the same direction it was moving on the earlier iteration. The utilization of inertia weight (w) has shown better performance in several applications. In the beginning when the method was developed, the value of w often reduced linearly from about 0.9 to 0.4 during a run. Proper selection of the inertia weight gives a balance between global and local exploration and exploitation, and results in less number of iteration on average to find a satisfactorily optimal solution. Generally, lower values of the inertial coefficient expedite the convergence of the swarm to the optimum point whereas higher values of the inertial coefficient promote exploration of the search space entirely. Declining inertia with time begins a shift from the exploratory (global search) to the exploitative (local search) mode.

To control the balance between the global and local exploration, to achieve quick convergence, and to attain an optimum, value of the inertia weight is decreasing according to the following equation

$$\omega^{t+1} = \omega_{max} - \left(\frac{\omega_{max} - \omega_{min}}{t_{max}}\right) t, \quad \omega_{max} > \omega_{min} \tag{9.9}$$

ω_{max} and ω_{min} are the initial and final values of the inertia weight respectively

t_{max} is the maximum iteration number and t is the current iteration number

Equations (9.11) and (9.12) express the velocity and position update equations integrated with an inertia weight. It can be observed that these equations are identical to Eqs (9.6) and (9.7) with the addition of the inertia weight ω as a multiplying factor of V_i^{t+1}.

$$X_i^t = f\left(\omega^t, X_i^{t-1}, V_i^{t-1}, P_{best}, G_{best}\right) \tag{9.10}$$

where

$$V_i^t = \omega^t V_i^{t+1} + c_1 r_1 \left(P_{best}^t - X_i^{t-1}\right) + c_2 r_2 \left(G_{best}^t - X_i^{t-1}\right) \tag{9.11}$$

$$X_i^{t+1} = X_i^t + V_i^{t+1} \tag{9.12}$$

9.2.4.2 PSO with constriction coefficient

Clerc [Clerc, (1999)] developed the PSO with Constriction Coefficient in 1999. Particle swarms interactions are represented by a set of complex linear equations. Utilization of constriction coefficient results in particle convergence over time. That is the amplitude of the particle's oscillations reduces as it focuses on the local and neighborhood previous best positions. Although the particle will converge to a point in due course, the constriction coefficient also prevents collapse if the right social circumstances are prepared. The particle oscillates around the weighted mean of P_{best} and G_{best}, when the previous best position and the neighborhood best position are close to each other the particle will conduct a local search. The particle will carry out a global search (more exploratory search) when the previous best position and the neighborhood best position are distant from each other. During this search process, the neighborhood best position and previous best position will be modified and the particle will be shifted from local search back to global search. Therefore, the constriction coefficient method makes a balance of the necessity for global and local search depending on whatever social circumstances are in place.

$$X_i^t = f\left(K, X_i^{t-1}, V_i^{t-1}, P_{best}, G_{best}\right) \tag{9.13}$$

$$K = \frac{2k}{\left|2 - \varphi - \sqrt{\varphi^2 - 4\varphi}\right|} \quad \text{where } \varphi = c_1 + c_2, \ \varphi > 4 \tag{9.14}$$

$$V_i^t = K\left[V_i^{t-1} + c_1\varphi_1\left(P_{best} - X_i^{t-1}\right) + c_2\varphi_2\left(G_{best} - X_i^{t-1}\right)\right] \tag{9.15}$$

$$X_i^t = X_i^{t-1} + V_i^t \tag{9.16}$$

By modifying φ, we can control the convergence characteristics of the system. Usually, $k = 1$ and $c_1 = c_2 = 2$, and φ is set to 4.1, thus, $K = 0.73$.

In PSO algorithm, all of these three parts of the velocity update equation (Eq. (9.15)) have different functions. The first part $\omega^t \times V_i^{t+1}$ is the inertia component, which is accountable for keeping the particle moving in the same direction it was originally traveling. The second part $c_1 r_1 \left(P_{best}^t - X_i^{t-1}\right)$, called the cognitive component, acts as memory of the particle, causes to be inclined to return to the regions of the search space within which it has experienced higher individual fitness. The third part $c_2 r_2 \left(G_{best}^t - X_i^{t-1}\right)$, which helps the particle to move toward the best region the swarm has encountered so far is called the social component.

9.2.4.3 Velocity clamping

The problem with basic PSO is that the particles velocities quickly explode to large values. Under some setting, the swarm can become scattered in a very wide area in the search domain, resulting in a very high particle velocity. In order to avoid such an explosion, a practice has been followed to limit the maximum particle velocity component as V_{max}. Eberhart and Kennedy first proposed the velocity clamping; it helps particles to stay inside the boundary and to take reasonable step size so as to examine through the search space. Without this velocity clamping in the searching space, the process will be likely to explode and positions of the particles change rapidly [Bratton and Kennedy, (2007)]. Whenever a search space is bounded by the range $[-X_{max}, X_{max}]$, velocity clamping confines the velocity to the range $[-V_{max}, V_{max}]$, where $V_{max} = k \times X_{max}$. The value k signifies a user-defined velocity clamping factor, $1.0 \geq k \geq 0.1$. During some optimization processes, the search space is not centered around 0 and therefore this $[-X_{max}, X_{max}]$ range is not an satisfactory definition of the search space. In such a case where the search space is bounded by the range, $[X_{min}, X_{min}]$ we define $V_{max} = k \times (X_{max} - X_{min})/2$.

If the maximum velocity V_{max} is too large, then the particles may move erratically and jump over the optimal solution. On the other hand, if V_{max} is too small, the particle's movement is limited and the swarm may not explore sufficiently or the swarm may become trapped in a local optimum.

The velocity update rule with this saturation is given by:

$$V_{i,j}^{t+1} = \begin{cases} V_{i,j}^{t+1} & \text{if } |V_{i,j}^{t+1}| < V_{max,j} \\ V_{max} & \text{if } |V_{i,j}^{t+1}| \geq V_{max,j} \end{cases} \tag{9.17}$$

Once, the velocity for each particle is determined, the position each particle is updated by applying the new velocity to the particle's previous position. This process is repeated until it satisfied some stopping criteria.

9.2.5 Stopping criteria

Stopping criteria is used to terminate the iterative search process. Some stopping criteria are discussed below:

1. The algorithm is terminated when a maximum number of iterations or function evaluations (FEs) have been reached. If this maximum number of iterations (or FEs) is very small, the search process may stop before a good result has been obtained.
2. The algorithm is terminated when there is no significant improvement over a number of iterations. This improvement can be measured in different ways. For instance, the process may be considered to have terminated whenever the average change of the particles' positions are too small or the average velocity of the particles is approximately zero over a number of iterations.
3. The algorithm is terminated when the normalized swarm radius is approximately zero. The normal swarm radius is defined as

$$R_{norm} = \frac{R_{max}}{\text{diameter}(S)} \tag{9.18}$$

where diameter(S) is the initial swarm's diameter and R_{max} is the maximum radius,

$$R_{max} = \|x_m - G_{best}\| \tag{9.19}$$

with

$$\|x_m - G_{best}\| \geq \|x_i - G_{best}\|, \quad m, \forall i = 1, 2, \ldots n \tag{9.20}$$

and $\|\cdot\|$ is a suitable distance norm.

The process will terminate when $R_{norm} < \varepsilon$. If ε is very large, the process can be terminated prematurely before a good solution has been reached while if ε is too small, the process may need more iterations.

9.2.6 Swarm communication topology

PSO algorithm relies on the social interaction among the particles in the entire swarm. A neighborhood of each particle is required to define properly because this neighborhood determines the extent of social interaction within the swarm and influences a particular particle's movement. Particles communicate with one another by exchanging information about the success of each particle in the swarm. When a particle in the whole swarm finds a better position, all particles move towards this particle. This performance of the particles is determined by the particles' neighborhood. When the neighborhoods in the swarm are small, less interaction occurs between individuals. For small neighborhood, the convergence is slower however, it improves the quality of solutions. Whereas for larger neighborhood, the convergence will be faster however, sometimes convergence may occurs earlier. Therefore, in practice, the search process starts with small neighborhoods size and then increases with time. The performance of PSO may be improved by designing different types of neighborhood structures. Some neighborhood structures or topologies are discussed below:

Five major neighborhood topologies have been used in PSO: Von Neumann, star, wheel, circle and pyramid [Valle et al. (2008)]. The selection for neighborhood topology decides which individual to use for G_{best}. In the circle topology, each individual in socially connected to its k nearest topological neighbors.

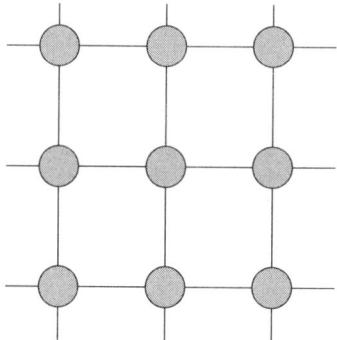

Fig. 9.7a   von Neumann topology

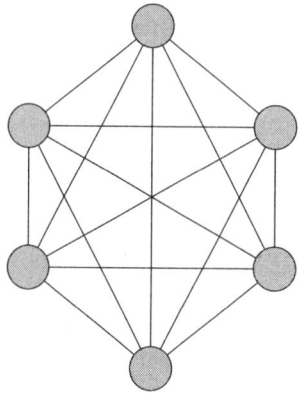

Fig. 9.7b   Star topology

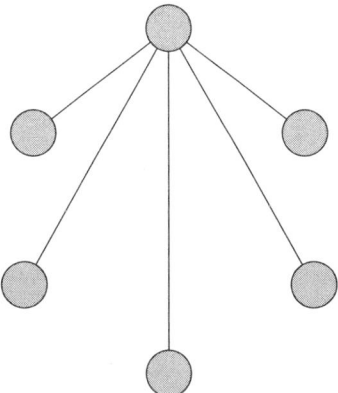

Fig. 9.7c Wheel topology

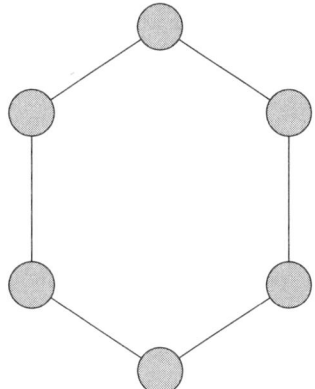

Fig. 9.7d Circle topology

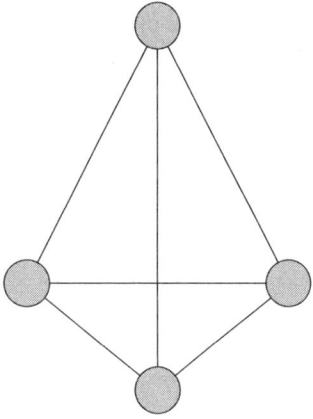

Fig. 9.7e Pyramid topology

238 Optimization in Chemical Engineering

Kennedy suggested that the G_{best} version [Fig. 9.7b] converges fast however, may be trapped in a local minimum, while the P_{best} network has more chances to find an optimal solution, although it converges slowly.

Example 9.3

Solve problem 3.3 using PSO, show only two iterations.

Solution

The objective function can be written as follows

$$W = 301.8\left[(x)^{0.286} + \left(\frac{10}{x}\right)^{0.286} - 2\right]$$

consider $0 \leq x \leq 12$

the initial population for iteration number $t = 0$

$x_1^0 = 1.0, \quad x_2^0 = 3.0, \quad x_3^0 = 5.0, \quad x_4^0 = 7.0, \quad x_5^0 = 9.0, \quad x_6^0 = 12.0$

$f_1^0 = 281.268038, \quad f_2^0 = 235.470634, \quad f_3^0 = 242.587797, \quad f_4^0 = 257.134263,$

$f_5^0 = 273.193037, \quad f_6^0 = 297.144363$

let $c_1 = c_2 = 2$

set the initial velocities of each particle to zero:

$v_i^0 = 0$, i.e., $v_1^0 = v_2^0 = v_3^0 = v_4^0 = v_5^0 = v_6^0 = 0$

Step 2   Set iteration number $t = 0 + 1 = 1$

Step 3   Find the personal best for each particle by

$$P_{best,i}^{t+1} = \begin{cases} P_{best,i}^t & \text{if } f_i^{t+1} > f_i^t \\ X_i^{t+1} & \text{if } f_i^{t+1} \leq f_i^t \end{cases}$$

so,

$P_{best,1}^1 = 1, \quad P_{best,2}^1 = 3, \quad P_{best,3}^1 = 5, \quad P_{best,4}^1 = 7, \quad P_{best,5}^1 = 9, \quad P_{best,6}^1 = 12$

Step 4   Find the global best by

$G_{best} = \min\{P_{best,i}^t\}$ where $i = 1, 2, 3, 4, 5, 6$

Since, the minimum personal best is $P^1_{best,2} = 3$, thus, $G_{best} = 3$

Step 5 Consider the random numbers in the range (0,1) as $r^1_1 = 0.512$, $r^1_2 = 0.796$, $\omega_{max} = 0.9$, $\omega_{min} = 0.4$ and maximum iteration number 10.

so, $\omega^1 = \omega_{max} = 0.9$

and find the velocities of the particles by

$$v_i^{t+1} = \omega^{t+1} v_i^t + c_1 r_1^t \left[P^t_{best,i} - X_i^t \right] + c_2 r_2^t \left[G^t_{best,i} - X_i^t \right]; i = 1, 2, \ldots, 6$$

so,

$v_1^1 = 0 \times 0.9 + 2 \times 0.512(1-1) + 2 \times 0.398(3-1) = 1.592$

$v_2^1 = 0 \times 0.9 + 2 \times 0.512(3-3) + 2 \times 0.398(3-3) = 0.0$

$v_3^1 = 0 \times 0.9 + 2 \times 0.512(5-5) + 2 \times 0.398(3-5) = -1.538$

$v_4^1 = 0 \times 0.9 + 2 \times 0.512(7-7) + 2 \times 0.398(3-7) = -3.184$

$v_5^1 = 0 \times 0.9 + 2 \times 0.512(9-9) + 2 \times 0.398(3-9) = -4.776$

$v_6^1 = 0 \times 0.9 + 2 \times 0.512(12-12) + 2 \times 0.398(3-12) = -7.164$

Step 6 Find the new values of x_i^1, $i = 1, \ldots, 6$

$x_1^1 = x_1^0 + v_1^1 = 1 + 1.592 = 2.592$

$x_2^1 = x_2^0 + v_2^1 = 3 + 0.0 = 3.0$

$x_3^1 = x_3^0 + v_3^1 = 5 - 1.538 = 3.462$

$x_4^1 = x_4^0 + v_4^1 = 7 - 3.184 = 3.816$

$x_5^1 = x_5^0 + v_5^1 = 9 - 4.776 = 4.224$

$x_6^1 = x_6^0 + v_6^1 = 12 - 7.164 = 4.836$

now calculate the values of the objection function (fitness function)

$f_1^1 = 236.732707$, $f_2^1 = 235.470634$, $f_3^1 = 235.656786$, $f_4^1 = 236.587280$,

$f_5^1 = 238.252585$, $f_6^1 = 241.574742$

Now we will move to next iteration $t = 1 + 1 = 2$

Step 7 Find the personal best for each particle by

$$P_{best,i}^{t+1} = \begin{cases} P_{best,i}^{t} & \text{if } f_i^{t+1} > f_i^{t} \\ X_i^{t+1} & \text{if } f_i^{t+1} \leq f_i^{t} \end{cases}$$

so,

$$P_{best,1}^{2} = 1.592,\ P_{best,2}^{2} = 3,\ P_{best,3}^{2} = 3.462,\ P_{best,4}^{2} = 3.816,\ P_{best,5}^{2} = 4.224,\ P_{best,6}^{2} = 4.836$$

The minimum personal best is $P_{best,2}^{2} = 3$, thus, $G_{best} = 3$

Consider the random numbers in the range (0,1) as $r_1^2 = 0.371$, $r_2^1 = 0.602$

$$\omega^2 = 0.9 - \frac{(0.9 - 0.4)}{10} \times 1 = 0.85$$

$$v_1^2 = 1.592 \times 0.85 + 2 \times 0.371(2.592 - 2.592) + 2 \times 0.602(3 - 2.592) = 1.7612$$

$$v_2^2 = 0.0 \times 0.85 + 2 \times 0.371(3.0 - 3.0) + 2 \times 0.602(3.0 - 3.0) = 0.0$$

$$v_3^2 = (-1.538) \times 0.85 + 2 \times 0.371(3.462 - 3.462) + 2 \times 0.602(3 - 3.462) = -1.86355$$

$$v_4^2 = (-3.184) \times 0.85 + 2 \times 0.371(3.816 - 3.816) + 2 \times 0.602(3 - 3.816) = -3.68886$$

$$v_5^2 = (-4.776) \times 0.85 + 2 \times 0.371(4.224 - 4.224) + 2 \times 0.602(3 - 4.224) = -5.533296$$

$$v_6^2 = (-7.164) \times 0.85 + 2 \times 0.371(4.836 - 4.836) + 2 \times 0.602(3 - 4.836) = -8.299944$$

Find the new values of x_i^2, $i = 1,\ldots,6$

$$x_1^2 = x_1^1 + v_1^2 = 2.592 + 1.7612 = 4.3532$$

$$x_2^2 = x_2^1 + v_2^2 = 3 + 0.0 = 3.0$$

$$x_3^2 = x_3^1 + v_3^2 = 3.462 - 1.86355 = 1.59845$$

$$x_4^2 = x_4^1 + v_4^2 = 3.816 - 3.68886 = 0.12714$$

$$x_5^2 = x_5^1 + v_5^2 = 4.224 - 5.533296 = -1.309296 \text{ (Beyond the range)}$$

$$x_6^1 = x_6^0 + v_6^1 = 4.836 - 7.164 = -2.328 \quad \text{(Beyond the range)}$$

Now we have to calculate f_i^2 at these new x_i^2 points. After sufficient number of iteration, the result will converge.

9.3 Differential Evolution

Differential evolution (DE) is a stochastic, population-based optimization technique which was introduced by Storn and Price in 1996 [Storn and Price (1997)]. DE is a powerful yet simple evolutionary algorithm (EA) for optimizing real-valued, multimodal functions [Price (1996)]. The EAs are encouraged by the natural evolution of species, have been effectively implemented in the field of optimization. However, user needs to choose the suitable parameters during the implementation of EAs. Improper selection of parameter setting may lead to high computational costs due to the time-consuming trial-and-error parameter and operator tuning process. The success of DE process to solve a particular problem significantly depends on correctly selection of trial vector generation strategies and their related control parameter values. Therefore, for solving a particular optimization problem at hand successfully, it is usually required to make a trial-and-error search for the most suitable strategy and to adjust the values of the related parameter. However, this trial-and-error searching process is time-consuming and needed high computational costs. Furthermore, as evolution continued, the population of DE may travel through different regions in the search space, within which some strategies related with specific parameter settings may be more effectual than others may. Therefore, it is required to adaptively determine a proper strategy and its related parameter values at different stages of evolution/search process [Qin *et al.* (2009)].

In DE, there are numerous trial vector generation strategies out of which a few may be suitable to solve a particular problem. The optimization performance of the DE will be significantly influenced by three crucial control parameters involved in DE namely population size (*NP*), scaling factor (*F*), and crossover rate (*CR*). The DE algorithm has an advantage over other EAs; since DE is inherently parallel, further significant acceleration can be accomplished if the algorithm is carried out on a parallel machine or a network of computers. This is especially correct for real world applications where calculating the objective function involves a considerable amount of time.

9.3.1 DE algorithm

DE algorithm attempts to evolve a population of *NP* *D*-dimensional parameter vectors commonly known as individuals, which encode the candidate solutions, i.e., $X_{i,G} = \{x_{i,G}^1, \ldots, x_{i,G}^D\}$, $i = 1, \ldots NP$ towards the global optimum. The initial population should cover the entire search space as much as possible by randomizing individuals uniformly within the search space constrained by the prescribed minimum and maximum parameter bounds $X_{\min} = \{x_{\min}^1, \ldots, x_{\min}^D\}$ and $X_{\max} = \{x_{\max}^1, \ldots, x_{\max}^D\}$. For example, the initial value of the *j*th parameter in the *i*th individual at the generation $G = 0$ is generated by

$$x_{i,0}^j = x_{\min}^j + \text{rand}(0,1)\left(x_{\max}^j - x_{\min}^j\right), \quad j = 1, 2, \ldots, D \tag{9.21}$$

where rand(0, 1) represents a uniformly distributed random variable within the range [0,1]

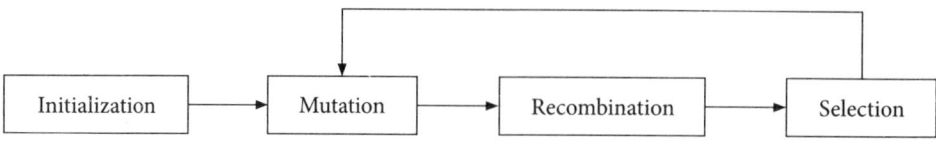

Fig. 9.8 Flowchart for differential algorithm

Figure 9.8 is the flowchart for DE method. This method starts with an initial population, which is generated randomly.

9.3.2 Initialization

Initialization is the first step of the DE algorithm. We need to define upper and lower bounds for each parameter:

$$x^j_{min} \leq x^j_{i,0} \leq x^j_{max} \tag{9.22}$$

After that, randomly select the initial parameter values uniformly on the intervals $\left[x^j_{min}, x^j_{max}\right]$ using the following equation.

$$x^j_{i,0} = x^j_{min} + \text{rand}(0,1)\left(x^j_{max} - x^j_{min}\right), \quad j = 1, 2, \ldots, D \tag{9.23}$$

Population Size is a crucial parameter for DE algorithm.

Price and Storn (1997) suggested a minimum population size of 4 times the dimensionality (D) of the problem and 5–10 times is considered as the most suitable population size. In simulating crystal structures via DE, a population of 40 for a 7-dimensional problem (a factor of 5.7) is used by Weber and Bürgi [Weber and Bürgi (2002)]. These values have surely been shown to work nicely in practice, confirming that adequate genetic material is contained in the populations [Storn and Price, (1997)]. In Example 9.4 a population of 20 has been considered for a 3-dimensional problem (a factor of 6.67). However, this could be using an excessive amount research with other evolutionary algorithms (also using real-value coding) has provided most excellent results with factors between 1.5 and 2 [Mayer, (2002)]. Values within this range could be more effective, by carrying only a sufficient number of population members.

Example 9.4

Create the initial population for the function given below

$$f(X) = 1200(2x_1x_3 + 2x_2x_3) + 2500x_1x_2 + 500\left(\frac{1000}{x_1x_2x_3}\right) + 100\left(\frac{1000}{10x_1x_2x_3}\right)$$

The limit for variables are as follows : $0 \leq x_1, x_x, x_3 \leq 10$

Solution

Here, we will consider population size 20.

Therefore, for first generation $G = 0$, we have factor $j = 1, 2, 3$ and individual $i = 1, 2, \ldots, 20$.

Consider any random number, rand $(0, 1)$

From Eq. (9.23), for $i = 1, j = 1, 2, 3$

$$x_{1,0}^1 = 0.0 + 0.638(10 - 0.0) = 6.38$$

$$x_{1,0}^2 = 0.0 + 0.285(10 - 0.0) = 2.85$$

$$x_{1,0}^3 = 0.0 + 0.819(10 - 0.0) = 8.19$$

Therefore, the first individual is $[6.38, 2.85, 8.19]$.

Now for $i = 2, j = 1, 2, 3$

$$x_{2,0}^1 = 0.0 + 0.122(10 - 0.0) = 1.22$$

$$x_{2,0}^2 = 0.0 + 0.493(10 - 0.0) = 4.93$$

$$x_{2,0}^3 = 0.0 + 0.720(10 - 0.0) = 7.20$$

Therefore, the second individual is $[1.22, 4.93, 7.20]$.

Similarly, we have to find 20 individuals for generation $G = 0$.

9.3.3 Mutation

After initialization, DE employs the mutation operation in the current population to generate a mutant vector $V_{i,G}$ with respect to each individual $X_{i,G}$, so-called target vector. For each target vector $X_{i,G}$, at the generation G, its associated mutant vector $V_{i,G} = \{v_{i,G}^1, v_{i,G}^2, \ldots, v_{i,G}^D\}$ can be generated via certain mutation strategy. For example, the five most frequently used mutation strategies implemented in the DE codes are listed as follows:

$$V_{i,G} = X_{r_1^i,G} + F\left(X_{r_2^i,G} - X_{r_3^i,G}\right) \tag{9.24a}$$

$$V_{i,G} = X_{best,G} + F\left(X_{r_1^i,G} - X_{r_2^i,G}\right) \tag{9.24b}$$

$$V_{i,G} = X_{i,G} + F\left(X_{best,G} - X_{i,G}\right) + F\left(X_{r_1^i,G} - X_{r_2^i,G}\right) \tag{9.24c}$$

$$V_{i,G} = X_{\text{best},G} + F\left(X_{r_1^i,G} - X_{r_2^i,G}\right) + F\left(X_{r_3^i,G} - X_{r_4^i,G}\right) \tag{9.24d}$$

$$V_{i,G} = X_{r_1^i,G} + F\left(X_{r_2^i,G} - X_{r_3^i,G}\right) + F\left(X_{r_4^i,G} - X_{r_5^i,G}\right) \tag{9.24e}$$

The indices $r_1^i, r_2^i, r_3^i, r_4^i, r_5^i$ are mutually exclusive integers generated randomly within the range [1, NP], which are also different from the index i. These indices are arbitrarily created once for each mutant vector. For scaling the difference vector, a positive control parameter F (scaling factor) is used. $X_{\text{best},G}$ is the best individual vector with the best fitness value in the population at generation G. A good initial choice of F was 0.5. The effective range of F values have been suggested between 0.4 and 1.0.

Note that the smaller the differences between parameters of parent, the smaller the difference vectors and therefore, the perturbation. That indicates whenever the population gets close to the optimum; the step length decreases automatically. This is similar to the automatic step size control found in the standard evolution strategies.

9.3.4 Crossover

After the mutation phase, crossover or recombination operation is applied to each pair of the target vector $X_{i,G}$ and its corresponding mutant vector $V_{i,G}$ to generate a trial vector: $U_{i,G} = \left(u_{i,G}^1, u_{i,G}^2, \ldots, u_{i,G}^D\right)$. In order to increase the diversity of the perturbed parameter vectors, crossover is introduced. In the basic version, DE employ the binomial (uniform) crossover defined as follows:

$$u_{i,G}^j = \begin{cases} u_{i,G}^j & \text{if } \left(\text{rand}_j[0,1] \leq CR\right) \text{ or } \left(j = j_{\text{rand}}\right) \\ x_{i,G}^j & \text{otherwise} \end{cases}, \quad j = 1, 2, \ldots, D \tag{9.25}$$

In Eq. (9.25), CR, the crossover rate is a user-defined constant in the range [0,1], which controls the fraction of parameter values copied from the mutant vector. The parameter j_{rand} is a randomly selected integer within the range [1,D]. The binomial crossover operator copies the jth parameter of the mutant vector $V_{i,G}$ to the corresponding element in the trial vector $U_{i,G}$ if rand$_j$ [0,1] ≤ CR or $j = j_{\text{rand}}$. Otherwise, it is copied from the corresponding target vector $X_{i,G}$. There is an another exponential crossover operator, in which the parameters of trial vector $U_{i,G}$ are inherited from the corresponding mutant vector $V_{i,G}$ starting from a randomly selected parameter index till the first time rand$_j$ [0,1] > CR. The remaining parameters of the trial vector $U_{i,G}$ are copied from the corresponding target vector $X_{i,G}$. The first reasonable attempt of selecting CR value can be 0.1. However, since the large CR value can accelerate convergence, the value of 0.9 for CR may also be a good initial choice if the problem is near unimodal or fast convergence is preferred. The condition $j = j_{\text{rand}}$ is introduced to make sure that the trial vector $U_{i,G}$ will be different from its corresponding target vector $X_{i,G}$ by at least one parameter. DE's exponential crossover operator is functionally equivalent to the circular two-point crossover operator.

9.3.5 Selection

The trial vector $U_{i,G+1}$ is compared to the target vector $X_{i,G}$ using the greedy criterion to decide whether the trial vector should become a member of generation $G + 1$ or not. Whenever the vector $U_{i,G+1}$ yields a smaller cost function value than $X_{i,G}$, then $X_{i,G+1}$ is set to $U_{i,G+1}$; or else, the old value $X_{i,G}$ is retained. The selection operation can be expressed as follows:

$$X_{i,G+1} = \begin{cases} U_{i,G} & \text{if } f(U_{i,G}) \leq f(X_{i,G}) \\ X_{i,G} & \text{otherwise} \end{cases}, \quad i = 1, 2, \ldots, N \tag{9.26}$$

Generation after generation the above 3 steps are repeated until some specific termination criteria are met.

9.4 Simulated Annealing

Since its inception in the early 80's, [Kirkpatrick *et al.* (1983)] simulated annealing (SA) has been applied to various combinatorial and numerical optimization problems. SA is a generic probabilistic metaheuristic algorithm for the global optimization problem. This method has been developed based on the thermal annealing of critically heated metals. A solid metal becomes molten at a high temperature and the atoms of the melted metal move freely with respect to each other. However, as the temperature is decreased, these movements of atoms get restricted. When the temperature decreases, the atoms tend to get ordered and finally form crystals with the minimum possible internal energy. This process of crystals formation essentially depends on the cooling rate; fast cooling may result defects inside the material. The material may not be able to get the crystalline shape if the temperature of the molten metal is diminished at a very fast rate; instead, it may achieve a polycrystalline shape with a higher energy state compared to that of the crystalline shape. Therefore, the temperature of the molten metal should be decreased at a sluggish and controlled rate to make sure proper solidification with a highly ordered crystalline shape that corresponds to the lowest energy state (internal energy). This process of slow rate cooling is known as annealing [Rao, (2009)]. SA is a random-search method, which exploits an analogy between the way in which a molten metal cools and freezes into a lowest energy crystalline state and the search for a minimum in a more general system. Using the cost function in place of the energy and defining configurations by a set of parameters $\{x_i\}$, it is simple by means of the Metropolis procedure to generate a population of configurations of a given optimization problem at some efficient temperature. This temperature is just a control parameter in the same units as the cost function. [Kirkpatrick *et al.* (1983)]. Iterative improvement, usually employed to these problems, is much like the microscopic rearrangement processes represented by statistical mechanics, where the cost function was playing the role of energy. Though, accepting only rearrangements that lower the cost function of the system is like extremely rapid quenching from high temperatures to $T = 0$, so it should not be surprising that resulting solutions are usually metastable. The Metropolis procedure from statistical mechanics gives a generalization of iterative upgrading in which controlled uphill steps can likewise be incorporated in the search for a better solution [Kirkpatrick *et al.* (1983)].

Local minimization is no guarantee of global minimization. Therefore, a fundamental concern in global minimization is to avoid being stuck in a local minimum. There are two categories of methods known to overcome this difficulty in stochastic minimization: The first class constitutes the so-called two-phases methods; the second class is based on SA [Dekkers and Aarts (1991)]. The main aspect of SA algorithm is its ability to avoid being trapped in a local minimum. This is done letting the algorithm to accept not only better solutions however, also worse solutions (by allowing hill-climbing moves) with a given probability. [Chibante *et al.* (2010)] From the mathematical point of view, SA can be considered as a randomization tool that permits wrong-way movements during the search for the optimum through an adaptive acceptance/rejection criterion. This mechanism is of significance for treating effectively non-convex problems [Wei-zhong and Xi-Gang (2009)]. Relatively it is slow compared to deterministic optimization techniques using gradient based searches, branch and bound techniques and/or decomposition. Its primary weaknesses is the computational time complexity. However, its robustness and capability of handling discontinuous functions and discrete variables, where deterministic techniques fail, make it an essential tool for single and multiple objective optimization problems [Suman *et al.* (2010)].

There are two critical parameters in the cooling process that decide how amorphous or firm will be the result for the metal in its frozen condition. The first parameter is the initial temperature from which the cooling starts and the second one is the rate at which the temperature is decreasing. For any stochastic local search algorithms, some control parameters like initial temperature (T_0), cooling rate, are subjective to some extent and must be defined from an empirical basis. This shows that the algorithm must be adjusted in order to maximize its performance. [Shojaee *et al.* (2010)].

9.4.1 Procedure

Simulated annealing method follows the steps similar to the cooling of molten metal. Annealing procedure involves first 'melting' the system at a high temperature, and then decreasing the temperature gradually taking enough steps at each temperature to keep the system close to equilibrium, until the system reaches the ground state [Nourani and Andresen (1998)]. A typical procedure of SA is given below:

9.4.1.1 Algorithm of SA

This simulation procedure starts with initialization of vector $X, (X_0)$ and temperature $T, (T_0)$. Various researches [citation] show that parameters like number of iteration, cooling rate, and termination criteria (freezing temperature) have great influence on the SA algorithm performance. At each temperature level, system should reach equilibrium with minimum energy. For this purpose, a large number of iteration is required at each temperature level. This section elucidates how to estimate these parameters. Figure 9.9 shows the flowchart of the SA.

Step 1 Estimate the initial temperature (T_0), define number of iteration in each temperature step (n), number of cycles (k), freezing or final temperature (T_f), constant for annealing schedule (c)

Step 2 Guess an initial value (X_0), Calculate the value of the objective function $f(X_0)$

Step 3 Generate a new design point (X_{i+1}) in the vicinity of X_i, calculate the value of $f(X_{i+1})$; Set iteration counter $i = 1$, cycles counter $k = 1$.

NONTRADITIONAL OPTIMIZATION 247

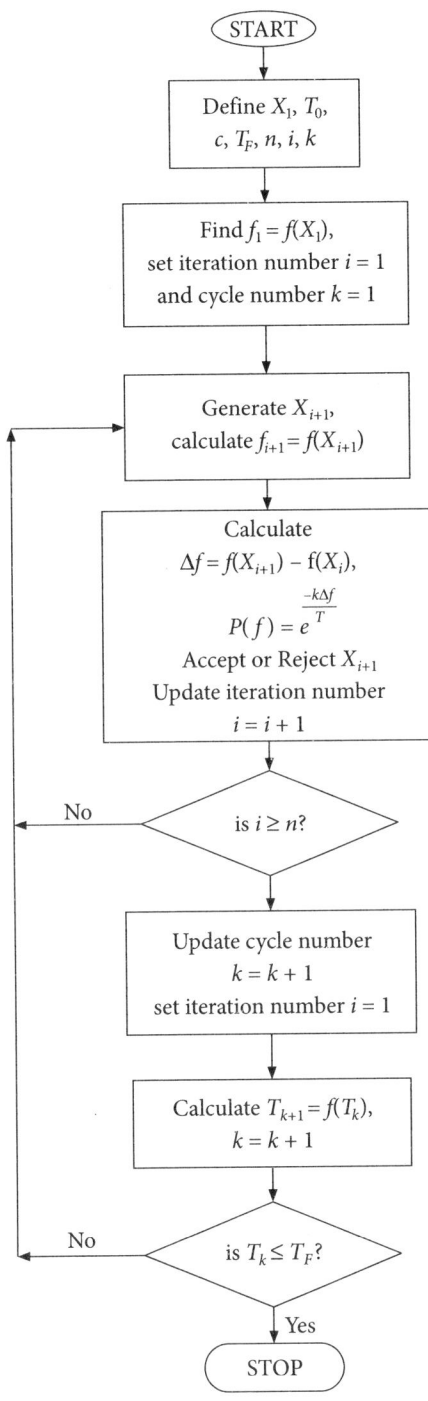

Fig. 9.9 Flowchart of simulated annealing

Step 4 Find the value of $Df = f(X_{i+1}) - f(X_i)$, use Metropolis criterion (Eq. (9.27)) to decide whether to accept or reject the new design point.

$$P(f) = \begin{cases} e^{\frac{-k\Delta f}{T}} & \Delta f > 0 \\ 1 & \Delta f \leq 0 \end{cases} \tag{9.27}$$

if probability $P(f)$ is within the acceptable range, accept the new design point; otherwise reject it and go back to the previous design point.

Step 5 Go to Step 3, with $i = i + 1$; Repeat the following sub-steps for n times (n is called the epoch length).

Step 6 If the given stop condition ($T_k \leq T_F$) is satisfied, stop. Otherwise, reduce the temperature $T_{k+1} = f(T_k)$ according to annealing schedule and $k = k + 1$; and go to Step 3.

Theoretically, only if the inner loop computing (Step 3 – Step 5) is saturated (n is sufficiently large) to let the system reach to an equilibrium state, the global optimal can be expected. However, in practical use, only a limited number of configurations can be created at each temperature level, and then there exist a trade-off between computational cost and the quality of the solution [Wei-zhong and Xi-Gang (2009)].

9.4.1.2 Estimation of initialization temperature (T_0)

The initial temperature (T_0) plays a crucial role during simulated annealing process. Because T_0 controls the acceptance rule defined by Eq. (9.27) as well as the convergence of the algorithm. If T_0 is too large, it requires a large number of temperature reduction steps to converge. Conversely, if the initial temperature is selected to be very small, the search procedure may be incomplete in the sense that it might fail to thoroughly explore the design space to locate the global minimum before convergence. If the initial temperature is not high enough, atoms of the molten metal would not have full freedom to rearrange their positions in a very regular minimum energy structure. In a typical execution of SA, the initial temperature is set sufficiently high to enable the algorithm to move off a local minimum. Research shows that there is not any deterministic criterion to set the initial pseudo-temperature in the literature; formulation of the initial temperature is characteristics of the particular problem [Shojaee *et al.* (2010)].

According to the fundamental concepts of SA, non-improver solutions are accepted in the primary iterations with high probability. Pao *et al.* considered an initial temperature so that the initial acceptance rate is about 70 per cent [Pao *et al.* (1999)]. Thompson and Bilbro set the probability of accepting to 0.75 for initial temperature setting of continuous problems. Then, the following probability distribution is solved to find T_0 [Thompson and Bilbro, (2005)]:

$$p = \exp(\Delta E/T) \tag{9.28}$$

where DE is the average cost of the random solutions plus its standard deviation. Hao Chen *et al.* set the initial temperature so that the initial acceptance probability for an average uphill move is 0.97 [Chen *et al.* (1998)]. Feng-Tse Lin, *et al.* proposed an Annealing-Genetic approach and use the following formula [Lin *et al.* (1993)]:

$$T_0 = \frac{\Delta E}{\left(\text{Population size}/2\right)} \tag{9.29}$$

where DE is the difference between the highest cost and the lowest cost found for the first generation of the randomly generated population.

Parameter setting for SA is problem dependent, and it is best accomplished using trial and error. Furthermore, researches show that SA algorithms are highly sensitive to parameters, and their performances are largely dependent on fine tuning of the parameters [Wong and Constantinides, (1998)].

For calculating the initial temperature of the example 9.5, we have followed a simple method. The average value of the cost function in the first generation was considered as initial temperature.

9.4.1.3 Perturbation mechanism

The perturbation mechanism is the process to create new solutions from the current solution. A randomly generated perturbation of the current system configuration is applied so that a trial configuration is obtained. In other words, it is a process to explore the neighborhood of the current solution creating small variations in the current solution. At the beginning of the process, the temperature is high and strong perturbations are applied to easily jump over high peaks in search of the global minima. Although applying too much perturbation is useless and should be avoided.

SA is normally used in combinatorial problems where the parameters being optimized are integer numbers. In an application where the parameters vary continuously, the exploration of neighborhood solutions can be made as discussed here. A solution s is defined as a vector $s = (x_1, x_2, \ldots, x_n)$ representing a point within the search space. A new solution is generated using a vector $\sigma = (\sigma_1, \sigma_2, \ldots, \sigma_n)$ of standard deviations to create a perturbation from the current solution. A neighbor solution is then generated from the current solution by:

$$x_{i+1} = x_i + N(0, \sigma_i) \tag{9.30}$$

where $N(0, \sigma_i)$ is a random Gaussian number with zero mean and σ_i standard deviation.

To improve the performance of simulated annealing method, it is usual to spend some time at each temperature by conducting a fixed number of iterations before reducing the temperature. Consider K be the number of iterations executed at each temperature, then Equation (1) can be written as

$$X_i = \{x_{1,i}, x_{2,i}, \ldots, x_{M,i}\}, i = 1, 2, \ldots \tag{9.31}$$

where i represents the number of perturbations applied to the solution, and $x_{1,r}$ is the value of x_1 after it is has been perturbed i-times. Hence, at the end of temperature T_1, the number of perturbations applied to the solution is K, and X_K represents the solution at the end of this temperature.

For finding the next design point (X_{i+1}) in the example 9.5, we have considered the range [$X_i - 5$, $X_i + 5$] and random numbers.

9.4.1.4 Cooling scheduling

The major drawback of SA is that, its massive requirement of computation time. This can be overcome using improved cooling schedule or annealing schedule. The sequence of temperatures and the number of rearrangements of the $\{x_i\}$ attempted to reach equilibrium at each temperature can be considered an annealing schedule. [Kirkpatrick *et al.* (1983)]. By controlling the temperature T, we control the probability of accepting a hill climbing move (a move that results in a positive Δx) and, therefore, the exploration of the state space. An annealing or cooling schedule controls how quickly the temperature T reduces from high to low values, as a function of time or iteration counts [Shojaee *et al.* (2010)]. The efficient annealing schedule, which lowers the temperature at every step and keeps the system in quasi-equilibrium at all times, is derived from a new quasi-equilibrium criterion.

The most common cooling schedule is the geometric rule for temperature variation:

$$T_{i+1} = sT_i \qquad (9.32)$$

with $s < 1$, good results have been reported in literature when s is in the range [0.8, 0.99] [Fouskakis and Draper, (2002)]

The problem dependent nature of parameter setting for SA and its sensitivity to parameters restrict the robustness and effectiveness of SA algorithms. SA possesses a formal proof of convergence to the global optima. This convergence proof relies on a very slow cooling schedule with an initial condition of a sufficiently large temperature and let it decay by following relation:

$$T_k = \frac{T_0}{\log(k)} \qquad (9.33)$$

where k is bound by the number of iterations [Ingber and Rosen, (1992)]. While this cooling schedule is impractical, it identifies a useful trade-off where longer cooling schedules tend to lead to better quality solutions.

A logarithmic cooling scheme introduced by Geman and Geman [Geman and Geman (1984)] is as follows:

$$T(t) = \frac{c}{\log(t+d)} \qquad (9.34)$$

where d is usually set equal to one. As the only existence theorem, it has been established that for c being greater than or equal to the largest energy barrier in the problem, this schedule will lead the system to the global minimum state in the limit of infinite time. Hajek (1988) [Hajek (1988)] proved that exponential cooling could not guarantee convergence.

9.4.1.5 Termination criteria

The process is to have converged when the current value of temperature T is very small or when changes in the function value (Δf) are observed to be sufficiently small. The stochastic search process is terminated whenever (a) a given total number of steps have been carried out, (b) the expected number of acceptances for a given number of trials has not been achieved, (c) the annealing temperature drops below the freezing point. In most the cases, condition (c) is used as termination criteria. The calculation is terminated and the best solution is taken as the optimal solution when annealing temperature becomes equal or smaller than T_{end}.

Example 9.5

Find the minimum of the function using simulated annealing.

$$f(X) = 1200(2x_1x_3 + 2x_2x_3) + 2500x_1x_2 + 500\left(\frac{1000}{x_1x_2x_3}\right) + 100\left(\frac{1000}{10x_1x_2x_3}\right)$$

The condition for solving this problem is $x_1, x_2, x_3 \geq 0$

Solution

We considered only two iterations for each temperature level. However, for practical application more iteration is required to reach thermal equilibrium.

The cooling schedule used for this problem is $T_{i+1} = 0.5 T_i$

$$X^{(1)} = \begin{bmatrix} 1 & 1 & 1 \end{bmatrix}^T, X^{(2)} = \begin{bmatrix} 1 & 5 & 10 \end{bmatrix}^T, X^{(3)} = \begin{bmatrix} 10 & 5 & 5 \end{bmatrix}^T, X^{(4)} = \begin{bmatrix} 7 & 6 & 5 \end{bmatrix}^T$$

$$f^{(1)} = 517.3, \; f^{(2)} = 166.7, \; f^{(3)} = 307.04, \; f^{(4)} = 263.43$$

the average value of $f^{(1)}, f^{(2)}, f^{(3)}$ and $f^{(4)}$ is 313.6175

so, we are considering 313.62 as initial temperature

Step 1 Initial design point $X_1 = [3\ 4\ 5]^T$ and $f_1 = 122.50$

Step 2 Random numbers generated using simulink MATLAB are as follows

$$r_1 = 0.38, \; r_2 = 0.57, \; r_3 = 0.51$$

Generate a new design point in the vicinity (± 5.00) of the current design point.
Choose the ranges of $x_1, x_2,$ and x_3; (–2,8), (–1,9), and (0,10)

now calculate $u_1 = -2 + 0.38[8 - (-2)] = 1.8$

$$u_2 = -1 + 0.57[9 - (-1)] = 4.7$$

$$u_3 = 0 + 0.51[10 - 0] = 5.1$$

which gives $X_2 = \begin{bmatrix} 1.8 \\ 4.7 \\ 5.1 \end{bmatrix}$ and $f_2 = 112.53$

value of for $p = 2$
$$\Delta f = f_2 - f_1 = 112.53 - 12250 = -9.97$$

The new temperature, $T_2 = 0.5 \times 313.62 = 156.81$

$$r_1 = 0.68, \ r_2 = 0.45, \ r_3 = 0.46$$

the new ranges of x_1, x_2, and x_3 (−3.2,6.8), (−0.3,9.7), and (0.1,10.1)

$$u_1 = -3.2 + 0.68[6.8 - (-3.2)] = 3.6$$

$$u_2 = -0.3 + 0.45[9.7 - (-0.3)] = 4.2$$

$$u_3 = 0.1 + 0.46[10.1 - 0.1] = 4.5$$

which gives $X_3 = \begin{bmatrix} 3.6 \\ 4.2 \\ 4.5 \end{bmatrix}$ and $f_3 = 129.534$

value of $\Delta f = f_3 - f_2 = 129.534 - 112.53 = 17.006$

with $k = 1, \ P[X_3] = e^{-\Delta f/T} = e^{-17.006/156.81} = 0.897$

so, we will accept the design point with a probability of 0.897
now we have to find X_4 in the neighbor of X_3 with random numbers

$$r_1 = 0.43, \ r_2 = 0.47, \ r_3 = 0.46$$

Calculated values for other iterations is given in the table below

Table 9.4 Calculated values of different iteration

P	i	T_i (c = 0.5)	r	u	X_i	$E_i = f_i = f(X_i)$	$\Delta E = E_{i+1} - E_i$	$P[E_{i+1}]$	accept/reject X_i
1	0	313.62			[3,4,5]	122.50			
	1		[0.38,0.57,0.51]	[1.8,4.7,5.1]	[1.8,4.7,5.1]	112.53	−9.97	1	accept
2	1	156.81	[0.68,0.45,0.46]	[3.6,4.2,4.5]	[3.6,4.2,4.5]	129.534	17.006	0.897	accept
	2		[0.43,0.47,0.46]	[2.9,3.9,4.1]	[2.9,3.9,4.1]	106.185	−23.349	1	accept
3	1	78.405	[0.48,0.42,0.44]	[2.7,3.1,3.5]	[2.7,3.1,3.5]	87.054	−19.131	1	accept
	2		[0.31,0.83,0.71]	[0.8,6.4,5.6]	[0.8,6.4,5.6]	127.355	40.301	0.598	accept
4	1	39.202	[0.63,0.22,0.31]	[2.1,3.6,3.7]	[2.1,3.6,3.7]	87.749	−39.606	1	accept
	2		[0.93,0.65,0.35]	[6.4,5.1,2.2]	[6.4,5.1,2.2]	149.422	61.673	0.207	reject
5	1	19.601	[0.61,0.56,0.36]	[3.2,4.2,2.3]	[3.2,4.2,2.3]	90.946	3.197	0.8495	accept
	2		[0.45,0.13,0.53]	[2.7,2.5,2.6]	[2.7,2.5,2.6]	78.383	−12.564	1	accept

The graph (Fig. 9.10) shows that how the probability of accepting the new design point changes with temperature. At low temperature, probability $P[E]$ is low. We reject the design point when probability is below certain value. In Example 9.5 (Table: 9.4) shows that design point [6.4,5.1,2.2] has been rejected as the $P[E]$ is 0.207.

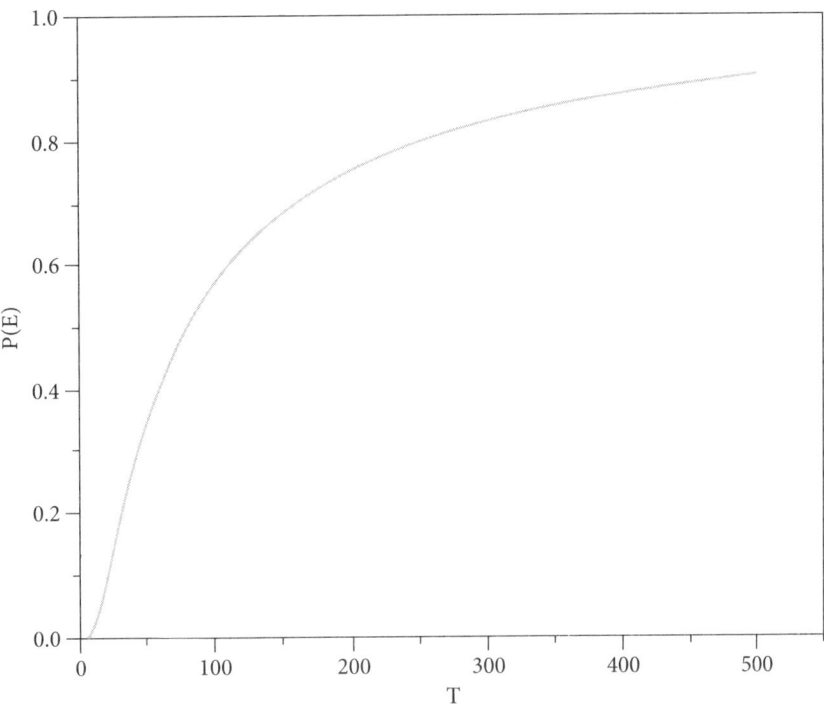

Fig. 9.10 Probability of accepting vs temperature plot

9.4.2 Applications of SA in chemical engineering

As simulated annealing is able to find the global optimum points, it is very useful in process engineering. Sankararao and Gupta [Sankararao and Gupta (2007)] used SA for optimizing fluidized bed catalytic cracking unit. They used multiobjective optimization for this problem. Some researchers [Kaczmarski and Antos (2006)] have used SA for chromatographic separation.

Summary

- This chapter discusses different non-traditional optimization techniques like genetic algorithm, particle swarm optimization, differential evolution, and simulated annealing. Genetic algorithm and differential evolution imitate the steps of genetic evolution, the particle swarm optimization method was developed based on the social behavior of a bird flocks, school of fish etc. Simulated annealing is a metaheuristic algorithm, which mimics the process of thermal annealing of critically heated metals. These methods are applicable for both continuous and discrete optimization. They are also able to find the global optimizer. However, we should be careful about the termination criteria of these algorithms.

Exercise

9.1 Covert these decimal number to binary
   a) 25   b) 3   c) 98

9.2 How do you calculate the fitness of a chromosomes during genetic reproduction?

9.3 Write the algorithm for solving the problem

$$\min f(X) = 80(x_1^2 + x_2)^2 + (1 - x_1)^2$$

subject to $X \in [0, 5]$

9.4 Compare the crossover and mutation operation.

9.5 solve the given problem using PSO method

$$f(X) = 1200(2x_1x_3 + 2x_2x_3) + 2500x_1x_2 + 500\left(\frac{1000}{x_1x_2x_3}\right) + 100\left(\frac{1000}{10x_1x_2x_3}\right)$$

9.6 Compare different neighborhood topologies for PSO method.

9.7 Write down the algorithm for differential evolution for solving the problem on 3.

9.8 Solve the optimization problem using SA

$$\min f(X) = 100(x_1^2 + x_2)^2 + (1 - x_1)^2$$

subject to $X \in [0, 7]$

9.9 Show 3 iterations for solving the problem below

$$\min f(X) = -12x_1 - 7x_2 + x_2^2$$

subject to $-2x_1^4 - x_2 + 2 = 0$

$0 \leq x_1 \leq 3, \ 0 \leq x_2 \leq 3$

9.10 Discuss the perturbation mechanism of simulated annealing.

9.11 Compare different cooling schedule of simulated annealing.

9.12 Discuss the termination criteria of simulated annealing. How the result depends on the final temperature?

9.13 What are the pros and cons of GA and SA? Can we conceive of a framework that combines the best of both worlds?

9.14 Are Genetic Algorithms useful if we do not have a full understanding of the objective function?

9.15 If two successive generations are identical then the Genetic Algorithm has found the optimal solution. Discuss if this statement is correct.

References

Babu, B. V. and Angira, R. 2006. *Modified Differential Evolution (MDE) for Optimization of Non-linear Chemical Processes,* Computers Chemical Engineering, 30(6–7): 989–1002.

Bhaskar, V., Gupta, S. K. and Ray, A. K. 2000. *Applications of Multi-objective Optimization in Chemical Engineering,* Rev Chemical Engineering, 16: 1–54.

Bratton, D. and Kennedy, J. 2007. *Defining a Standard for Particle Swarm Optimization,* IEEE Swarm Intelligence Symposium, pp. 120–27.

Chen, H., Flann, N. S. and Watson, D. W. 1998. *Parallel Genetic Simulated Annealing: A Massively Parallel SIMD Algorithm,* IEEE Transactions on Parallel and Distributed Systems, 9(2):126–136.

Chibante, R. Araújo A. and Carvalho A. 2010. *Parameter Identification of Power Semiconductor Device Models using Metaheuristics,* Simulated Annealing Theory with Applications (edited by Rui Chibante), Chapter 1.

Clerc, M. 1999. *The Swarm and the Queen: Towards a Determininistic and Adaptive Particle Swarm Optimization,* 'Congress on Evolutionary Computation' (CEC99), pp. 1951–57.

Clerc, M. and Kennedy, J. 2002. *The Particle Swarm – Explosion, Stability, and Convergence in a Multidimensional Complex Space,* 'IEEE Transactions on Evolutionary Computation', 6: 58–73.

Curteanu, S. Leon, F. 2008. *Optimization Strategy Based on Genetic Algorithms and Neural Networks Applied to a Polymerization Process,* 'International Journal of Quantum Chemistry', vol. 108, 617–30.

Davis, L. 1991. *Handbook of Genetic Algorithms,* New York: Van Nostrand Reinhold, pp. 11–96.

Dekkers, A. and Aarts, E. *Global Optimization and Simulated Annealing;* Mathematical Programming 50(1991): 367–93.

Edgar, T. F., Himmelblau, D. M. and Lasdon, L. S. 2002. *Optimization of Chemical Processes,* McGraw-Hill Inc.

Fouskakis, D. and Draper, D. 2002. *Stochastic Optimization: A Review,* International Statistical Review, 70(3): 315–49.

Garrard, A. and Fraga, E. S. 1998. *Mass Exchange Network Synthesis using Genetic Algorithms,* Computers Chemical Engineering, 22, pp. 1837.

Geman, S. and Geman, D. 1984. *Stochastic Relaxation, Gibbs Distributions, and Bayesian Restoration of Images IEEE Trans.* Pattern Anal. Mach. Intell., PAMI-66: 721–41.

Ghalia, M. B. *Particle Swarm Optimization with an Improved Exploration-Exploitation Balance,* IEEE, vol. 978-1-4244-2167-1/08/ 2008.

Goldberg, D. E. 1989. *Genetic Algorithms in Search Optimization and Machine Learning,* Addison-Wesley Pub. Co., pp. 147–260.

Hajek, B. 1988. *Cooling Schedules for Optimal Annealing Math Oper Res* 13: 311–29 op cit Azencott.

Handbook of Evolutionary Computation (1997). IOP Publishing Ltd. and Oxford University Press, release 97/1.

Heppner, F. and Grenander, U. A. 1990. *Stochastic Nonlinear Model for Coordinated Bird Flocks.* In S. Krasner, Ed., The Ubiquity of Chaos, Washington. DC: AAAS Publications.

Herrera, F., Lozano, M., Verdegay, J. L. 1998. *Tackling Real Coded Genetic Algorithms: Operators and Tools for Behavioural Analysis.* Artificial Intelligence Review, 12: 265.

Holland, J. H. 1975. *Adaptation in Natural and Artificial Systems,* University of Michigan Press, Ann Arbor, MI.

Ingber, L. and Rosen, B. 1992. *Genetic Algorithms and Very Fast Simulated Reannealing:* 'A Comparison, Mathematical Computer Modeling', vol. 16, No. 11, pp. 87–100.

Jung, J. H., Lee, C. H. and Lee, I-B. 1998. *A Genetic Algorithm for Scheduling of Multiproduct Batch Processes,* Comp. Chem. Eng., 22: 1725.

Kaczmarski, K. and Antos, D. 2006. *Use of Simulated Annealing for Optimization of Chromatographic Separations;* Acta Chromatographica, No. 17: 20–45.

Kennedy, J. and Eberhart, R. 1995. *Particle Swarm Optimization.* 'In Proceedings of IEEE International Conference on Neural Networks', vol. IV, pp. 1942–48, Perth, Australia.

Kirkpatrick, S., Gelatt, C. D. Vecchi, M. P. Jr. 1983. *Optimization by Simulated Annealing; Science, New Series,* vol. 220, No. 4598. pp. 671–80.

Lin, F-T., Kao, C-Y. and Hsu C-C. 1993. *Applying the Genetic Approach to Simulated Annealing in Solving Some NP-Hard Problems,* IEEE Transactions on Systems, Man, and Cybernetics, 23(6).

Löeh, T., Schulz, C. and Engell, S. 1998. *Sequencing of Batch Operations for Highly Coupled Production Process: Genetic Algorithms vs. Mathematical Programming,* Comp. Chem. Eng., 22: S579.

Mayer, D. G. 2002. *Evolutionary algorithms and agricultural systems,* Boston: Kluwer Academic Publishers, pp. 107.

Nandasana, A. D. Ray, A. K. and Gupta, S. K. 2003. *Application of the Non-dominated Sorting Genetic Algorithm (NSGA) in Chemical Engineering,* Int J Chem Reactor Eng 1: 1.

Nourani, Y. and Andresen, B. *A Comparison of Simulated Annealing Cooling Strategies;* J. Phys. A: Math. Gen., 31(1998): 8373–85.

Pao, D. C. W. Lam, S. P. and Fong, A. S. 1999. *Parallel Implementation of Simulated Annealing Using Transaction Processing,* IEE Proc-Comput. Digit. Tech., 146(2): 107–13.

Price, K. V. Differential Evolution: *A Fast and Simple Numerical Optimizer,* 0–7803–3225–3–6/96 1996 IEEE.

Qin, A. K. Huang, V. L. and Suganthan, P. N. *Differential Evolution Algorithm With Strategy Adaptation for Global Numerical Optimization,* IEEE Transactions on Evolutionary Computation, 13(2): 2009.

Rao, S. S. 2009. *Engineering Optimization: Theory and Practice* (4e), John Wiley and Sons.

Reynolds, C. W., *Flocks, Herds, and Schools: A Distributed Behavioral Model.* Computer Graphics, vol. 21(1987): 25–34.

Sankararao, B. Gupta, S. K. *Multi-objective Optimization of an Industrial Fluidized-bed Catalytic Cracking Unit (FCCU) using Two Jumping Gene Adaptations of Simulated Annealing;* Computers and Chemical Engineering, 31(2007): 1496–1515.

Shojaee, K., Shakouri, H. G. and Taghadosi, M. B. 2010. *Importance of the Initial Conditions and the Time Schedule in the Simulated Annealing a Mushy State SA for TSP,* Simulated Annealing Theory with Applications (edited by Rui Chibante); Chapter 12.

Shopova and Vaklieva-Bancheva (2006) Shopova, E. G.; Vaklieva-Bancheva N. G. Comp Chem Eng 2006, 30: 1293.

Storn, R. Price, K. 1997. *Differential Evolution – a Simple and Efficient Heuristic for Global Optimization Over Continuous Spaces.* Journal of Global Optimization 11: 341–359.

Storn R. *On the Usage of Differential Evolution for Function Optimization.* Biennial Conference of the North American Fuzzy Information Processing Society (NAFIPS). (1996): 519–523.

Suman, B. Hoda, N. and Jha, S. *Orthogonal Simulated Annealing for Multiobjective Optimization;* Computers and Chemical Engineering 34(2010): 1618–31.

Thompson, D. R. and Bilbro, G. L. 2005. *Sample-Sort Simulated Annealing,* IEEE Transactions on Systems, Man, and Cybernetics—PART B: Cybernetics, 35(3): 625–632.

Trelea, I. C. 2003. *The Particle Swarm Optimization Algorithm: Convergence Analysis and Parameter Selection.* Information Processing Letters, 85: 317–25.

Valle, Y. del, Venayagamoorthy G. K., Mohagheghi S., Hernandez J–C., and Harley R. G. 2008. *Particle Swarm Optimization: Basic Concepts, Variants and Applications in Power Systems,* IEEE Transactions on Evolutionary Computation, 12(2).

Weber, T. and Bürgi, H. B. 2002. *Determination and Refinement of Disordered Crystal Structures Using Evolutionary Algorithms in Combination with Monte Carlo Methods;* Acta Cryst. A58, pp. 526–40.

Wei-zhong A., Xi-Gang Y., *A Simulated Annealing-based Approach to the Optimal Synthesis of Heat-Integrated Distillation Sequences,* Computers and Chemical Engineering, 33(2009): 199–212.

Wong, K. L., Constantinides A. G. 1998. *Speculative Parallel Simulated Annealing with Acceptance Prediction,* Electronics Letters, 34(3): 312–13.

Chapter 10

Optimization of Various Chemical and Biochemical Processes

10.1 Heat Exchanger Network Optimization

An important task during the design of chemical processes is to changing the process streams from their available temperatures to the required temperatures without additional cost. Heat recovery is an important approach for reducing the cost by using the heat available from streams to be cooled to the streams to be heated. The distinguish between hot streams and cold streams are made based on whether the stream is cooled or heated not on the basis of temperature. Developing an efficient energy system is very essential in process industry to reduce waste heat available. This waste heat from one process can be recovered and reused for another process. Heat Exchanger Network (HEN) is used to optimize this heat recovery consequently the investment cost and energy consumption in the process plant. Synthesis of HEN is one of the most commonly discussed problems in a process industry. This method has importance during determination of energy expenditure for a process and improving the recovery of energy in industry. The first systematic method was introduced in 1970s to heat recovery with the concept of pinch analysis. Hohmann (1971), and Linnhoff and Flower (1978) introduced the pinch analysis for synthesis of HEN. A single task is decomposed into three different subtasks (i.e., targets) like minimum utility cost, minimum number of units, and minimum investment cost network configurations. The major benefit of decomposing this HEN synthesis problem is that handling of these sub-problems are much simpler than the original single-task problem. The decomposed sub-problems are given below:

1. **Minimum utility cost**   This subtask or target is related to the amount of maximum energy recovery that can be attained in a feasible HEN with a constant heat recovery approach temperature (HRAT), that allows to eliminate various non-energy efficient HEN structures. Hohmann in 1971, first introduced the minimum utility cost and then Linnhoff and Flower [Linnhoff and Flower, (1978)] discussed this technique. Cerda *et al.* (1983) discussed this as an LP transportation model and is improved as LP transshipment model by Papoulias and Grossmann [Papoulias and Grossmann (1983)].

2. **Minimum number of units** For a fixed utility cost, this target finds the match combination with the minimum number of units and distribution of their load. The overall cost of the HEN depends on the number of units. There are two popular models namely MILP transportation and MILP transshipment developed by various researchers (Cerda and Westerberg (1983), Papoulias and Grossmann (1983)). The vertical heat transfer formulation can also be employed to optimize HEN (Gundersen and Grossmann (1990) and Gundersen, Duvold and Hashemi-Ahmady (1996)).

3. **Minimum investment cost network configurations** This optimization is done based on the match information and heat load obtained from preceding targets. The NLP problem is formulated and optimization is done to minimize the overall cost of the network using the superstructure-based model developed by Floudas *et al.* [Floudas *et al.* (1986)]. In this model, the objective function is the investment cost of the heat exchangers (i.e., area of heat transfer only as matches and utility loads are fixed) that are proposed as a superstructure. The objective function is expressed as a function of temperatures, considering logarithmic mean temperature difference (LMTD) as driving temperature forces. This objective function is nonlinear and convex in nature. In addition, the constraints energy balance equations for the mixers and heat exchangers are nonlinear as they possess bilinear products of unknown flow rates times corresponding to unknown temperatures. The NLP problem formulation is nonconvex due to the bilinear energy balance equalities, which implies that use of local NLP solvers yield local solutions (Floudas and Ciric (1989)). Heat Exchanger Networks (HENs) synthesis is intrinsically a Mixed Integer and Nonlinear Programming (MINLP) problem. The theories of MINLP has been discussed in chapter 7 (section 7.2).

10.1.1 Superstructure

A simple HEN superstructure has been shown in Fig. 10.1. This superstructure comprised of four heat exchangers (HE1, HE2, HE3, and HE4) with two hot fluid and one cold fluid inlet.

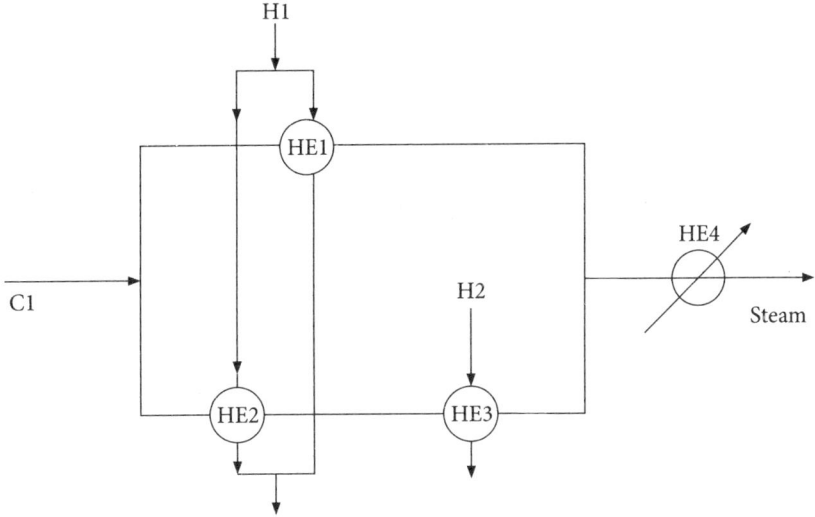

Fig. 10.1 HEN superstructure

10.1.2 Problem statement

The problem discussed in this section can be declared as follows, the condition given:

i. a set of hot process streams and a set of cold process streams for heat exchange with their specified flow rates, inlet and target temperatures, and physical properties (thermal conductivity, density, viscosity, heat capacity, and fouling factor)
ii. a single heating utility and a single cooling utility to fulfill the requirements of energy with their inlet and target temperatures as well as their physical properties
iii. capital and operating cost data involved in the network installation and operation.

During HEN design, determine the minimum overall annual cost target for these problems. We have to consider the target for the optimum power cost as well as traditional HEN design targets for minimum utilities, number of units and the heat transfer area.

The following assumptions were used in this work:

- The film heat-transfer coefficient of a stream is match independent.
- Considering the heat transfer in a single phase only.
- Only counter current and multi-pass shell-and-tube heat exchangers are considered.

The objective of HEN synthesis is to find out:

i. Optimal process flowsheet and design parameters.
ii. Values of temperatures and stream flow rates through the HEN and the process.
iii. Optimal configuration of the HEN including values for heat transfer area, number of units and requirement of utility.

10.1.3 Model formulation

Serna-González et al. [Serna-González et al. (2010)] have developed a HEN model, which considers both the fixed and operating costs.

The objective function is given by

$$\min TAC = A_f CC + C_U + C_P \tag{10.1}$$

where A_f is the annualized factor for investment, CC is the capital cost for heat exchange units, C_U is the hot and cold utilities costs, and C_P is the power cost.

The annual utility cost of a network is given by:

$$C_U = H_Y \left(C_H Q_{H_{\min}} + C_C Q_{C_{\min}} \right) \tag{10.2}$$

where $Q_{H_{\min}}, Q_{C_{\min}}$ are the minimum hot and cold utility target respectively.

C_H, C_C are unit cost of hot and cold utility respectively and H_Y represents the hour of operation of network per year.

Considering an equal distribution of area among all heat exchangers in the HEN, the network capital cost is given as

$$C_{CAP} = N_{u,mer}\left[a + b\left(A_{min}/N_{u,mer}\right)^c\right] \quad (10.3)$$

where $N_{u,mer}$ is the minimum number of units for a maximum energy recovery network, A_{min} is the network area target, and a, b and c are cost law coefficients that depend on the type of heat exchanger, construction materials, and pressure rating.

In order to evaluate A_{min}, we develop a mathematical model that is based on the spaghetti design model provided by the composite enthalpy-temperature curves. With a specified value of ΔT_{min}, for developing the spaghetti design of a process, the first step is to split the balanced composite curves into $k = 1,\ldots, K$ enthalpy intervals defined wherever a slope change occurs in either composite profile [Linnhoff and Ahmad (1990)]. The spaghetti design assumes that each hot stream i is integrated with all cold streams j within each enthalpy interval k. This generates a vertical structure for heat transfer, featuring parallel stream splitting and isothermal mixing. Under the assumptions that each stream match of the spaghetti design represents one and only one heat exchanger and that the film heat transfer coefficients of streams are match independent, the minimum area can be expressed as the sum of the stream contact areas:

$$A_{min} = \sum_i A_{ci} = \sum_j A_{cj} \quad (10.4)$$

$$A_{ci} = \sum_{j=1}^{J}\left(\frac{1}{h_i} + \frac{1}{h_j}\right)(UA_{i,j}) \quad \text{for} \quad i \in I \quad (10.5)$$

$$A_{cj} = \sum_{i=1}^{I}\left(\frac{1}{h_i} + \frac{1}{h_j}\right)(UA_{i,j}) \quad \text{for} \quad j \in J \quad (10.6)$$

$$UA_{i,j} = \sum_{k=1}^{K} UA_{i,j,k} \quad \text{for} \quad i \in I, j \in J \quad (10.7)$$

$$UA_{i,j,k} = \frac{q_{i,j,k}}{F_{Tk}\Delta T_{MLk}} \quad \text{for} \quad i \in I, j \in J, k \in K \quad (10.8)$$

$$q_{i,j,k} = \frac{CP_{i,k} CP_{j,k} \Delta T_{i,k}}{\sum_j CP_{j,k}} \quad \text{for} \quad i \in I, j \in J, k \in K \quad (10.9)$$

where A_{ci} is the contract area of the hot stream i and A_{cj} is the contract area of the cold stream j. For each enthalpy interval k, $U_{i,j,k}$ is the UA value for the match between streams i and j, $q_{i,j,k}$ is the amount of heat transferred from hot stream i to cold stream j, $CP_{i,k}$ is the heat capacity flow rate of hot stream i and $CP_{j,k}$ is the heat capacity flow rate of cold stream j. $\Delta T_{i,k}$ is the temperature change of any hot stream i, F_{Tk} is the correction factor for the log mean temperature difference, ΔT_{MLk} is the log mean temperature difference for any pair of streams.

If hot stream i is allocated in the shell side, the total stream pressure drop may be evaluated by the following equation [Linnhoff and Ahmad (1990)], [Serna-Gonzalez et al. (2004)].

$$\Delta P_i = K_i A_{ci} \left(\frac{1}{h_i} - R_{di} \right)^{-5.109} \quad \text{for} \quad i \in I \tag{10.10}$$

whereas if cold stream j is in the tube side,

$$\Delta P_j = K_j A_{cj} \left(\frac{D_{ti}}{D_t h_j} - R_{dj} \right)^{-3.5} \quad \text{for} \quad j \in J \tag{10.11}$$

where h_i and h_j are the dirt film heat transfer coefficient.

R_{di} is the fouling factor for hot stream i and R_{dj} is the fouling factor for hot stream j. Constants K_i and K_j depend on the stream physical properties and volumetric flow rate as well as the geometry of the heat exchanger.

The total cost of power, which is necessary for pumping the fluid streams, is given by:

$$C_P = \sum_i C_{P,i} Q_i \Delta P_i + \sum_j C_{P,j} Q_j \Delta P_j \tag{10.12}$$

where Q_i and Q_j are volumetric flow rate of the process streams i and j respectively. $C_{P,i}$ and $C_{P,j}$ are unit cost of power used by the streams i and j respectively.

For a given value of ΔT_{min}, Eqs (10.1)–(10.6) combined with Eqs (10.10)–(10.12) to develop a nonlinear programming (NLP) problem for minimization of the total annual cost of a HEN. The independent variables over which the minimization is carried out are the minimum network area, the stream contact areas, pressure drops of the streams and the film heat transfer coefficients. The values of the fixed parameters $q_{i,j,k}$, $UA_{i,j,k}$ and $UA_{i,j}$ have been calculated from Eqs (10.9), (10.8), and (10.7) respectively. Also, it should be noted that $Q_{H_{min}}$, $Q_{C_{min}}$, and $N_{u,mer}$ are calculated before the NLP problem is solved.

By varying the value of ΔT_{min} used for utility targeting and recalculating the unit, area and power targets, the trade-off between capital and operating cost can be predicted to find the cost-optimal value of ΔT_{min} for HEN design. The heat exchanger network synthesis (HENS) problem generally solved through sequential optimization [Floudas et al. (1986)] is incomplete, i.e., it is solved sub-optimally. In fact, minimizing the utilities consumption in the first step, then the number of matches required, and lastly the overall costs including the cost of the heat transfer area is definitely sub-optimal because the efficiency of each decision level depends on the quality of the solution found in the preceding decision level. In addition, the different economic issues involved are conjugated during the economic analysis of the energetic integration problem, which is not accomplished using the described series of partial sub-problems. The optimization problems developed for HEN synthesis are mostly MINLP. In mathematical programming methods, MINLP problem can be solved by deterministic, stochastic or coupling of them. Deterministic methods like GBD, OA etc are sometimes failed to converge owing to mixed nature of binary and continuous variables. Stochastic methods such as Simulated Annealing (SA), and Genetic Algorithm (GA) can solve these problems efficiently.

10.2 Distillation System Optimization

Continuous distillation is one of the most widely used separation techniques in the chemical process industry. As large amount of energy is required, the optimum operation of continuous distillation columns has economic importance for process industries. During optimization of continuous distillation columns, for a specified degree of separation the problem of finding the optimal values of: (i) the number of stages, (ii) reflux ratio, (iii) feed location(s), have been addressed. In a typical distillation column, the choice of the operating parameters include fixing of the feed location and reflux ratio, whereas the number of stages becomes the design variable. The total (minimum) reflux condition, gives the minimum (infinite) number of stages [Ramanathan et al. (2001)].

The objective of such an optimization is the determination of the optimal operating and design parameters for achieving the desired degree of separation at the lowest total cost. The total cost of a distillation operation is made up of two major components. In the first component, the annual running cost of utilities i.e., the heating media to produce vapours, and the cooling media for vapour condensation, are accounted for. The second cost component refers to the annual fixed charges that take into account the interest and depreciation on the installation cost of the column, condenser, and reboiler; this component also includes maintenance of the installed equipment.

For optimizing the continuous distillation columns, the mixed integer non-linear programming (MINLP) seems to be an attractive approach [Viswanathan and Grossmann (1993)]. The MINLP formalism can be entrapped into a locally optimum solution instead of the desired globally optimum one. Moreover, MINLP methods are complex, computationally intensive and many simplifications become necessary to make them affordable [Wang et al. (1998)]. It is noticed, from the above discussion, that a majority of studies on continuous distillation optimization include applications of the deterministic optimization methods (MINLP, Powell's method, tunneling method, etc.). These formalisms are mostly calculus-based involving direct computation of the gradient. The gradient-based techniques invariably require the objective function to be smooth, continuous and differentiable (well-defined slope values). When the objective function is multimodal, noisy, and fraught with discontinuities, simultaneous fulfillment of these criteria cannot be guaranteed, thus, leading to suboptimal solutions. For instance, if the search space includes mixed integer (e.g., the number of distillation stages) and continuous (e.g., reflux ratio) variables, then the objective function could be non-monotonic and possess multiple local minima.

Continuous Simple Distillation

A simple continuous distillation column is given Fig. 10.2. This distillation column consists of feed (F), bottom product or residue (D), and distillate (D).

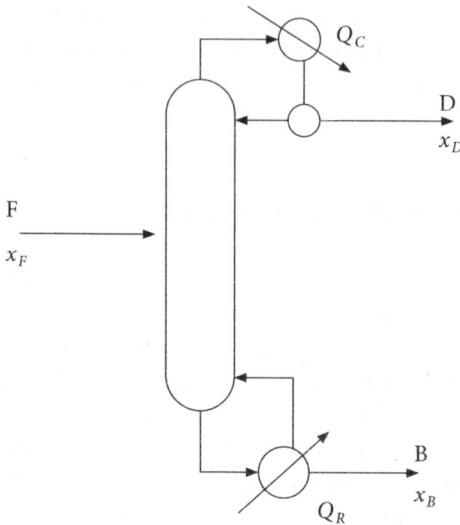

Fig. 10.2 Continuous distillation column

Problem Formulation

The objective function used in this study is representative of the total annual cost (C_T) that is made-up of two components, namely, the operating cost (C_1), and the fixed cost (C_2). While C_1 accounts for the energy cost pertaining to the reflux ratio and reboiler duty, the cost component C_2 accounts for the number of stages. The overall optimization objective is expressed as:

Minimize $C_T(x)$ $x_k^L \leq x_k \leq x_k^U$ (10.13)

where C_T is a function of the K-dimensional decision variable vector, $x = [x_1, x_2, ... x_k, ... x_K]^T$ and x_k^L and x_k^U respectively, refer to the lower and upper bounds on x_k. The three decision variables ($K = 3$) considered for optimization are: (i) the total number of stages (N), which is a function of real-valued x_1, (ii) reflux ratio (x_2), (iii) the feed location f_l (a function of x_3 and x_1). The evaluation procedure for the cost components C_1 and C_2 is discussed in Appendix I.

Computation of Total Cost

The objective function (C_T) for both, continuous simple and continuous azeotropic distillations represents the total annual cost ($). This cost comprises two additive components:

Total Cost, C_T = Energy Cost, C_1 + (depreciation + interest + maintenance) × (Fixed Cost, C_2)

(10.14)

where the energy cost, C_1, which is directly proportional to the heating cost is calculated according to:

$$C_1 = \frac{Q_R \times C_s \times N_D \times 24}{\lambda_{steam}} \tag{10.15}$$

where Q_R is the boiler duty; λ_{steam} is the latent heat of steam vaporization; C_s refers to the steam cost and N_D denotes the number of years working days.

The fixed cost, C_2 ($\$yr^{-1}$), consists of packing ($C_{pack}$) and column ($C_{col}$) cost where C_{pack} is computed as:

$$C_{pack} = A_c \times N \times HETP \times C_{pack}^o \tag{10.16}$$

Here, A_c representing the column area is calculated from the total vapor load on the basis of vapor velocity corresponding to the top temperature and capacity factor (C_f) of the packing ($C_f = 1.5$); C_{pack}^o denotes the packing cost per unit volume ($\$m^{-3}$) and the $HETP$ value for the packing is in meter. The second cost component of C_2 i.e., C_{col}, is calculated on the basis of internals from the following correlation.

$$C_{col} = 3.14 \times 1.4 \times d_c \times N_{st} \times HETP \times W_s \times \rho_s \times C_{steel} \tag{10.17}$$

In this expression, d_c, W_s, ρ_s and C_{steel}, refer to the column diameter, column thickness (m), density (kgm^{-3}) of the column material (steel) and its cost ($\$kg^{-1}$). Assuming depreciation, interest, and maintenance costs of d (%), I (%) and m (%), respectively, the total annual cost to be minimized is evaluated as:

$$C_T = C_1 + (d + I + m)C_2 \tag{10.18}$$

In the operation of an azeotropic distillation column, a small quantity of entrainer is lost through the vent condenser and bottom product. Thus, the cost of entrainer loss approximately amounting to 3 per cent of the total entrainer quantity, must be additionally considered while evaluating C_T:

$$C_T = C_1 + (d + I + m)C_2 + C_3 \tag{10.19}$$

where, C_3 ($\$kg^{-1}$) refers to the cost of entrainer, i.e., benzene.

The solution to the minimization problem defined in Eq. (10.13) should satisfy the following constraints:

Purity constraints:

$$x_d^{spec} - x_d^{simu} \leq 0 \tag{10.20a}$$

$$x_b^{spec} - x_b^{simu} \leq 0 \tag{10.20b}$$

where x_d^{spec} and x_b^{spec}, represent the desired top and bottom concentrations (mole %) of the more volatile component, and x_d^{simu} and x_b^{simu} refer to the optimized (simulated) values of the top and bottom concentrations of the more volatile component.

Equality constraints:

As defined by the material balance, equilibrium, summation of the mole fraction and heat balance (MESH) equations.

Component material balance:

$$M_{ij} = \left(1 + \frac{S_i}{V_i}\right) \times v_{ij} + \left(1 + \frac{s_i}{L_i}\right) \times l_{ij} - f_{ij} - l_{i+1,j} - v_{i-1,j} \tag{10.21}$$

$i = 2,\ldots,N-1;\ j = 1,2,\ldots,C$

Material balance for the reboiler and condencer:

$$M_{1j} = \left(1 + \frac{S_1}{V_1}\right) \times v_{1j} + \left(1 + \frac{s_1}{L_1}\right) \times l_{1j} - f_{1j} - l_{2,j} \tag{10.22}$$

$$M_{Nj} = \left(1 + \frac{S_N}{V_N}\right) \times v_{Nj} + \left(1 + \frac{s_N}{L_N}\right) \times l_{Nj} - f_{Nj} - v_{N-1,j} \tag{10.23}$$

where M_{ij} represents the discrepancy function expressed in terms of moles. hr^{-1}; S_i and s_i are the vapor and liquid side-streams, and f_{ij} refers to the feed flow.

Equilibrium relationship:

$$Q_{ij} = \frac{\eta_i \times m_{ij} \times V_i \times l_{ij}}{L_j} + \frac{(1+\eta_i) \times v_{i-1,j} \times V_i}{V_{i-1}} - V_{ij} \tag{10.24}$$

where Q_{ij} represents the discrepancy function (moles.hr^{-1}); η_i is the Murphree stage efficiency, and m_{ij} represents the equilibrium constant for the component j on the ith stage. The UNIQUAC method was used to calculate m_{ij}.

Energy balance equation:

$$E_i = \left(1 + \frac{S_i}{V_i}\right) \times H_i + \left(1 + \frac{s_i}{L_i}\right) \times h_i - h_{fi} - H_{i-1} - h_{i+1} \tag{10.25}$$

$i = 2,\ldots,N-1$

Energy balance equations for the reboiler and condenser:

$$E_1 = \left(1 + \frac{S_1}{V_1}\right) \times H_1 + \left(1 + \frac{s_1}{L_1}\right) \times h_1 - h_{f1} - h_2 \tag{10.26}$$

$$E_N = \left(1 + \frac{S_N}{V_N}\right) \times H_N + \left(1 + \frac{s_N}{L_N}\right) \times h_N - h_{fN} - H_{N-1} \tag{10.27}$$

where E_i represents the discrepancy function (kcal.hr^{-1}); H_i and h_i are the enthalpy values for vapor and liquid respectively. To solve these equations a linear pressure and temperature profile was assumed. The pressure on the ith stage (p_i) is given by:

$$p_i = P_1 - i \times \Delta p \tag{10.28}$$

where P_1 is the bottom pressure and Δp refers to the pressure drop across the stage. The initial guess value for the temperature at each stage is given by:

$$T_i = T_1 + \frac{(i-1) \times (T_N - T_1)}{(N-1)} \tag{10.29}$$

where T_N and T_1 are the temperature of the condenser and reboiler respectively, that assume value of the boiling points of the more volatile component and less volatile component.

Napthali and Sandholm method [Napthali and Sandholm (1971)] has been utilized for the simulation of simple multi-component distillation to solve the steady-state MESH equations for each plate. The Napthali and Sandholm (NS) method uses the Newton–Raphson technique to simultaneously solve all the variables in the MESH equations. For simulation purposes, the full NS matrix method for the continuous distillation simulation is combined with the UNIQUAC method for predicting the vapour-liquid equilibrium (VLE).

10.3 Reactor Network Optimization

The reactor network is a very important part of many process industries. Efficient conversion of raw materials to desired products in the reactor division can greatly influence the process energy use, separation requirements consequently the overall economics. A reactor network contains different types of reactor with different streams (i.e., inlet, outlet, recycle). Economic feasibility of a chemical process strongly depends on the efficiency of the reaction system. Reactor network optimization aims to identify the most effective conceptual reactor design in terms of mixing and feeding strategies [Ashley and Linke, (2004)]. The objective of reactor network synthesis is to determine the types, sizes, and operating conditions of the reactor units in addition to the interconnections between the reactors, which convert the given raw materials into the desired products [Schweiger and Floudas (1999)]. The basis for the conceptual approach is to look closely at the fundamental relationships of the reaction process. For reactor networks, performance is generally defined by objectives such as maximum yield, selectivity and conversion. These are all dependent on reaction rates that are based on component concentrations and temperature as described by the reaction kinetics.

The objective of reactor network synthesis is to determine the optimal mixing and feeding patterns that lead to the best possible performance attainable by a reactor network. The simplified models and representations are used to allow the quick identification of the optimal reactor network layout at a conceptual level. However, the optimization results need to be interpreted and translated into practical schemes by the engineer. Common features amongst optimal designs are serial arrangements of PFRs. This results from the discretization of PFRs by cascades of equal-volume CSTRs. Serial arrangements of PFRs allow for a finer discretization which leads to a better approximation of plug flow behaviour. Such structures should be interpreted by the design engineer as single PFRs.

Under the situation of known feed concentration and reaction kinetics, the job of reactor network synthesis (RNS) is to choose appropriate types of reactor, structures of the process and significant design parameters, which will optimize a particular objective function. Two important mathematical programming approaches for reactor networks synthesis are superstructure optimization and targeting [Jin *et al.* (2012)]

The superstructure based methods first introduce by Jackson in 1968 [Jackson, (1968)]. He considered a network composed of parallel plug flow reactors (PFRs) interconnected with side streams. Superstructure optimization involves generalized representations of ideal reactor units from which the optimal reactor mixing and feeding structure is to be extracted.

The superstructure representation consists of all possible combinations of reactor units. Reactor types are modelled as ideal CSTR, PFR and distributed side stream reactors (DSSR). Plug flow behaviour is approximated by a cascade of equal volume sub-CSTRs (Kokossis and Floudas, 1990). During optimization, it is ensured that there is always one reactor in existence, and that active reactors feature sequential connections with a minimum flow from the previous reactor. This maintains the expected sequential structure and allows optimization results to remain simple and interpretable.

Initially a network structure is recommended in superstructure optimization methods. Then from this initial network, an optimal sub-network that optimizes a preferred variable is obtained. The major advantages of superstructure optimization are that constraints can be included directly and conveniently, it is possible to modify the objective function and the optimal value of objective function and the reactor network can also be derived simultaneously. Although, the mathematical model of superstructure network is mostly a complicated non-linear programming problem (Schweiger and Floudas, 1999). This problem is very difficult to solve using the conventional optimization techniques. The topic is considered as an optimization problem where the objective function to be optimized is the yield or selectivity of a desired product, based on the complex reactions considered. The constraint problems are the material balances in the nodes of superstructure and the design equations of the reactors used. Kokossis and Floudas (1990) applied deterministic mixed integer nonlinear programming (MINLP) techniques to optimize superstructures comprising all possible combinations of continuous stirred tank reactors (CSTRs) and plug flow reactors (PFRs). More recently, stochastic techniques in the form of simulated annealing have been applied to the optimization of reactor network superstructures (Marcoulaki and Kokossis, 1996), with extensions to non-isothermal and multiphase systems by Mehta and Kokossis (2000).

Proposed superstructure

A simple superstructure is proposed [Silva *et al.* (2008)] to discuss the reactor network synthesis. Distinct possibilities of the typical CSTR-PFR arrangement have been considered. The most arrangement used in the literature are composed by a unique CSTR, a PFR solely, a CSTR followed by a PFR, a PFR followed by a CSTR or a CSTR and a PFR operating in parallel. Figure 10.3 shows us the proposed superstructure where all these combinations are available. The mathematical formulation is presented below.

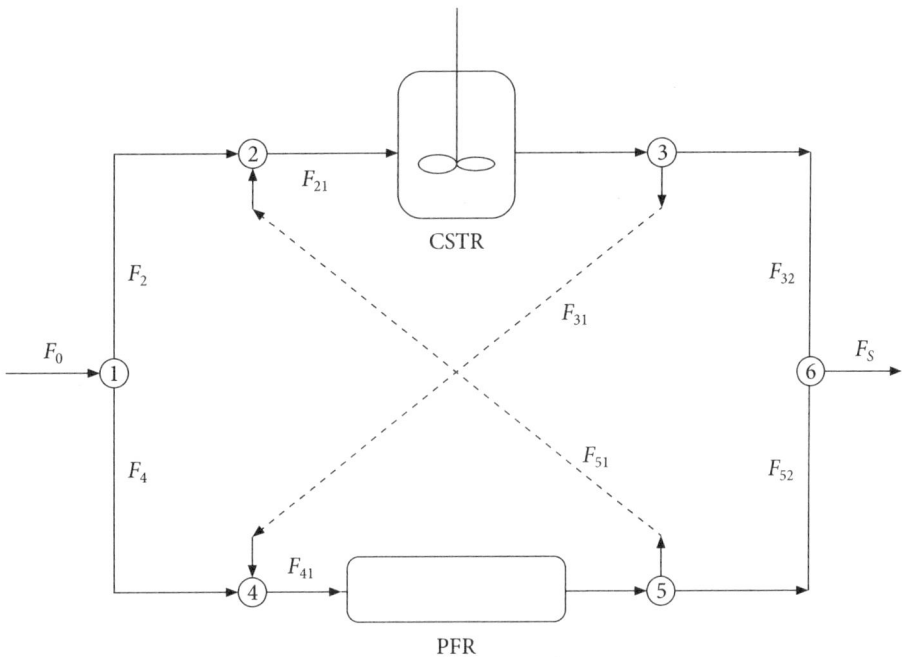

Fig. 10.3 Reactor network superstructure

Mathematical formulation

The formulation of mathematical model considers the mass balances in the nodes 1 to 6 of the superstructure. The feed introduced to the superstructure through the node 1. F_0 is the amount of initial feed, and it can be split randomly into two parts, F_2 and F_4. This can be written as:

$$F_2 = F_0 \times RAN1 \tag{10.30}$$

$$F_4 = F_0 - F_2 \tag{10.31}$$

Node 2 near the CSTR inlet is a mixing node. The inlet flow rate F_{21} is formed after the mixing the PFR outlet F_{51} and the branched flow rate F_2:

$$F_{21} = F_2 + F_{51} \tag{10.32}$$

Node 3 after the CSTR outlet is a splitting node. The CSTR outlet flow is F_3, which is equal to F_{21}, and can be split randomly into two parts, F_{31} and F_{32}:

$$F_{31} = F_3 \times RAN2 \tag{10.33}$$

$$F_{32} = F_3 - F_{31} \tag{10.34}$$

Again, node 4 near the PFR inlet is a mixing node. The inlet flow rate F_{41} is formed after the mixing of the CSTR outlet F_{31} and the branched flow rate F_4:

$$F_{41} = F_4 + F_{31} \tag{10.35}$$

And node 5 after the PFR outlet is a splitting node. The FPR outlet flow rate is F_5, which is equal to F_{41}, and can be split randomly into two parts, F_{51} and F_{52}:

$$F_{51} = F_5 \times RAN3 \tag{10.36}$$

$$F_{52} = F_5 - F_{51} \tag{10.37}$$

where random numbers $RAN1$, $RAN2$ and $RAN3$ are in between 0 and 1.

Node 6 is a splitting node and corresponds to the superstructure outlet. The outlet flow rate F_6 is represented by:

$$F_6 = F_{32} + F_{52} \tag{10.38}$$

These equations are the constraints for the superstructure. These constraints are used to formulate the optimization problem. The design equations developed for reactors used in this superstructure have to be considered as constraint equations. The design equation for the CSTR is:

$$\frac{V}{F} = \frac{x}{(-r)} \tag{10.39}$$

For the PFR, the design equation is:

$$F\frac{dx}{dV} = (-r) \tag{10.40}$$

where:

V = volume of reactor;

F = molar feed flow rate;

x = conversion;

$(-r)$ = rate of reaction.

The developed objective function depends on the reaction system used and it is solved to maximize yield and/or selectivity to a desired product. The superstructure equations and the algebraic and differential equations from the design of the reactors are used as constraints problem.

10.4 Parameter Estimation in Chemical Engineering

One of the fundamental tasks of engineering and science, and indeed of mankind in general, is the extraction of information from data. Parameter estimation is a subject that provides tools for the effective utilization of data in the evaluation of constants appearing in mathematical models and for assisting in modeling of phenomena [Beck and Arnold, (1977)]. The models may be in the form of algebraic, differential, or integral equations and their associated initial and boundary conditions.

Parameter estimation is a powerful tool for various engineering applications such as the design of process equipments. Actually, this technique offers the ability to estimate unknown parameters, which are often vital for the design and optimization. These methods use experimental data for determination of unknown parameters related to the model equations. During parameter estimation, we need to fit the experimental data with the model equation with minimum error. The name "parameter estimation" is not universally used. Other terms are non-linear least squares, non-linear estimation, non-linear regression.

10.4.1 Derivation of objective function

Mathematical models are normally utilized for many purposes such as analysis of experimental data, understanding the behavior of process, for process design, process optimization and process control. In most of the cases, we use the least square method for finding the best-fit model. The results found from the parameter estimation are somewhat uncertain because the data obtained from experiments are uncertain due to the presence of experimental errors. Characterization of this uncertainty is our primary concern for proper estimation of the ultimate results. Definition of the maximum likelihood function may be convenient to interpret the parameter estimation process [Schwaab et al. (2008)].

The experimental data is considered as random variables, whose joint probability distribution can be described by the following equation

$$P(z^e, z^*, V) \qquad (10.41)$$

which expresses the probability to get the experimental values z^e, with the unknown real values z^* and a measure of the experimental errors V. The maximum likelihood estimation consist in maximizing Eq. (10.41), considering the model constraints

$$g(z^*, \theta) = 0 \qquad (10.42)$$

where g is a vector of model functions and θ is a vector of model parameters. If we assume that the model is perfect and that the experiments have been performed properly, it is fair to acknowledge that the experimental outcomes are the most probable ones. Thus, efforts should be made for maximizing the probability of getting these experimental results (Brad, 1974). While, the experimental errors follow the normal distribution, maximization of the likelihood function can be written as the minimization of the following function

$$S(\theta) = \left(z^* - z^e\right)^T V^{-1} \left(z^* - z^e\right) \tag{10.43}$$

where z is a vector that contains the independent x and dependent y variables and V is the covariance of measurements. When the objective function defined in Eq. (10.43) is used, the process is usually called data reconciliation. Considering that the independent variables x are known with high precision, then the objective function becomes

$$S(\theta) = \left(y^* - y^e\right)^T V_y^{-1} \left(y^* - y^e\right) \tag{10.44}$$

and we can rewrite the model equations as

$$y^* = f\left(x^*, \theta\right) \tag{10.45}$$

it is assumed that the dependent variables can be estimated (experimentally or numerically) as function of the model parameters and of the independent variables.

Whenever, the experimental measurements of the dependent variables are uncorrelated, the matrix Vy is diagonal and Eq. (10.44) takes the form of the known weighted least squares function

$$S(\theta) = \sum_{i=1}^{NE} \sum_{j=1}^{NY} \frac{\left(y_{ij}^* - y_{ij}^e\right)^2}{\sigma_{ij}^2} \tag{10.46}$$

where σ_{ij}^2 is the variance of the experimental fluctuations of the dependent variable j in the experiment i. The number of experiments and the number of dependent variables are represented by NE and NY respectively. When all variances are equal (this is generally assumed for models that have only one dependent variable), then Eq. (10.46) can be simplified to take the shape of the well-known least square function

$$S(\theta) = \sum_{i=1}^{NE} \left(y_i^* - y_i^e\right)^2 \tag{10.47}$$

The sum over different dependent variables was eliminated as the role of the least squares function is normally unsuitable for multi-response models. This is due to variations among the magnitudes of the responses and because of the involvement of various physical units. Matrix V provides the proper normalization and dimensionalization of each term in the sum.

After specifying the objective function, various numerical methods can be employed to minimize the objective function. The conventional derivative-based methods are used very frequently. These methods say that the minimization is done along a direction that combines gradient vector (vector of 1st derivatives with respect to model parameters) and the Hessian matrix (matrix of 2nd derivatives with respect to model parameters) of the objective function. A class of methods called direct search methods (section 5.2) conduct the objective function minimization based on only the evaluation of the objective function, without the estimation of derivatives. Though, the concept of minimization without the computation of derivatives is very attractive, Bard [Bard (1974)] describes that gradient methods do better than direct search methods both in reliability and in speed of convergence. The direct and gradient search methods both are considered as local search methods as the search starts from an initial guess value and then results in a minimum. An excellent compilation of direct and derivative-based search processes has been discussed by Bard (1970, 1974).

For the application in chemical engineering field, minimization of the objective function in parameter estimation problems is very difficult and it may lead to severe numerical complication. These complexities arise due to many reasons such as large number of model parameters, a high correlation among the model parameters and multimodal nature of the objective function. The heuristic optimization methods such as SA, GA, and PSO (chapter 9) may be used to overcome these challenges. These algorithms are characterized by random search with the huge number of function evaluations. These techniques ensure a higher probability for finding the global minima, when compared to direct search and derivative-based search techniques. These algorithms do not need initial guess values for model parameters and do not utilize derivatives of the objective function.

These methods have been discussed in chapter 9. Pedro Mendes and Douglas B. Kell [Mendes and Kell (1998)] used to estimate the parameters for biochemical reaction. Their study shows that the Simulated Annealing and Levenberg-Marquardt methods (section 5.4) are most suitable for solving the optimization problem.

10.4.2 Parameter estimation of dynamic system

Problems on parameter estimation for nonlinear dynamic systems are described in this section [Moles et al. (2003)]. Here we have considered minimization of a cost function that determines the goodness of the fitted model with respect to a known experimental data set, subject to the system dynamics (as a set of differential equality constraints) in addition to other algebraic constraints. The mathematical representation of a nonlinear programming problem (NLP) with differential-algebraic constraints is as follows:

Find p to minimize

$$J = \int_0^{T_f} \left(y_{\text{msd}}(t) - y(p,t)\right)^T w(t) \left(y_{\text{msd}}(t) - y(p,t)\right) dt \tag{10.48a}$$

subject to

$$f\left(\frac{dx}{dt}, x, y, p, v, t\right) = 0 \tag{10.48b}$$

$$x(t_0) = x_0 \tag{10.48c}$$

$$h(x, y, p, v) = 0 \tag{10.48d}$$

$$g(x, y, p, v) = 0 \tag{10.48e}$$

$$p^L \leq p \leq p^U \tag{10.48f}$$

where J denotes the cost function to be minimized, p represents the vector of decision variables of the optimization problem; we have to estimate this set of parameters, y_{msd} is the experimental measure of a subset of the output state variables, $y(p, t)$ represents the predicted values for outputs from the model, $w(t)$ represents a matrix of scaling or weighting factors, x is the differential state variables, v denotes a vector of other parameters (usually time-invariant) that are not calculated, f is the set of algebraic and differential equality constraints explaining the dynamics of the system (i.e., the nonlinear process model), and h and g are the possible equality and inequality path and point constraints that represent additional requirements for the system performance. The vector p is subject to lower and upper bounds that can be considered as inequality constraints.

The formulated problem in Eq. (10.48) is a nonlinear programming problem (NLP) which has differential-algebraic (DAEs) constraints. Very often, these problems are multimodal (non-convex) due to the nonlinear and constrained nature of the system dynamics. Therefore, if these NLP-DAEs are solved using typical local optimization methods, for example the standard Levenberg–Marquardt method, it is obvious that the found solution will be of local nature, as explained by Mendes and Kell (1998). The simplest and earliest effort to overcome this non-convexity of many optimization problems have been developed based on the concept of repeatedly use of a local optimization algorithm, starting from a number of different starting point.

Example 10.1

A parameter estimation study has been given by Moles *et al.* [Moles *et al.* (2003)]. They estimated the parameters for biochemical pathways that consists of 36 kinetic parameters.

The mathematical formulation of this nonlinear dynamic model is:

$$\frac{dG_1}{dt} = \frac{V_1}{1 + \left(\frac{p}{Ki_1}\right)^{ni_1} + \left(\frac{Ka_1}{S}\right)^{na_1}} - k_1 G_1 \tag{10.49}$$

$$\frac{dG_2}{dt} = \frac{V_2}{1 + \left(\frac{p}{Ki_2}\right)^{ni_2} + \left(\frac{Ka_2}{M_1}\right)^{na_2}} - k_2 G_2 \tag{10.50}$$

$$\frac{dG_3}{dt} = \frac{V_3}{1 + \left(\frac{p}{Ki_3}\right)^{ni_3} + \left(\frac{Ka_3}{M_2}\right)^{na_3}} - k_3 G_3 \tag{10.51}$$

$$\frac{dE_1}{dt} = \frac{V_4 G_1}{K_4 + G_1} - k_4 E_1 \tag{10.52}$$

$$\frac{dE_2}{dt} = \frac{V_5 G_2}{K_5 + G_2} - k_5 E_2 \quad (10.53)$$

$$\frac{dE_3}{dt} = \frac{V_6 G_2}{K_6 + G_3} - k_6 E_3 \quad (10.54)$$

$$\frac{dM_1}{dt} = \frac{kcat_1 E_1 \left(\frac{1}{Km_1}\right)(S - M_1)}{1 + \frac{S}{Km_1} + \frac{M_1}{Km_2}} - \frac{kcat_2 E_2 \left(\frac{1}{Km_3}\right)(M_1 - M_2)}{1 + \frac{M_1}{Km_3} + \frac{M_2}{Km_4}} \quad (10.55)$$

$$\frac{dM_2}{dt} = \frac{kcat_2 E_2 \left(\frac{1}{Km_3}\right)(M_1 - M_2)}{1 + \frac{M_1}{Km_3} + \frac{M_2}{Km_4}} - \frac{kcat_3 E_3 \left(\frac{1}{Km_5}\right)(M_2 - P)}{1 + \frac{M_2}{Km_5} + \frac{P}{Km_6}} \quad (10.56)$$

where M_1, M_2, E_1, E_2, E_3, G_1, G_2 and G_3 represent the concentration of the component involved in the various biochemical reactions and S and P keep fixed initial values for each experiment (i.e., parameter under control). The remaining 36 parameters are considered to construct the optimization problem. These parameters are divided into two different categories: Hill coefficients, allowed to vary in the range $(0.1, 10)$ and all the others, allowed to vary in the range $(10^{-12}, 10^{+12})$.

The global optimization problem is represented as the minimization of a weighted distance measure, J, between predicted and experimental values of the 8 state variables, given as the vector y

$$J = \sum_{i=1}^{n} \sum_{j=1}^{m} w_{ij} \left\{ \left[y_{\text{pred}}(i) - y_{\text{exp}}(i) \right]_j \right\}^2 \quad (10.57)$$

where n and m are the number of data for each experiment and the number of experiments respectively, y_{exp} is the known experimental data, and y_{pred} represents the vector of states that corresponds to the predicted theoretical evolution using the model with a given set of 36 parameters. Additionally, w_{ij} represents the different weights considered to normalize the contributions of each term.

$$w_{ij} = \left(\frac{1}{\max\left[y_{\text{exp}}(i) \right]_j} \right)^2 \quad (10.58)$$

This problem (10.57–10.58) can be solved using nonlinear programming.

10.5 Environmental Application

In recent times, environmental pollution is the main issue for process industries. We need to reduce the quantity of effluent from industries and maintain the effluent standard as per environmental regulation. Optimization methods have been implemented successfully in the field of environmental pollution control. There are various ways by which environmental issues can be tackled. Upper limits of concentrations or pollutant flows in waste streams are set based on regulatory requirements. The designs that satisfy these constraints are evaluated in terms of economic indicators where we can include the cost of waste treatment and disposal in the economic objective function (Papalexandri and Pistikopoulos, 1996; Diwekar *et al.* 1992). The environmental issues can be implemented during process design of a chemical industry as constraints with the economic optimization, as objectives or as trading off the environmental objectives against other design objectives [García and Caballero (2012)]. Zhang *et al.* [Zhang *et al.* (2014)] have developed a multi-objective optimization method for municipal wastewater treatment. They simultaneously optimized the treatment cost and multiple effluent quality indexes (including effluent COD, $NH_4^+ - N$, $NO_3 - N$) of the municipal wastewater treatment plant (WWTP).

Air pollution control

Controlling the air pollution emissions is required to keep our environmental clean. There are various control methods that can be utilized to control air pollution emissions. Shaban *et al.* [Shaban *et al.* (1997)] formulated a mixed integer linear programming model that determines the best selection strategy. The objective of the program is to minimize the total control cost consisting of operating and investment costs. The development of the model has been discussed in following section. There are many constraints that are inflicted on the model including allocation constraints, time limits for utilization of each control process, maximum funds available for investment, and a pollution abatement level prescribed by regulatory body.

Model development

For developing the model, we are considering various emission sources; a number of pollutants are emitted from each source. There are many control options/technologies available for a given emission source. Updated and retrofit emission control devices in addition to improved operating procedures are among these pollution control technologies. A certain cost and reduction capacity is associated with each of these options. It is preferred to choose the best combination of control devices in an attempt to decrease emissions to a desired limit and, as a result, the total cost of the pollution control system is minimized. A mixed integer linear programming (MILP) model has been developed that will decide the optimal selection strategy. Different variables and parameters used to develop the model will be defined as:

Sets

I_i set of control options, which can be utilized on source i

J_i set of pollutants discharged from source i

K_j set of sources on which control j can be employed

Variables

T_j set-up time of control j (at which time the control device j is being made available).

T_{ij} time-length for which control j is utilized on source i.

x_{ij} binary variable that indicates whether control process j is used on source i ($x_{ij} = 1$) or not ($x_{ij} = 0$).

Parameters

C_j^0 installation or set-up cost of control device j.

C_j^{max} maximum set-up cost of control j (cost when control j is made available by its earliest possible availability time).

C_{ij} operating cost per unit time when control j is being used on source i.

B budget available for development of the control processes.

T_i start-up time for pollution source i.

T_j^{min} earliest time control j can be made accessible.

T_j^{max} latest time control j can be made accessible.

T length of time horizon of interest.

R_{ikj} total reduction per year of pollutant k in source i when control j is employed on source i.

K desired overall reduction of pollutant k.

P total number of pollutants from the various sources

Here, i represents the pollution source, j denotes the pollution control device, and k represents single pollutant.

Formulation of objective function

For each pollution source i, it is required to choose one control j in such a way that the total pollution cost will be minimized. The total cost is composed of the initial investment cost and the operating cost. The operating cost of all control processes over all pollution sources is represented by the following equation:

$$\text{Operating cost} = \sum_j \sum_{i \in K_j} C_{ij} T_{ij} \quad (10.59)$$

The investment cost is divided into two parts. A minimum fixed cost due to installation or set-up and an acceleration cost that is incurred when the set-up cost of a given control is speeded up. The set-up cost is given by:

$$\text{Set-up cost} = \sum_j C_j^0 \left(\sum_{i \in K_j} x_{ij} \right) \quad (10.60)$$

Whenever a control device j is not utilized on any of the pollution sources i, then in Eq. (10.60)

$$\sum_{i \in K_j} x_{ij} = 0 \quad (10.61)$$

Subsequently, the set-up cost associated with control j is reduced to zero as it should be.

The accelerated cost of a control j depends on how fast the set-up of the control is. For example, this cost is equal to its maximum C_j^{max} when the set-up time of control j is equal to its earliest possible set-up time T_j^{min}. Also, the acceleration cost attains a minimum when the set-up time of control j is at its latest.

If we consider a linear relationship between set-up time and acceleration cost, then it can be expressed as:

$$\text{Acceleration cost} = aT_j + b \tag{10.62}$$

where

$$a = -\frac{C_j^{max} - C_j^{min}}{T_j^{max} - T_j^{min}} \tag{10.63}$$

$$b = -\frac{T_j^{max} C_j^{max} - T_j^{min} C_j^{min}}{T_j^{max} - T_j^{min}} \tag{10.64}$$

The decision variable T_j is as given before, the set-up time or time at which control j is being made accessible.

In order to make sure that the acceleration cost is always zero in the situation when control device j is not in use, Eq. (10.62) is modified as:

$$\text{Acceleration cost} = aT_j + b\sum_{i \in K_j} x_{ij} \tag{10.65}$$

If control device j is not selected, the 2nd term in the Eq. (10.65) is equal to zero

$$\left(\sum_{i \in K_j} x_{ij} = 0\right) \tag{10.66}$$

In such a case, the 1st term to be equal to zero, a condition should be imposed on the availability time T_j. This condition will be dealt with as constraints as discussed below. The objective function becomes

$$\min \sum_j \left\{ \left(aT_j + b\sum_{i \in K_j} x_{ij} + C_j^0 \sum_{i \in K_j} x_{ij} \right) + \sum_{i \in K_j} C_{ij} T_{ij} \right\} \tag{10.67}$$

Constraints

i. For each control device j to be considered for use on a source i; T_i, the start-up time of source i must be greater than T_j^{min}, the earliest availability time of control j:

$$x_{ij}\left(T_j^{min} - T_i\right) \leq 0 \quad \forall j, i \in K_j \tag{10.68}$$

ii. For each source i, at most one control j can be used:

$$\sum_{j \in I_i} x_{ij} \leq 1 \quad \forall i \qquad (10.69)$$

iii. Each control process j can be used at most once:

$$\sum_{i \in K_j} x_{ij} \leq 1 \quad \forall j \qquad (10.70)$$

iv. Whenever, a control process j is not used for any of the pollution sources, then, the time T_j at which control j is being made available must be set to zero with the intention that control j will not have any effect on the overall cost. We can write this in equation form:

$$M \sum_{i \in K_j} x_{ij} \geq T_j \quad \forall j \qquad (10.71)$$

where the parameter M signifies a large positive number. M should be set to the upper bound on T_j (i.e., T_j^{max}) in order to have a tight constraint set. As given above, whenever control j is not utilized on any of the sources $i \in K_j$, then constraints in Eq. (10.71) ensure that $T_j = 0$. However, if control j is used on any of the sources i, then the constraint is rendered redundant and is always satisfied due to the presence of the large number M.

v. When a control process j is selected to utilize on source i (i.e., $x_{ij} = 1$), then the availability time of control j should be larger than the start-up time for source i (i.e., $T_j \geq T_i$). This can be given as the following equation:

$$T_j - x_{ix} T_i \geq 0 \quad \forall j, i \in K_j \qquad (10.72)$$

vi. For each control process j, the availability time T_j is bounded between a minimum time T_j^{min} and a maximum time T_j^{max} (i.e., $T_j^{min} \leq T_j \leq T_j^{max}$). However, the lower and upper bounds on j should be relaxed to zero when the control process j is not selected for use. This can be accomplished by the following:

$$\left(\sum_{i \in K_j} x_{ij}\right) T_j^{min} \leq T_j \leq \left(\sum_{i \in K_j} x_{ij}\right) T_j^{max} \quad \forall j \qquad (10.73)$$

vii. When control j is used on source i, then $T_j \geq T_i$ and the time increment that j is used on i is $T_{ij} = T - T_j$. If j is not used on i, then $T_{ij} = 0$. The time period for use by each control process j on pollution source i is given below:

$$T_{ij} = (T - T_j) x_{ij} \quad \forall j, i \in K_j \qquad (10.74)$$

The above set of constraints is nonlinear. Since, linear model equations are much easier to solve than nonlinear models, it is always advantageous to rearrange nonlinear constraints in the form of linear equations if it is possible [Glover, (1975); Ashford and Daniel, (1992)]. The method described by Raman and Grossman (1991) is used to rearrange the constraint given by Eq. (10.74) into linear form. For this reason, x_{ij} is associated with $T_{ij} \geq 0$ and $(1 - x_{ij})$ with $T_{ij} \leq 0$ and the constraint set is rewritten as:

$$L_1(1 - x_{ij}) \leq T_{ij} - T + T_j \leq U_1(1 - x_{ij}) \tag{10.75}$$

and

$$L_2 x_{ij} \leq T_{ij} \leq U_2 x_{ij} \quad \forall j, i \in K_j \tag{10.76}$$

where L_1 = lower bound on $T_{ij} - T + T_j$, U_1 = upper bound on $T_{ij} - T + T_j$, L_2 = lower bound on T_{ij}, and U_2 = upper bound on T_{ij}.

Note that in the case when $x_{ij} = 1$, constraints 10.75, 10.76 reduce to:

$$0 \leq T_{ij} - T + T_j \leq 0 \tag{10.77}$$

and

$$L_2 \leq T_{ij} \leq U_2 \tag{10.78}$$

or simply $T_{ij} - T - T_j$, which is the desired result.

If, however, $x_{ij} = 0$, then constraints 10.75, 10.76 are rendered redundant due to proper use of the lower and upper bounds L_1, L_2, U_1, and U_2. Suitable choices for these bounds are discussed in Shaban *et al.* (1997).

viii. There is a certain budget B for process development; the investment to develop the control processes should not exceed the budget:

$$\sum_j \left\{ \left(aT_j + b \sum_{i \in K_j} x_{ij} \right) + C_j^0 \sum_{i \in K_j} x_{ij} \right\} \leq B \tag{10.79}$$

ix. It is required to reduce a given pollutant released to a designated reduction level:

$$\sum_j \sum_{i \in K_j} R_{ijk} T_{ij} \geq K \quad \forall k = 1, 2, \ldots, P \tag{10.80}$$

x. Non-negativity and integrality:

$$T_{ij} \geq 0 \quad \forall j, i \in K_j \tag{10.81a}$$

$$T_j \geq 0 \quad \forall j \tag{10.81b}$$

$$x_{ij} = 0 \text{ or } 1 \quad \forall j, i \in K_j \tag{10.81c}$$

This problem is a MILP which can be solved by Branch and Bound method discussed in chapter 6.

Summary

- This chapter explains different optimization methods applied to the chemical and biochemical engineering. Heat exchanger network (HEN) and reactor network (RN) are the integrated part of any chemical process. HEN and RN have been optimized to get maximum profit. Distillation system is also optimized in this chapter. An optimized multi-component distillation system has been developed, which can be used in process industry. Parameter estimation is a very familiar technique to find the correlation between theoretical model and experimental data. Optimization methods can be used to find the best-fit model for a process. Beside these economic considerations, environmental pollution is required to consider during process development. We can incorporate this environmental issue with the design equation during its implementation.

Review Questions

10.1 What are the important factors we need to consider during heat exchanger network design?

10.2 Develop an optimization problem, which minimizes the operating cost a heat exchanger.

10.3 Draw the HEN superstructure with four heat exchangers.

10.4 Develop the model equation for a continuous distillation column with two feed inlet. How do you select the feed inlet position?

10.5 Discuss the algorithm of Napthali and Sandholm for solving MESH equations.

10.6 Develop a reactor network superstructure as shown in Fig. 10.3, considering both CSTR with volume V_1 and V_2.

10.7 Data is given below; use least square method to find the linear model.

x	3	5	7	9	10	15	20	22	25
y	12	18	22	32	38	53	67	72	78

10.8 Develop a model for wastewater treatment plant using MOO methods.

References

Ashford, R. W. and Daniel, R. C. 1992. *Some Lessons in Solving Practical Integer Programs.* J. Opl. Res. Soc. 43, 425–33.

Ashley, V. and P. Linke. 2004. *A Novel Approach to Reactor Network Synthesis using Knowledge Discovery and Optimisation Techniques.* Chemical Engineering Research and Design, 82(8): 952–60.

Brad, Y. 1974. Nonlinear Parameter Estimation, Academic, Orlando, Fla.

Brad, Y. 1970. *Comparision of Gradient Methods for the Solution of Nonlinear Parameter Estimation Problems.* SIAM J. Numer. Anal., 7: 157.

Beck, J. V., Arnold, K. J. 1977. *Parameter Estimation in Engineering and Science,* John Wiley and Sons.

Cerda, J. and A. W. Westerberg. 1983. *Synthesizing Heat Exchanger Networks Having Restricted Stream/Stream Match Using Transportation Problem Formulations.* Chem. Engng. Sci., 38: 1723–40.

Cerda, J., A. W. Westerberg, D. Masonand, B. Linnhoff. 1983. *Minimum Utility Usagein Heat Exchanger Network Synthesis a Transportation Problem.* Chem. Engng Sci., 38: 373–87.

da Silva, L. K., M. A. da Silva Sá Ravagnani. 2008. *Giovani Pissinati Menoci and Marcia Marcondes Altimari Samed; Reactor network synthesis for isothermal conditions;* Maringá, vol. 30, No. 2, pp. 199–207, Acta Sci. Technol.

Diwekar, U. M., Frey, H. C., Rubin, E. S. 1992. *Synthesizing Optimal Flowsheets: Applications to IGCC System Environmental Control,* Ind. Eng. Chem. Res., 31(8): pp. 1927–36.

Floudas, C. A., Ciric, A. R., Grossmann, I. E. 1986. *Automatic Synthesis of Optimum Heat Exchanger Network Configurations,* AICHE Journal, 32(2): 276–90.

García, N., José, A., Caballero. 2012. *How to Implement Environmental Considerations in Chemical Process Design: An Approach to Multi-Objective Optimization for Undergraduate Students,* Education for Chemical Engineers, 7: pp. 56–67.

Glover, F. 1975. *Improved Linear Integer Programming Formulation of Nonlinear Integer Problems.* Manage. Sci., 22: 455–60.

Gundersen, T. and Grossmann, I. E. 1990. *Improved Optimization Strategies for Automated Heat Exchanger Network Synthesis through Physical Insights.* Computers and Chemical Engineering, 14(9): 925–44.

Gundersen, T., Duvold, S. and Hashemi-Ahmady, A. 1996. *An Extended Vertical MILP Model for Heat Exchanger Network Synthesis,* Computers and Chemical Engineering, 20(Suppl.): S97–S102.

Hohmann. 1971. *Optimum Networks for Heat Exchange,* PhD Thesis, University of S. California.

Jackson, R. 1968. *Optimization of Chemical Reactors with Respect to Flow Configuration.* J. Opt. Theory Applic. 2: 240–59.

Jin, S., Li X., Tao S. *Globally Optimal Reactor Network Synthesis Via the Combination of Linear Programming and Stochastic Optimization Approach*, Chemical Engineering Research and Design 90(2012): 808–13.

Kokossis, A. C. and Floudas, C. A. 1990. *Optimization of Complex Reactor Networks-I, Isothermal Operation,* Chem. Eng. Sci., 45: 595–614.

Linnhoff, B. and J. R. Flower. 1978. *Synthesis of Heat Exchangers Networks, Systematic Generation of Energy Optimal Networks,* AIChE J., 24: 633–42.

Linnhoff, B. and Ahmad S. *Cost Optimum Heat Exchanger Networks: I. Minimum Energy and Capital Using Simple Models for Capital Costs.* Computers and Chemical Engineering, 14(1990): 729.

Marcoulaki, E., Kokossis, A. 1996. *Stochastic Optimization of Complex Reaction Systems.* New York: Comput. Chem. Eng., vol. 20, Suppl., pp. S231–36.

Mehta, V. L. and Kokossis, A. C. 2000, *Nonisothermal Synthesis of Homogenous and Multiphase Reactor Networks,* AIChE J, 46(11): 2256–73.

Mendes, P. and Kell, D. B. 1998. *Non-linear Optimization of Biochemical Pathways: Application to Metabolic Engineering and Parameter Estimation,* Bioinformatics 14(10): 869–83.

Moles, C. G., Mendes, P. and Banga J. R. *Parameter Estimation in Biochemical Pathways: A Comparison of Global Optimization Methods,* Genome Research, 13(2003): 2467–74.

Napthali, L. M. and Sandholm, D. P. 1971. *Multicomponent Separation Calculations by Linearization,* AIChE J, 17: 1148.

Papalexandri K. P., Pistikopoulos E. N. 1996. *Generalized Modular Representation Framework for Process Synthesis,* AIChE J., 42(4): 1010–32.

Papoulias, S. A., Grossmann, I. E. *A Structural Optimization Approach in Process Synthesis–II,* Heat Recovery Networks Computers and Chemical Engineering, 76(1983): 707–21.

Ramanathan, S. P., Mukherjee S., Dahule R. K., et al. *Optimization of Continuous Distillation Columns Using Stochastic Optimization Approaches,* Trans IChemE, vol. 79, Part A, April 2001.

Raman, R. and Grossmann, I. E. 1991. *Relation between MILP Modeling and Logical Inference for Chemical Process Synthesis,* Comp. Chem. Eng., 15: 73–84.

Schwaab, M., Evaristo Chalbaud Biscaia, Jr., José Luiz Monteiro, José Carlos Pinto. *Nonlinear Parameter Estimation through Particle Swarm Optimization;* Chemical Engineering Science, 63(2008): 1542–52.

Schweiger, C. A. and Floudas C. A. 1999. *Synthesis of Optimal Chemical Reactor Networks,* Computers and Chemical Engineering Supplement, pp. S47–S50.

Schweiger, C. and Floudas, C. 1999. *Optimization Framework for the Synthesis of Chemical Reactor Networks.* Industrial Engineering and Chemical Research, 38(3): 744–66.

Shaban, H. I., A. Elkamel, R. Gharbi. 1997. *An Optimization Model for Air Pollution Control Decision Making;* Environmental Modelling and Software, vol. 12, No. 1, pp. 51–58

Serna-Gonzalez, M., Ponce-Ortega, J. M. and Jiménez-Gutiérrez A. *Two Level Optimization Algorithm for Heat Exchanger Networks Including Pressure Drop Considerations,* Industrial and Engineering Chemistry Research, 43(2004): 6766.

Serna-González, M., José María Ponce-Ortega. 2010. *Oscar Burgara-Montero; Total Cost Targeting for Heat Exchanger Networks Including Pumping Costs,* 20th European Symposium on Computer Aided Process Engineering – ESCAPE20.

Viswanathan, J. and Grossmann, I. E. 1993. *Optimal Feed Locations and Number of Trays for Distillation Columns with Multiple Feeds,* Ind. Eng. Chem. Res., 32: 2943–49.

Wang, K., Qian, Y., Yuan, Y. and Yao, P. 1998. *Synthesis and Optimization of Heat Integrated Distillation Systems Using an Improved Genetic Algorithm,* Comput. Chem. Eng., 23: 125–36.

Zhang, R., Wen-Ming Xie, Han-Qing Yu, Wen-Wei Li. *Optimizing Municipal Wastewater Treatment Plants Using an Improved Multi-Objective Optimization Method,* Bioresource Technology, 157(2014): 161–65.

Chapter 11

Statistical Optimization

Large number of experimental run is required to develop the input-output relationship of any process. Utilization of statistical methods can reduce the number of the experimental run. This chapter elucidates the design of experiment for optimize use of experimental data. Response surface methodology is used to optimize the process output. A process plant can be optimized efficiently using these methods.

11.1 Design of Experiment

A huge number of experiments are required in research, development and optimization of any system. This research is carried out in labs, pilot plants or full-scale plants. An experiment may be a model based experiment or experiment with the physical system [Lazić, (2004)]. When the experimentation cost is very high, it is very difficult for the researcher to examine the numerous factors that have an effect on the processes using trial and error methods within a short time and limited resources. As an alternative, we need a technique that identifies some key factors in the most effective way. Then this leads the process to the best setting to satisfy the increasing demand for increased productivity with improved quality.

A one-factor-at-a-time experiment is used to study the effect of various factors during experimentation. This method involves changing a single factor at a time to analyze the impact of that factor on the product or process output. The main advantage of "one-factor-at-a-time" experiment is that we can perform the experiment easily. Despite this fact, these methods do not permit us to investigate of how a factor influences the process or a product in the presence of other factors. When the response of a process is changed due to the existence of one or more other factors, that relationship is called an interaction. Sometimes the effects of interaction terms are more significant compared to the individual effects. Because in the application environment of the process or product, most of the factors present together rather than the isolated incidents of single factors at different times. A chemical reacting system can be considered as an example of interaction between two factors. The reaction rate slightly increases when the temperature is increased and there is no effect of pressure on reaction rate. However, the reaction rate changes rapidly when both the temperature and pressure are changing simultaneously. In this system, an interaction does exist between the two factors that affect the chemical reaction. The major disadvantage of the one-factor-at-a-time strategy is that it is not able to consider any probable interaction between various factors. An interaction is the failure of the one factor to provide the same influence on the response at various levels of another factor [Montgomery (2001)]. The DOE technique guarantees that all factors and their interactions are investigated methodically. The one-factor-at-a-time experiments overlook the interactions and may direct to misleading outcomes. Therefore, information collected from the analysis of DOE is much more complete and reliable.

As the experiment is considered an essential part of science and technology, the question of efficiency of utilizing an experimental data is therefore, necessary. It is found that very less percentage of experimental data is used efficiently. To enhance the efficiency of research, it is essential to introduce something novel into conventional experimental research. The degree of precision and completeness of experimental data and information about the system that is being tested are used to determine the efficiency of experimental research. Fisher [Fisher 1926,1958,1960] methodically introduced the idea and theories of statistics into designing experimental investigations, which includes the concept of factorial design and the analysis of variance.

Even though applications of statistical design started in1930s, it was accelerated by the development of response surface methodology (RSM) by Box and Wilson (1951).

For the design of experiment, the main purposes are:

i. Decrease or minimization of total number of experiments
ii. Varying of all factors simultaneously that formalizes the activities of experimenter
iii. Selection of a proper strategy that facilitates reliable solutions to be found after each series of experiments

Therefore, the following factors should be considered before a design can be chosen

a. Choice of factor, levels, and range
b. Selection of the response variable
c. Selection of experimental design
d. Performing the experiment efficiently
e. Statistical analysis of the data

11.1.1 Stages of DOE

The DOE method consist five steps namely planning, screening, optimization, robustness testing and verification.

i. **Planning**

   Before starting the process of data collection and testing, it is essential to plan the experimentation very carefully. The usefulness of any experimental data depends on how it is designed. During planning, we have to keep in mind some of these considerations i) a thorough and precise objective identifying is required to perform the investigation, ii) estimation of resources and time available to accomplish the objective and iii) collection of prior experience with the experimentation method.

We have to create a team that will identify probable factors for investigation and the most suitable response(s) to measure. This team should be comprised of individuals from various disciplines related to the process or product. The knowledge from various disciplines can be gathered from those team members. The teamwork strategy encourages synergy, which provides a richer set of factors for analyzing and consequently a more comprehensive experiment. We will get better understanding of the process or product when the experiment is planned carefully. During implementation, a well-planned experiment is easy to perform and evaluate. Conversely, botched experiments may result in inconclusive data sets that may not be possible to analyze even with an excellent statistical tool (www.weibull.com).

ii. **Screening**

The DOE method starts with the screening of variables. Screening experiments are required to distinguish the significant factors that influence the process under consideration out of the large pool of potential factors. Prior knowledge of the process is required to perform the experiments that eliminate insignificant factors and focus interest on the key factors that require further thorough analyses. This procedure use a fractional factorial design, since only a fraction of the possible values of the corners are investigated. Very few executions are required as screening experiments are usually efficient designs. Our focus is not on studying the effect of interactions but on identifying the vital few factors.

iii. **Optimization**

After proper selection of the relevant factors that are influencing the process, the next step is to decide the best setting of these factors to produce the preferred objective. The objective of the study depends on the process or product under investigation, it may be either to increase yield or to decrease variability or to get settings that will give us both at the same time.

iv. **Robustness Testing**

Robustness is defined as the degree to which a system operates correctly in the presence of exceptional inputs or stressful environmental conditions [IEEE Std 24765: 2010]. After deciding the optimal settings of the factors, it is required to make the process or product insensitive to variations that may arise in the application environment. These variations occur due to the fluctuations in factors that influence the process, but they are beyond our control. These factors (e.g., humidity in atmosphere, ambient temperature, variation in raw material, etc.) are called the uncontrollable factors or noise. It is necessary to recognize the sources of these variations and take actions to assure that the process or product is made insensitive (or robust) to these factors.

v. **Verification**

Validation of the best settings is required to verify that whether the process functions as desired and all objectives are satisfied. The final stage involves this validation by performing a small number of follow-up experimental runs.

11.1.2 Principle of DOE

The three basic stages of experimental design are replication, randomization, and blocking. The basic experiment is repeated by using replication. There are two essential properties of replication. The first one is that it helps the experimenter to get an assessment of the experimental error. Secondly, whenever the sample mean ($\bar{y}$) is employed to evaluate the effect of a factor in the experiment, replication allows the experimenter to get a more accurate estimation of this effect. For example; if σ^2 is the variance of an individual observation and n is the number of replicates, the variance of the sample mean is

$$\sigma_{\bar{y}}^2 = \frac{\sigma^2}{n} \tag{11.1}$$

Randomization is the fundamental issue for using any statistical methods in experimental design. Using randomization process, we randomly determine both the distribution of the experimental material and the order in which the individual runs or trials of the experiment are to be executed. In statistical methods, the observations (or errors) should be independently distributed random variables. By randomization, we make this assumption valid. We also help in "average out" the effects of extraneous factors that may be present by properly randomizing the experiment [Montgomery, (2001)].

Blocking is a procedure that is employed to increase the accuracy with which comparisons are made among the factors of interest. Blocking is also used to diminish or remove the variability contributed by nuisance factors; that the factors may affect the experimental response but in which we are not interested directly. For instance, to complete all experimental runs in a chemical process industry it may require more than one batch of raw material. Though, there might be variations between those batches due to source to source variability. Whenever we are not interested to study particularly this effect, we can consider these differences between the batches of raw material as a nuisance factor. Usually, a set of relatively homogeneous experimental conditions is considered as a block. The variability within the same batch of raw material would be expected smaller than the variability between different batches. Therefore, each batch of raw material would form a block in the chemical process industries.

Computer software programs are extensively used for assisting experimenters to select and construct experimental designs. Commercial software like MINITAB (chapter 12) is available to design the experiment.

Example 11.1

A chemical engineer is interested to investigate the reaction rate as a function of pH.

This is a single factor experiment with 5 level of the factor and 5 replicates. We conducted the experiment with five different pH level and repeated 5 times in each pH level.

Table 11.1 Single factor experiment with 5 level of the factor and 5 replicates

pH	Experimental run number				
3	1	2	3	4	5
5	6	7	8	9	10
7	11	12	13	14	15
9	16	17	18	19	20
11	21	22	23	24	25

Now we arrange the experimental number randomly (randomization). After randomization, the 5th observation comes first. The randomized data is given in Table 11.2.

Table 11.2 Arrangement of experimental run after randomization

Test sequence	Run number	pH
1	5	3
2	7	5
3	21	11
4	16	9
5	3	3
6	19	9
7	20	9
8	25	11
9	2	3
10	15	7
11	4	3
12	9	5
13	18	9
14	10	5
15	14	7
16	1	3
17	17	9
18	6	5
19	23	11
20	12	7
21	8	5
22	13	7
23	22	11
24	11	7
25	24	11

the data from reaction rate (mol × 10^{-2}/L.s) is given below

Table 11.3 Reaction rate data at different pH

pH	Observations (Reaction rate, mol×10^{-2}/L.s)					Total	Average
	1	2	3	4	5		
3	2	3	2	5	3	15	3
5	3	4	3	5	4	19	3.8

7	5	5	4	6	5	26	5.2
9	4	3	3	4	4	18	3.6
11	2	3	2	3	3	13	2.6
						91	3.64

11.1.3 ANOVA study

"Analysis of Variance" (ANOVA) is a technique that employs tests based on variance ratios of several groups of observations. In this analysis, it is decided if significant differences are there among the means of these groups of observations wherein each group follows a normal distribution. Whenever three or more means are there, the analysis of variance method extends the t-test applied to determine whether there is a difference between two means.

Decomposition of the Total Sum of Squares

Consider, we have 'a' treatments or different levels of a single factor that we would like to compare. The observed response from each of the 'a' treatments is a random variable. The data can be represented as in Table 11.3. This name Analysis of Variance is obtained from a partitioning of the total variability into its component parts. The total corrected sum of squares can be written as

$$SS_T = \sum_{i=1}^{a}\sum_{j=1}^{n}\left(y_{ij} - \bar{y}\right)^2 \tag{11.2}$$

the overall variability in the data can be measured by using Eq. (11.2).

The total corrected sum of squares SS_T can be represented as

$$\sum_{i=1}^{a}\sum_{j=1}^{n}\left(y_{ij} - \bar{y}\right)^2 = \sum_{i=1}^{a}\sum_{j=1}^{n}\left[\left(\bar{y}_i - \bar{y}\right) + \left(y_{ij} - \bar{y}_i\right)\right]^2 \tag{11.3}$$

or

$$\sum_{i=1}^{a}\sum_{j=1}^{n}\left(y_{ij} - \bar{y}\right)^2 = n\sum_{i=1}^{a}\left(\bar{y}_i - \bar{y}\right)^2 + \sum_{i=1}^{a}\sum_{j=1}^{n}\left(y_{ij} - \bar{y}_i\right)^2 + 2\sum_{i=1}^{a}\sum_{j=1}^{n}\left(\bar{y}_i - \bar{y}\right)\left(y_{ij} - \bar{y}_i\right) \tag{11.4}$$

In Eq. (11.4), the cross-product term is zero, because

$$\sum_{j=1}^{n}\left(y_{ij} - \bar{y}_i\right) = y_i - n\bar{y}_i = y_i - n\left(y_i/n\right) = 0 \tag{11.5}$$

Therefore, we have

$$\sum_{i=1}^{a}\sum_{j=1}^{n}\left(y_{ij} - \bar{y}\right)^2 = n\sum_{i=1}^{a}\left(\bar{y}_i - \bar{y}\right)^2 + \sum_{i=1}^{a}\sum_{j=1}^{n}\left(y_{ij} - \bar{y}_i\right)^2 \tag{11.6}$$

The total variability in the data as shown in the Eq. (11.6), is measured by the total corrected sum of squares, can be splitted into a sum of squares of the differences between the treatment averages and the grand average (1st term on the right side of Eq. (11.6)), plus a sum of squares of the differences of observations within treatments from the treatment average (2nd term on the right side of Eq. (11.6)). Now, the differences between treatment means represent the difference between the observed treatment averages and the grand average, while the differences of observations within a treatment from the treatment average can be due only to random error. Therefore, we can write Eq. (11.6) as

$$SS_T = SS_{Treatments} + SS_E \qquad (11.7)$$

Example 11.2

To discuss the analysis of variance, we will consider the reaction rate data given in Table 11.3. Calculate the value of SS_T from Eq. (11.2)

$$SS_T = \sum_{i=1}^{a}\sum_{j=1}^{n}(y_{ij} - \bar{y})^2 \qquad (11.2)$$

$$SS_T = \sum_{i=1}^{5}\sum_{j=1}^{5} y_{ij}^2 - \frac{y^2}{N} \qquad (11.8)$$

$$= (2)^2 + (3)^2 + \cdots + (3)^2 + (3)^2 - \frac{(91)^2}{25}$$

$$= 354 - 331.24 = 22.76$$

and calculate the value of $SS_{Treaments}$ from Eq. (11.9)

$$SS_{Treaments} = \frac{1}{n}\sum_{i=1}^{5} y_i^2 - \frac{y^2}{N} \qquad (11.9)$$

$$= \frac{1}{5}\left[(15)^2 + (19)^2 + (26)^2 + (18)^2 + (13)^2\right] - \frac{(91)^2}{25}$$

$$= 351 - 331.24 = 19.76$$

from Eq. (11.7), we get

$$SS_E = SS_T - SS_{\text{Treatments}} \tag{11.10}$$

$$SS_E = 22.76 - 19.76 = 3.0$$

Where, $SS_{\text{Treaments}}$ is called the sum of squares due to treatments and SS_E is called the sum of squares due to error.

The value of F_0 is calculated using the Eq. (11.11)

$$F_0 = \frac{SS_{\text{Treatments}}/(a-1)}{SS_E/(N-a)} = \frac{MS_{\text{Treatments}}}{MS_E} \tag{11.11}$$

$$= 32.933$$

11.1.4 Types of experimental design

The experimental designs are classified based on different characteristic like geometric configuration, level of experiments, etc.

11.1.4.1 First-order designs

The most familiar first-order designs are 2^k factorial (k is the number of control variables), Plackett–Burman, and simplex designs.

Factorial Designs

A factorial design is a type of designed experiment that allows us to study the effects that various factors can have on a response. A factorial experiment is an experimental approach wherein the design variables are varied simultaneously, rather than one at a time. When we are conducting an experiment, varying the levels of all factors at the same time rather than one at a time that allows us to study the interactions between the factors. The factorial designs are broadly classified into two groups namely full factorial design and fractional factorial design.

Full Factorial Designs

A full factorial design is a design strategy, where researchers measure responses at all combinations of the factor levels. Figure 11.1 represents a two level design of three variables (X_1, X_2, and X_3). The number of available combinations is eight.

Fractional Factorial Designs

A fractional design is a design strategy in which experiments are conducted only for a selected subset or "fraction" of the runs in the full factorial design. When resources are limited or the number of factors in the design is large, the fractional factorial designs are a good choice because

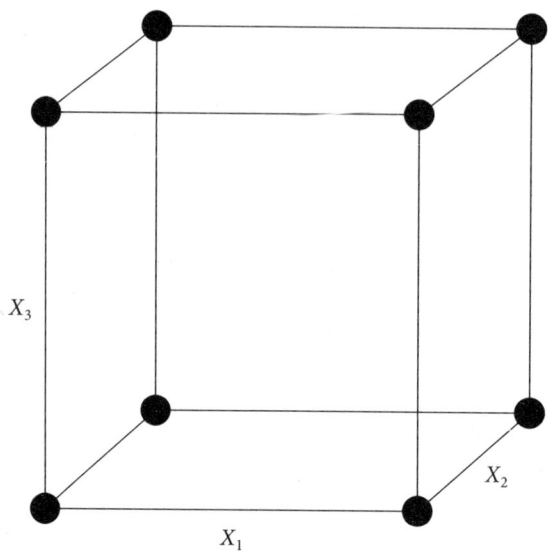

Fig. 11.1 Full factorial design with three variables

they use fewer runs than the full factorial designs. As fractional factorial design uses a subset of a full factorial design, some of the main effects and 2-way interactions are confounded and cannot be partitioned from the other higher-order interaction effects. Usually experimenters are assuming that the higher-order effects are negligible in order to get information about major effects and considering the lower-order interactions with fewer runs. In Fig. 11.2, a ½ fractional factorial design with 3 factors and 2 levels has been shown. Only four points are considered as design points.

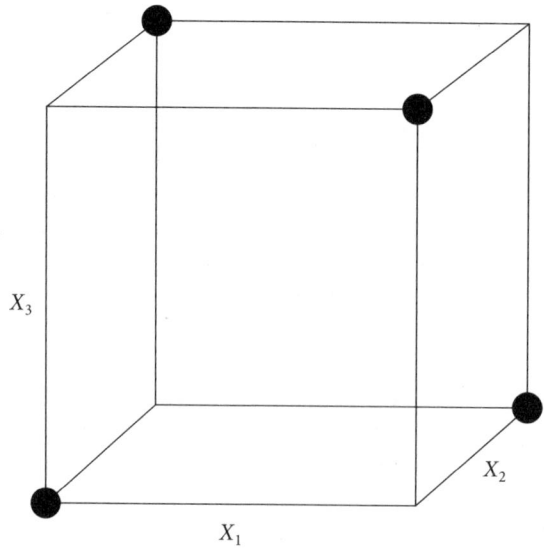

Fig. 11.2 Fractional factorial design with three variables

Plackett-Burman Designs

The Plackett-Burman design is a standard two-level design as active screening design, which offers the option of 4, 8, 16, 32 or more trials-runs, but only the power of two. When the number of variables is large and the aim is to choose the most significant variables for further experimentation, usually only the main effects are of interest. In such cases, the most cost effective alternative is to use the designs that have as many experiments as there are parameters. This is a popular and economical approach that gives information only on the effects of single factors, but not on interactions. The Plackett–Burman designs (1946) are two level fractional factorial designs for studying $k = N - 1$ variables in N number of runs, where N is a multiple of 4. For N = 12, 20, 24, 28 and 36, sometimes the Plackett–Burman designs are significant, as they fill gaps in the standard designs. Unfortunately, the structure of these particular Plackett–Burman designs are very messy alias. For instance, the very popular 11th factor in the 12-runs choice causes each main effect to be partially aliased with 45 two-factor interactions. Theoretically, we can escape from this if absolutely there exist no interactions, although this assumption is a very dangerous. The unexpected aliasing takes place with many Plackett–Burman designs. Therefore, it is suggested to keep away from them in favor of the standard two-level designs [Lazic, (2004)].

Since these designs cannot be represented as cubes, sometimes they are called non-geometric design. Some of the software packages (e.g., MINITAB) for design of experiments are available with Plackett–Burman designs.

11.1.4.2 Second-order designs

The most popular second-order designs are Central Composite Design and Box-Behnken Design. Following section discusses these methods.

Central composite design

The Central Composite Design (CCD) has been extensively utilized as the experimental design. For fitting a quadratic surface, the CCD is a very efficient. It helps us to optimize the effective parameters with a minimum number of experiments, and also to analyze the interaction between the parameters. The CCD consists of a 2^k factorial runs with $2k$ axial runs and n_0 center runs. In CCD, each variable is investigated at two levels and as the number of factors, k, increases the number of runs for a complete replicate of the design increases rapidly.

The total number of experiment require is given by

$$N = 2^k + 2k + n_0 \tag{11.12}$$

The different level for each experiment is $-\alpha$, -1, 0, $+1$, and $+\alpha$
where

$$\alpha = 2^{k/4} \tag{11.13}$$

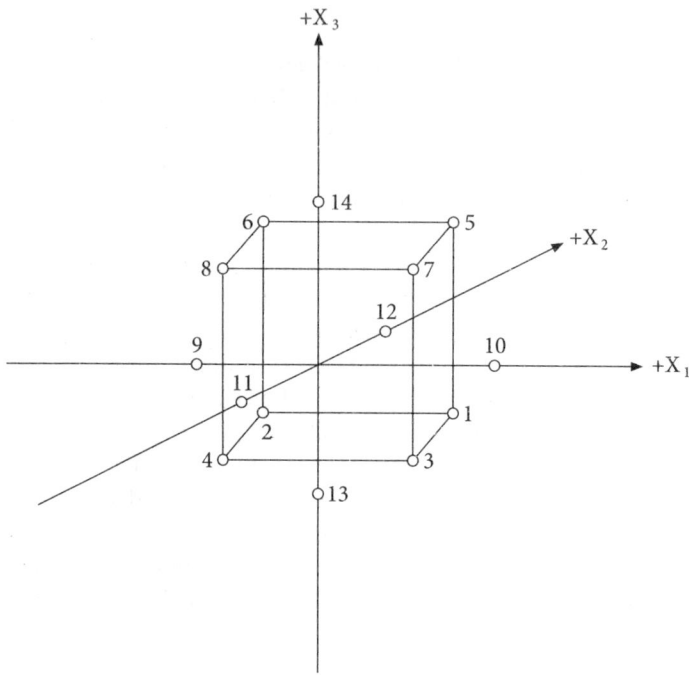

Fig. 11.3 Central composite design

Rotatability

It is important for the second order model to provide good predictions throughout the region of interest. One way to define "good" is require that at points of interest **x**, the model comprise a reasonably consistent and stable variance of the predicted response. The variance of the predicted response at some point x is

$$V[\hat{y}(\mathbf{x})] = \sigma^2 \mathbf{x}'(\mathbf{X}'\mathbf{X})^{-1}\mathbf{x} \tag{11.14}$$

Box and Hunter (1957) suggested that a second order response surface design should be rotatable. This means that the $V[\hat{y}(\mathbf{x})]$ is the same at all points **x** that are the same distance from the centre of the design. That is, the variance of predicted response is constant on spheres. A design with this property will leave the variance of $\hat{y}$ unaffected when we rotate the design about the centre (0, 0,…,0), hence, the name rotatable design. Rotatability is a reasonable basis for the selection of a response surface design. Because RSM is used for optimization and the position of the optimum point is unknown to us before conducting the experiment, it makes sense to use a design that gives equal precision of estimation in all directions (it can be shown that any first order orthogonal design is rotatable). A central composite design is made rotatable with proper selection of α. The value of α for rotatability depends on the number of points in the factorial portion of the design; actually, $\alpha = (n_F)^{1/4}$ give us a rotatable central composite design where n_F is the number of points used in the factorial portion of the design.

Box–Behnken design

Box and Behnken (1960) have proposed some three-level designs for fitting response surfaces. The main advantage of Box–Behnken design over CCD is that it requires less number of experimental design points which reduces the experimental cost, consequently cost of optimization process. Figure 11.4 shows the geometric representation of Box–Behnken (B–B) design. This design is a spherical in shape where all points are lying on a sphere of radius $\sqrt{2}$. The B–B design does not have any points at the vertices of the cubic region constructed by the lower and upper limits for each variable. Sometimes this design could be helpful particularly when the points on the corners of the cube represent factor-level combinations that are extremely expensive or sometimes impossible to test owing to physical process constraints.

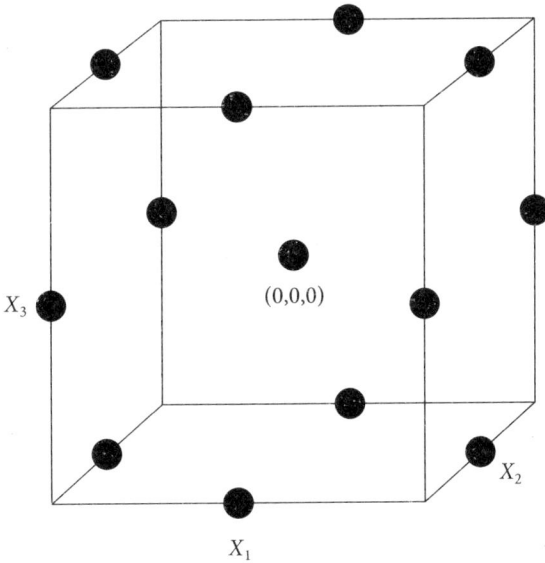

Fig. 11.4 Box–Behnken design

Table 11.4 Three variable B–B design

Run	x_1	x_2	x_3
1	−1	−1	0
2	1	−1	0
3	−1	1	0
4	1	1	0
5	−1	0	−1
6	1	0	−1
7	−1	0	1
8	1	0	1

9	0	-1	-1
10	0	1	-1
11	0	-1	1
12	0	1	1
13	0	0	0
14	0	0	0
15	0	0	0

The value in table 11.4 is in coded form, which is given by the following equation

$$\text{Coded value} = X_i = \frac{(x_i - \bar{x}_i)}{\Delta x} \quad (11.15)$$

where Δx is called step change.

11.1.4.3 D-optimal design

There are many popular design optimality criteria; probably the most widely used is the D-optimality criterion. A design is called D-optimal if $\left|(X'X)^{-1}\right|$ is minimized. It turns out that a D-optimal design minimizes the volume of the joint confidence region on the vector of regression coefficients. A measure of the relative efficiency of design 1 to design 2 according to the D-criterion is represented by

$$D_e = \left(\frac{\left|(X_2'X_2)^{-1}\right|}{\left|(X_1'X_1)^{-1}\right|}\right)^{\frac{1}{p}} \quad (11.16)$$

where X_1 and X_2 are the X matrices for the two designs and p is the number of model parameters.

11.2 Response Surface Methodology

Response surface methodology (RSM) is a compilation of mathematical and statistical techniques. RSM is a method for designing experiments, building models, evaluating the effects of various factors and reaching the optimum conditions with a limited number of planned experiments [Khuri and Cornell (1996)]. By designing the experiments carefully, our aim is to optimize a response (output variable) which is influenced by various independent variables (input variables). RSM shows how a given set of input variables affect a particular response within the specified region of interest, and what values of input variables will yield an optimizer for a particular response. Initially RSM was developed to find out optimum operating conditions in the chemical process industry, but now it is employed in a variety of fields and applications, not only in the engineering and physical sciences, but also in clinical, social, and biological sciences [Khuri (2001)]. The main aim of RSM is the optimization of process. This could imply, for instance, the operating cost minimization of a production process, the minimization of the variability of a quality characteristic, maximization of the yield of a chemical process, or achieving the desired specifications for a response. It is frequently considered the multiple responses of interest in practical problems in chemical industry [Castillo, (2007)]. The main advantage of RSM is that, it formulate the optimization problem with very less cost of analysis methods and their associated numerical noise. For optimization using RSM, the first step is design the experiment and develop the response surface based on experimental data. Design of experiment is discussed in the previous section (11.2).

For optimizing an industrial process, RSM techniques suggest to construct a parametric model for the expected response using designed experiments. Different model can be considered to find the relationship between response and variables. The response function can be represented as

$$y = f(x_1, x_2, x_3) + \varepsilon \tag{11.17}$$

where ε is the error or noise observed in the response y. If the expected response is represented by $E(y) = f(x_1, x_2, x_3) = \eta$, then the surface represented by

$$\eta = f(x_1, x_2, x_3) \tag{11.18}$$

The linear model is

$$y = \beta_0 + \sum_{i=1}^{k} \beta_i x_i + \varepsilon \tag{11.19}$$

and second order model is

$$y = \beta_0 + \sum_{i=1}^{k} \beta_i x_i + \sum_{i}\sum_{j} \beta_{ij} x_i x_j + \sum_{i=1}^{k} \beta_{ii} x_i^2 + \varepsilon \tag{11.20}$$

Usually the response surface is presented graphically as shown in Figs 11.5a, 11.6a, and 11.7a. We often plot the contours of the response surface for better visualization of the shape of response surface. Contours, also helps us to determine the optimum point.

Example 11.3

Anupam *et al.* [Anupam *et al.* (2011)] have discussed Cr(VI) from water using activated carbon depends on three factors pH, powdered activated carbon dose (PAC) and time of adsorption. They found the correlation for chromium removal as

$$\text{Cr(VI) removal}(\%) = f(\text{pH}, \text{PAC}, \text{time})$$

or

$$\begin{aligned}\text{Cr(VI) removal}(\%) = {} & 77.92 - 8.46(\text{pH}) + 9.41(\text{PAC}) + 3.86(\text{time}) \\ & + 0.68(\text{pH})^2 - 3.53(\text{PAC})^2 - 7.33(\text{time})^2 \\ & - 1.08(\text{pH})(\text{PAC}) + 1.08(\text{pH})(\text{time}) - 0.452(\text{time})(\text{PAC})\end{aligned} \quad (11.21)$$

Three surfaces can be constructed as shown in Figs 11.5a, 11.6a, and 11.7a. The corresponding contours can be represented by Fig. 11.5b, 11.6b, and 11.7b.

$$\text{Cr(VI) removal}(\%) = f(\text{PAC, time}) \text{ at pH} = 0. \quad (11.22a)$$

$$\text{Cr(VI) removal}(\%) = f(\text{pH, time}) \text{ at PAC} = 0 \quad (11.22b)$$

$$\text{Cr(VI) removal}(\%) = f(\text{pH, PAC}) \text{ at time} = 0 \quad (11.22c)$$

These surfaces are called response surfaces.

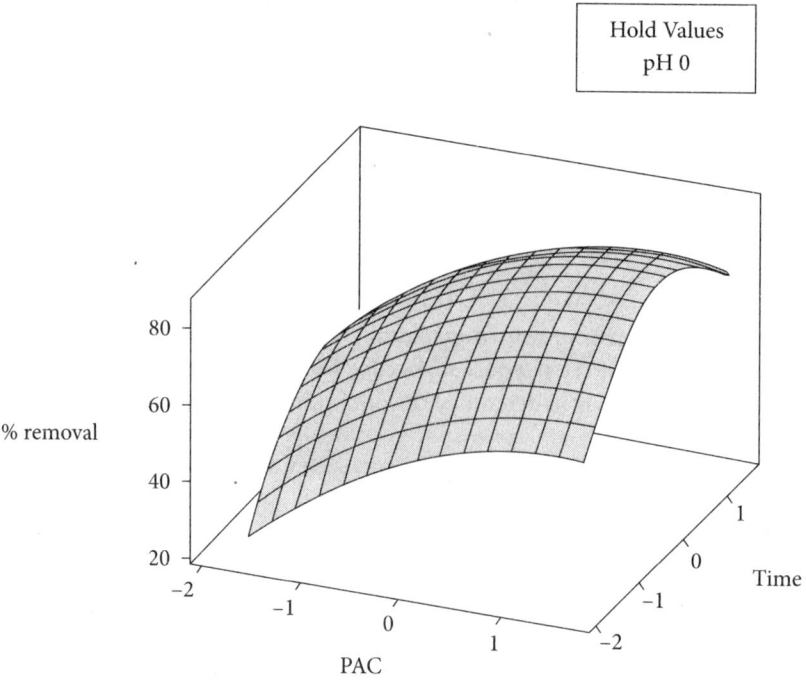

Fig. 11.5a Response surface formed from Eq. (11.22a)

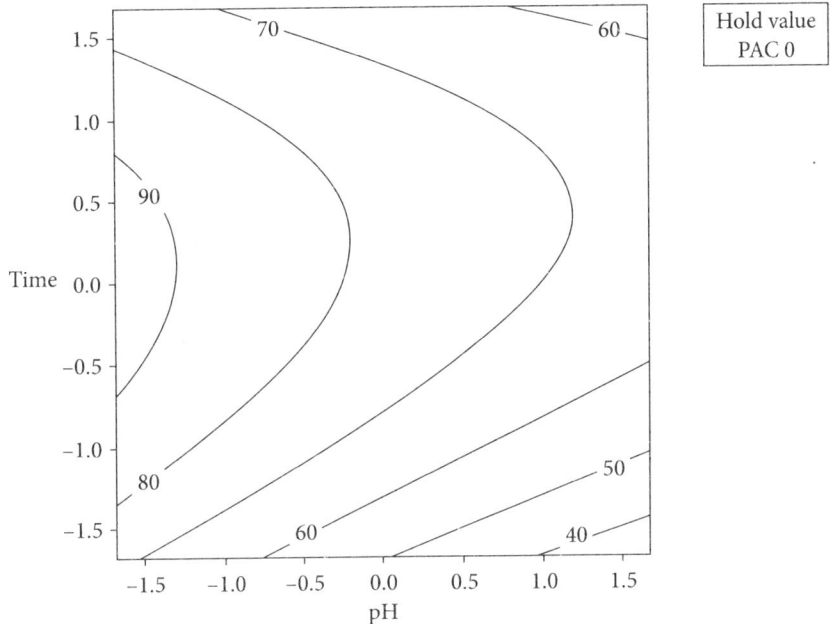

Fig. 11.5b Contour representation of Fig. 11.5a

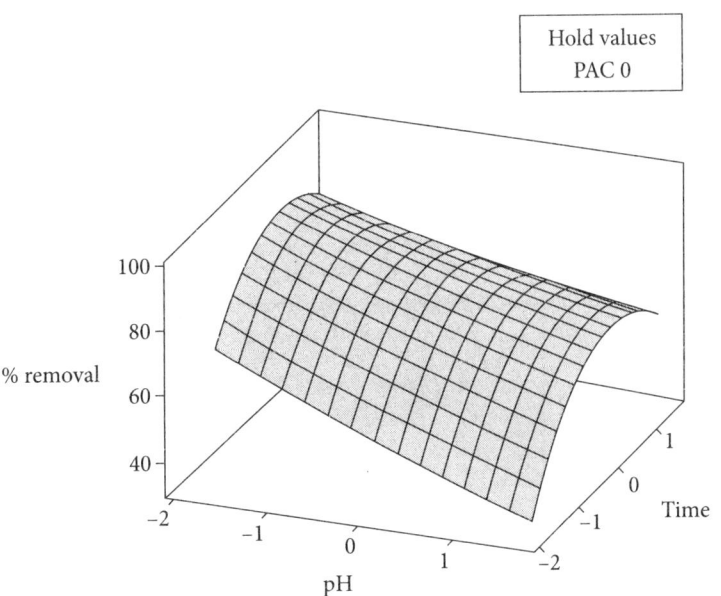

Fig. 11.6a Response surface formed from Eq. (11.22b)

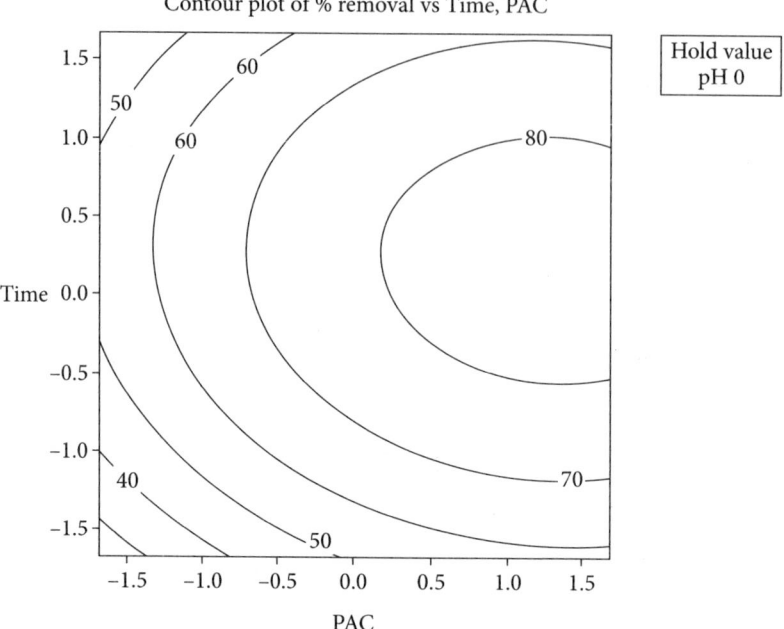

Fig. 11.6b Contour representation of Fig. 11.6a

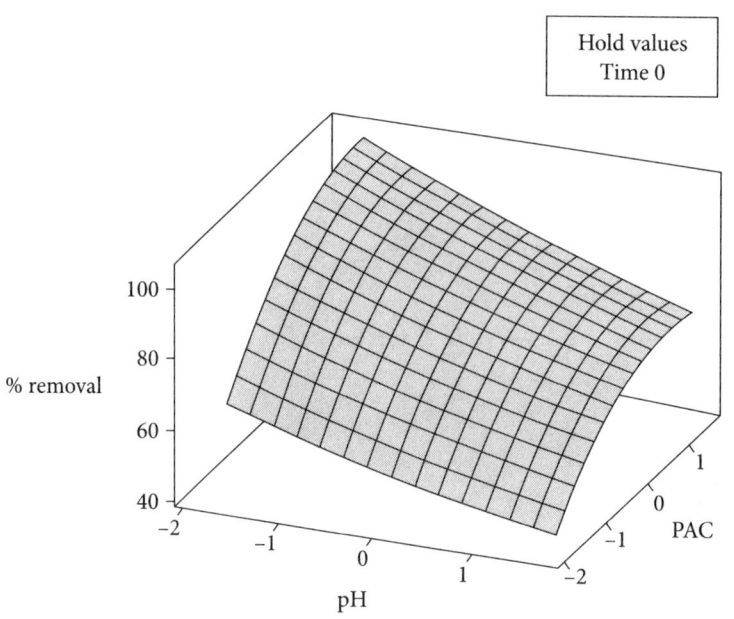

Fig. 11.7a Response surface formed from Eq. (11.22c)

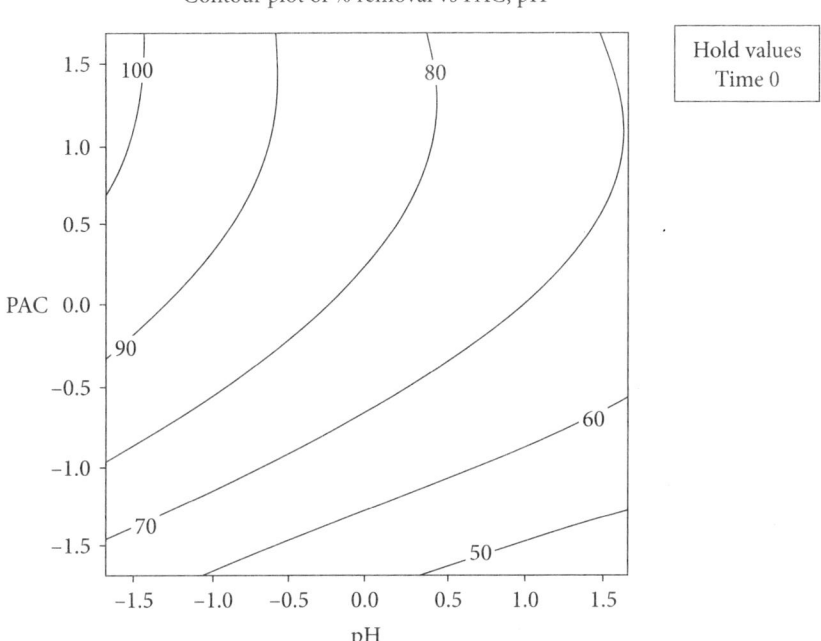

Fig. 11.7b Contour representation of Fig. 11.7a

For a conventional DOE when it is expected that many of the design variables initially considered have small or no effect on the response, screening experiments are performed in the early stages of the process. The purpose is to recognize the design variables that have large effects and need further investigation. Genetic Programming can be used efficiently for screening, genetic algorithm has been demonstrated in Section 9.1.

11.2.1 Analysis of a second order response surface

Suppose we are interested to find the location of the stationary point. We may obtain a general mathematical solution for the same. Writing the second order model in matrix form, we have

$$\hat{y} = \hat{\beta}_0 + \mathbf{x}'\mathbf{b} + \mathbf{x}'\mathbf{B}\mathbf{x} \tag{11.23}$$

where

$$\mathbf{x} = \begin{bmatrix} x_1 \\ x_2 \\ \vdots \\ x_k \end{bmatrix}, \mathbf{b} = \begin{bmatrix} \hat{\beta}_1 \\ \hat{\beta}_2 \\ \vdots \\ \hat{\beta}_k \end{bmatrix} \text{ and } \mathbf{B} = \begin{bmatrix} \hat{\beta}_{11} & \hat{\beta}_{12}/2 & \cdots & \hat{\beta}_{1k}/2 \\ & \hat{\beta}_{22} & \cdots & \hat{\beta}_{2k}/2 \\ & & \ddots & \vdots \\ \text{sym} & & & \hat{\beta}_{kk} \end{bmatrix}$$

Here, **b** is a $(k \times 1)$ vector of the first order regression coefficients and **B** is a $(k \times k)$ symmetric matrix whose main diagonal elements are the pure quadratic coefficients $(\hat{\beta}_{ii})$ and whose off-diagonal elements are one-half the mixed quadratic coefficients $(\hat{\beta}_{ij}, i \neq j)$. The derivative of $\hat{y}$ with respect to the elements of the vector x equated to 0.

$$\frac{\partial \hat{y}}{\partial \mathbf{x}} = \mathbf{b} + 2\mathbf{B}\mathbf{x} = 0 \qquad (11.24)$$

The stationary point is the solution of this equation.

$$\mathbf{x}_s = -\frac{1}{2}\mathbf{B}^{-1}\mathbf{b} \qquad (11.25)$$

Furthermore, by substituting Eq. (11.25) into Eq. (11.23), we can establish the predicted response at the stationary point as

$$\hat{y}_s = \hat{\beta}_0 + \frac{1}{2}\mathbf{x}_s'\mathbf{b} \qquad (11.26)$$

Example 11.4

Find the stationary point for the problem formulated by Anupam *et al.*

Table 11.5 Values of the coefficient (Anupam *et al.*)

Term	Coefficients
Constant	77.92
pH	−8.46
PAC	9.41
time	3.86
pH × pH	0.68
PAC × PAC	−3.53
time × time	−7.33
pH × PAC	−1.08
pH × time	1.88
PAC × time	−0.45

$$\mathbf{b} = \begin{bmatrix} -8.46 \\ 9.41 \\ 3.86 \end{bmatrix} \quad \mathbf{B} = \begin{bmatrix} 0.68 & -0.54 & 0.94 \\ -0.54 & -3.53 & -0.225 \\ 0.94 & -0.225 & -7.33 \end{bmatrix}$$

$$\mathbf{B}^{-1} = \begin{bmatrix} 1.121 & -0.181 & 0.149 \\ -0.181 & -0.255 & -0.015 \\ 0.149 & -0.015 & -0.117 \end{bmatrix}$$

which gives the stationary point (using Eq. (11.25))

$$\mathbf{x}_s = \begin{bmatrix} -5.303 \\ -0.462 \\ -0.929 \end{bmatrix}$$

Characterizing the response surface

Characterization of response surface in the vicinity of this stationary point is necessary. By characterize, we mean determine whether the stationary point is a point of minimum or maximum response or saddle point. We also interested to study the relative sensitivity of the response to the variables $x_1, x_2, \ldots, x_k$.

The most straightforward way to do this is to examine a contour plot of the fitted model. If there are two or three process variables, the construction and interpretation of this contour plot is quite simple. However, even when there are relatively few variables, a more formal analysis, called the canonical analysis, can be useful.

11.2.2 Optimization of multiple response processes

Most of the processes in real life required the optimization with respect to various criteria simultaneously. Sometimes the operating conditions required to satisfy various conditions or constraints on m responses, $y_1, \ldots, y_m$. For example, simultaneous optimization of selectivity as well as production rate for a reacting system.

Desirability Approach

The desirability approach was initially recommended by Harrington [Harrington, (1965)] and later modified by Derringer and Suich [Derringer and Suich, (1980)] to its most familiar form that is in practical use nowadays. The desirability function approach is a well-accepted method in industry for dealing with the multiple response processes optimization. The quality of a process or product that has multiple quality characteristics, with one of them out of some desired limits and that is totally unacceptable. This technique finds the operating conditions x that offer the "most desirable" values of the response [Castillo, (2007)].

For each response $y_i(x)$, a desirability function $d_i(y_i)$ assigns numbers between 0 and 1 to the probable values of y_i. The value $d_i(y_i) = 0$ represents a absolutely undesirable value of y_i and $d_i(y_i) = 1$ represents a completely desirable or ideal response value. To obtain the overall desirability D, the individual desirabilities are combined using the geometric mean as Eq. (11.27):

$$D = \left[d_1(y_1) \times d_2(y_2) \times \ldots \times d_m(y_m) \right]^{1/m} \qquad (11.27)$$

where the number of responses is represented by m. It is obvious that the overall desirability is zero whenever any of these responses i is completely undesirable ($d_i(y_i) = 0$). In practical application of this method, the fitted response models $\hat{y}_i$ are used.

Various desirability functions $d_i(y_i)$ can be used depending on whether a particular response y_i is to be minimized, maximized, or assigned to a target value. Derringer and Suich [Derringer and Suich, (1980)] proposed a useful class of desirability functions. Suppose L_i, U_i and T_i be the lower, upper, and target values desired for response i, where $L_i \leq T_i \leq U_i$. If a response is of the "target is best" type, then its individual desirability function is

$$d_i(\hat{y}_i) = \begin{cases} 0 & \text{if } \hat{y}_i(x) < L_i \\ \left(\dfrac{\hat{y}_i(x) - L_i}{T_i - L_i}\right)^s & \text{if } L_i \leq \hat{y}_i(x) \leq T_i \\ \left(\dfrac{\hat{y}_i(x) - U_i}{T_i - U_i}\right)^t & \text{if } T_i \leq \hat{y}_i(x) \leq U_i \\ 0 & \text{if } \hat{y}_i(x) > U_i \end{cases} \quad (11.28)$$

where the exponents s and t decide how strictly the target value is desired. When $s = t = 1$, the desirability function increases linearly towards T_i, the function is convex when $s < 1$, $t < 1$ and the function is concave for $s > 1$, $t > 1$.

If as an alternative a response is to be maximized, the individual desirability is instead defined as

$$d_i(\hat{y}_i) = \begin{cases} 0 & \text{if } \hat{y}_i(x) < L_i \\ \left(\dfrac{\hat{y}_i(x) - L_i}{T_i - L_i}\right)^s & \text{if } L_i \leq \hat{y}_i(x) \leq T_i \\ 1.0 & \text{if } \hat{y}_i(x) > T_i \end{cases} \quad (11.29)$$

where in this case T_i is considered as a large enough value for the response.

Finally, if we are interested to minimize a response, we could use

$$d_i(\hat{y}_i) = \begin{cases} 1.0 & \text{if } \hat{y}_i(x) < T_i \\ \left(\dfrac{\hat{y}_i(x) - U_i}{T_i - U_i}\right)^s & \text{if } T_i \leq \hat{y}_i(x) \leq U_i \\ 0 & \text{if } \hat{y}_i(x) > U_i \end{cases} \quad (11.30)$$

where T_i signifies a small enough value for the response.

Summary

- Properly designed experiment saves both our time and money. DOE method is used to plan the experiment before it started. Various experimental designs such as Box–Behnken, Central Composite etc. have been discussed in this chapter. The experimental data is utilized to find the model equation. Some statistical analysis is required for this purpose. The best-fit model is used for optimization using response surface methodology. Contours are also useful for finding the optimum point.

Review Questions

11.1 Why DOE is necessary for chemical process design?

11.2 What are the different stages used for DOE?

11.3 What is meant by replication? Why it is required?

11.4 Show that the Box–Behnken design of 3 factors is not a rotatable design.

11.5 Calculate the number of experiment required for B–B design with 4 factors.

11.7 When fractional factorial design is useful?

11.8 Show that a CCD is rotatable when $\alpha = (n_F)^{1/4}$ where n_F is the number of points used in the factorial portion of the design.

11.9 Show that using the relation $\alpha = \sqrt{k}$ results in a rotatable CCD for $k = 2$ and $k = 4$, but not for $k = 3$.

11.10. Consider the following response surface of two factors:

$$\hat{y}_i = 70 + 0.1x_1 + 0.3x_2 + 0.2x_1^2 + 0.1x_2^2 + x_1 x_2$$

a. Find the coordinates of the stationary point.

b. What type of response function is this?

11.11 What do you mean by D-Optimal design?

References

Anupam, K., Dutta, S., Bhattacharjee, C., and Datta S. *Adsorptive Removal of Chromium (VI) from Aqueous Solution Over Powdered Activated Carbon: Optimization Through Response Surface Methodology,* 'Chemical Engineering Journal', 173(2011): 135–43.

Box, G. and Behnken, D. 1960. *Some New Three Level Designs for the Study of Quantitative Factors.* Technometrics, 2, 4, pp. 455–75.

Box, G. E. P. and Wilson, K. B. 1951. *On the Experimental Attainment of Optimum Conditions,* 'Journal of the Royal Statistical Society', Series B, 13, 1–45.

Castillo, E. D. 2007. *Process Optimization: A Statistical Approach*: Springer.

Derringer, G. C. and Suich, R. 1980. "Simultaneous Optimization of Several Response Variables," Journal of Quality Technology, 12, pp. 214–19.

Fisher, R. A. 1926. *The Arrangement of Field Experiments. Journal of Ministry of Agriculture,* England, 33: 503–13.

Fisher, R. A. 1958. *Statistical Methods for Research Workers,* 13th edition. Edinburgh: Oliver and Boyd.

Fisher, R. A. 1960. *The Design of Experiments,* 7th edition. Edinburgh: Oliver and Boyd.

Harrington, E. C. 1965. *The Desirability Function,* Industrial Quality Control, 21, pp. 494–98.

Khuri, A. I. 2001. *An Overview of the Use of Generalized Linear Models in Response Surface Methodology.* Nonlinear Anal. 47: 2023–34.

Khuri, A. I. and Cornell, J. A. 1996. *Responses surfaces: Design and Analyses.* 2nd ed. New York: Marcel Dekker.

Lazić, Ž. R. 2004. *Design of Experiments in Chemical Engineering,* Weinheim: Wiley-VCH Verlag GmbH and Co. KGaA.

Montgomery, D. C. 2001. *Design and Analysis of Experiments,* 5th Edition, John Wiley and Sons, Inc.

Plackett, R. L. and Burman, J. P. 1946. *The Design of Optimal Multifactorial Experiments,* Biometrika 33: 305–25.

Chapter 12

Software Tools for Optimization Processes

Optimization problem developed for different processes are very complicated. Mostly, the number of variables and constraints are very large. These complicated problems are difficult to solve by hand. Therefore, we take help of computer to solve those problems with very less effort that save our time and money. Computer programming solves these optimization problems using some numerical techniques. The algorithms are discussed in previous chapters. Some commercial software/software tools are available that can be used for optimization purpose. The most popular software/software tools are LINGO, MATLAB, MINITAB®, GAMS etc. In this chapter, we will discuss these software/software tools with some examples.

12.1 LINGO

Introduction to LINGO

LINGO is modeling language and optimizer developed by LINDO Systems Inc. It is a simple tool for formulating large problems concisely, solve them, and analyze the solution using the power of linear and nonlinear optimization.

The outer window shown in Fig. 12.1, labeled **Lingo 14.0**, is the LINGO main frame window. All other sub-windows will be enclosed within this main frame window. The entire command menu and the command toolbar are placed at the top of the frame window. The status bar at the lower edge of the main frame window gives us various information related to the current state of LINGO. Both the status bar and the toolbar can be suppressed using the *LINGO|Options* command. The main window consists of a sub-window labeled **Lingo Model-Lingo1** is used for Lingo Model. Here we will demonstrate how to enter and solve a small model in Windows platform. The text of the model's equations identical on all platforms, it is platform independent. Though, the procedure for entering a model is somewhat different on non-Windows platforms.

308 Optimization in Chemical Engineering

When we start LINGO in Windows platform, screen should look like the following:

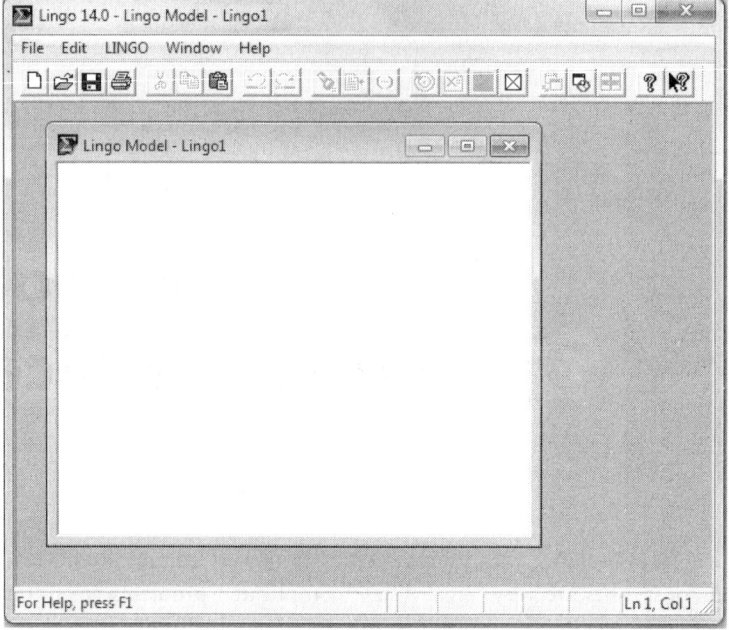

Fig. 12.1 LINGO main window

Entering the Model

Example 12.1

We will now solve the problem 6.1 using LINGO.

The problem formulated in chapter 6 is given below:

$$\text{Max } F = 14x_{P_1} + 11x_{P_2} \tag{6.12a}$$

subject to

$$3x_{P1} + 2x_{P2} \leq 36 \tag{6.12b}$$

$$x_{P1} + x_{P2} \leq 14 \tag{6.12c}$$

$$x_{P1}, x_{P2} \geq 0 \tag{6.12d}$$

After entering the above model within the **Lingo Model-Lingo1** window, our model window should look like this:

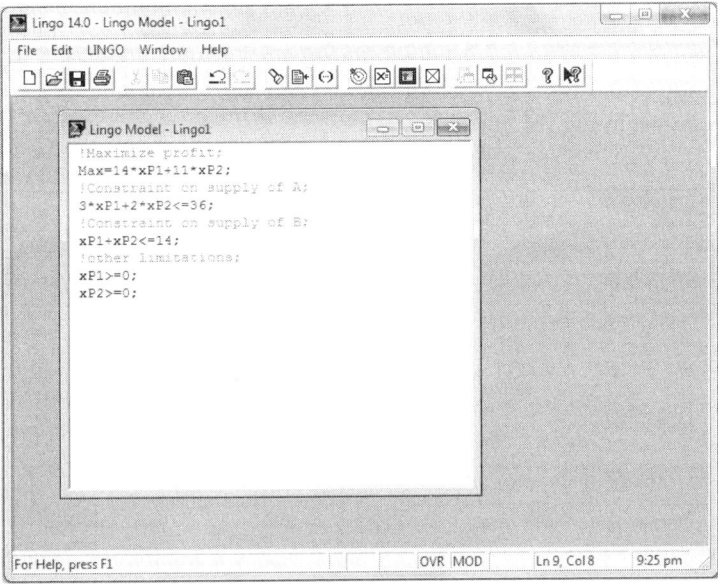

Fig. 12.2 Lingo Model-Lingo1 window with a maximization model

Solving a LINGO Model

After entering the LINGO model into the LINGO Model window, the model can be solved by clicking the *Solve* button on the toolbar, by choosing LINGO|Solve from the menus, by using the keyboard shortcut "ctrl + U" or by pressing the [icon] button from toolbar menu as shown below

Fig. 12.3 Toolbar of LINGO

If there is any syntax error, you will get an error message as given below:

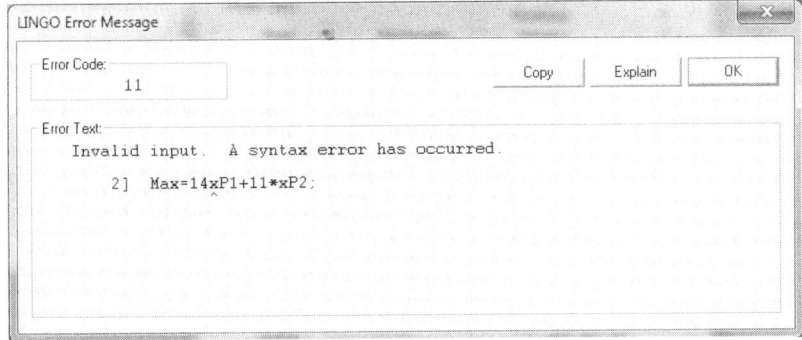

Fig. 12.4 LINGO error message box

Solver Status Window

If there are no syntax errors, LINGO will call up the suitable internal solver to start searching for the optimal solution to our model. When the solver starts working, a *solver status* window appears on the screen as shown in Fig. 12.5:

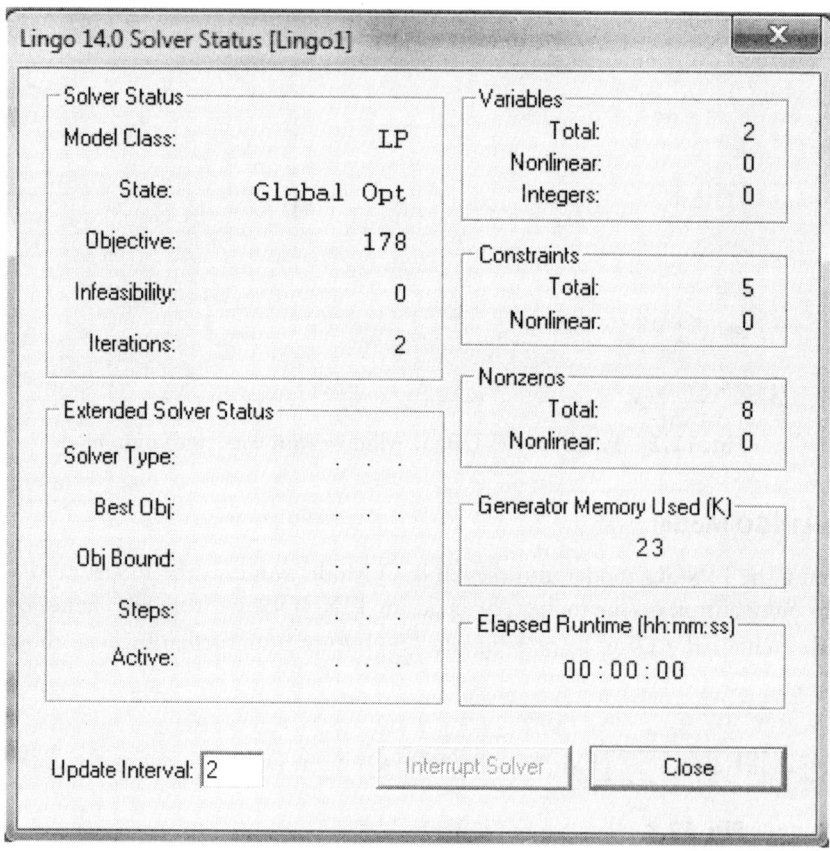

Fig. 12.5 Solver status window

This solver status window gives us information on the number of nonlinear, integer, and total variables in our model; the nonlinear and the total number of constraints used in the model; and the number of nonlinear and total nonzero variable coefficients used. The details of the model classification like LP, QP, ILP, IQP, NLP, etc is given in the Solver Status box within this window. This also provides the status of the current solution (local or global optimum, feasible or infeasible, etc.), the value of the objective function, the infeasibility of the model (amount constraints are violated by), and the number of iterations required to solve the model. Similar information is available in the Extended Solver Status box for more advanced branch-and-bound, global, and multistart solvers.

LINGO solution report window (Fig. 12.6) gives many parameters like reduced cost, dual price, slack or surplus etc.

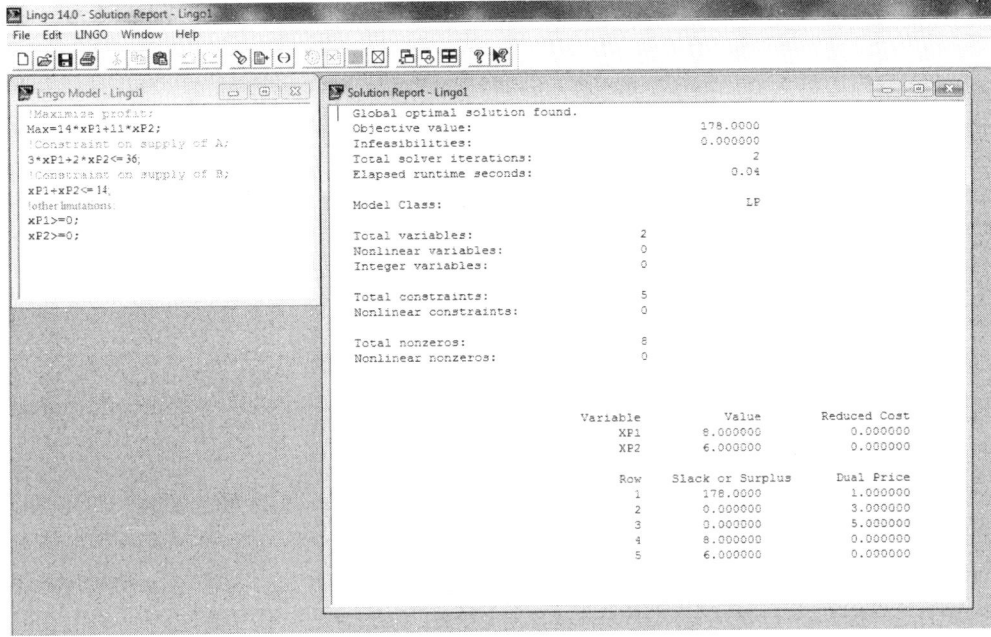

Fig. 12.6 LINGO solution report window

For variables, which are not included in the optimal solution, the reduced cost gives us an idea about how much the objective function value would decrease (for maximization) or increase (for minimization) if one unit of that variable were to be included in the solution. For instance, if for a certain variable the reduced cost 4, then the optimal value of the maximization problem would decrease by 4 units if 1 unit of that variable is to be added. The value of the reduced cost for any variable, which is included in the optimal solution is always zero. We can get an idea about how tight the constraint is, from the Slack or Surplus column in the Solution Report window. The value of slack or surplus is zero when a constraint is completely satisfied as an equality. The positive of slack/surplus indicates how much of the variable could be added to the optimal solution to make the constraint equality. When the slack/surplus is negative, the constraint has been neglected. The Dual Price column explains the improvement in the objective function if the constraint is relaxed by one unit.

General LINGO Syntax

i. An expression in LINGO may be broken up into several lines as we want, but the expression must be completed with a semicolon. In the example, we have been used two lines rather than just one to write the objective function:

Max: 14*xP1

+ 11*xP2;

ii. We can also enter some comments for better understanding of the readers (Fig. 12.2). Comments start with an exclamation point (!) and terminate with a semicolon (;). When LINGO solves the model, it overlooks all the text between an exclamation point and terminating semicolon.

Comments can be written in more than one line and can share lines with other expressions in LINGO. For example:

Max: 14*xP1 !Profit from product P1; + 11*xP2; !Profit from product P2;

iii. In LINGO, the uppercase and lowercase in variable names have the same meaning they are not distinguishable. Therefore, the variable names given below would all be considered equivalent:

PROFIT

Profit

profit

When creating variable names in LINGO, all names must start with an alphabetic character (A–Z). Subsequent characters may be either alphabetic, numeric (0–9), or the underscore (_). The length of the variable names may be up to 64 characters.

Using Sets in LINGO

LINGO allows us to group many occurrences of the same variable into sets. When the number of variables is large, we can represent it as a set. Sets are either primitive or derived. A primitive set is one that consists of distinct members only. Whereas a derived set, contains other sets as its members.

Example 12.2

The warehouse problem can be written easily using sets in LINGO.

Consider, a Wireless Widget (WW) Company has five warehouses supplying their widgets to six vendors. Each warehouse has a limit of widgets supply that cannot be exceeded, and each vendor has a demand for widgets that must be fulfilled. The company wants to decide how many widgets to distribute from each warehouse to each vendor to minimize the total transportation cost.

Solution

This problem can be represented as follows:

$$\text{Minimize } f = \sum_{i=1}^{5} \sum_{j=1}^{6} c_{ij} x_{ij} \tag{12.1a}$$

$$\text{subject to } \sum_{i=1}^{5} \sum_{j=1}^{6} x_{ij} \leq b_j \tag{12.1b}$$

$$\sum_{i=1}^{5} \sum_{j=1}^{6} x_{ij} \leq d_i \tag{12.1c}$$

$$x_{ij} \geq 0, \ (i = 1, 2, \ldots, 5; j = 1, 2, \ldots, 6) \tag{12.1d}$$

The corresponding dual problem can be given by

The values of the c_{ij}, d_i, and b_j are given in Table 12.1, 12.2, and 12.3

Table 12.1 Widget capacity data

Warehouse	Widget on hand
1	60
2	55
3	51
4	41
5	52

Table 12.2 Vendor widget demand

Vendor	Widget demand
1	35
2	37
3	22
4	43
5	38
6	32

Table 12.3 Shipping cost per widget ($)

	V1	V2	V3	V4	V5	V6
WW1	6	2	7	4	2	9
WW2	4	9	5	3	8	6
WW3	7	6	7	3	9	2
WW4	2	3	9	7	5	3
WW5	5	6	2	3	8	1

The corresponding LINGO code:

```
MODEL:
! A Transportation Problem with 5 Warehouses and 6 Vendors;
SETS:
WAREHOUSES: CAPACITY;
VENDORS: DEMAND;
```

```
LINKS( WAREHOUSES, VENDORS): COST, VOLUME;
ENDSETS
! Data is given here;
DATA:
!set members;
WAREHOUSES = WH1 WH2 WH3 WH4 WH5;
VENDORS = V1 V2 V3 V4 V5 V6;
!attribute values;
CAPACITY = 60 55 51 41 52;
DEMAND = 35 37 22 43 38 32;
COST = 6 2 7 4 2 9
4 9 5 3 8 6
7 6 7 3 9 2
2 3 9 7 5 3
5 6 2 3 8 1;
ENDDATA
! The objective function;
MIN = @SUM( LINKS( I, J):
COST( I, J) * VOLUME( I, J));
! The demand constraints;
@FOR( VENDORS( J):
@SUM( WAREHOUSES( I): VOLUME( I, J)) =
DEMAND( J));
! The capacity constraints;
@FOR( WAREHOUSES( I):
@SUM( VENDORS( J): VOLUME( I, J)) <=
CAPACITY( I));
END
```

Variable Types in LINGO

All LINGO variables model are considered to be continuous and non-negative unless otherwise specified. The default domain for given variables can be override by using LINGO's four variable domain functions. These variable domain functions are given below:

i. @FREE – any positive or negative real value

ii. @GIN – any positive integer value

iii. @BIN – a binary value (i.e., 0 or 1)

iv. @BND – any value within the specified bounds

Similar syntax is used for the @GIN, @FREE, and @BIN variable domain functions. The general form for the declaration of a variable x using any of these functions is @FUNCTION(X); (see example 12.3)

The @BND function has a somewhat different syntax, which consists of the lower and upper bounds for the acceptable variable values. We can declare a variable x between a lower bound and an upper bound is given by @BND(lower bound, X, upper bound).

Example 12.3

Solve the problem given in Example 6.6.

Solution

Maximize: $Z = 5x_1 + 7x_2$

subject to constraints

$x_1 + x_2 \leq 6$

$5x_1 + 9x_2 \leq 43$

$x_1, x_2 \geq 0$ and Integer

The given integer programming problem can be solved using LINGO with the simple code

```
!Code for Example 6.6;
max=5*x1+7*x2;
x1+x2<=6;
5*x1+9*x2<=43;
x1>=0;
x2>=0;
@GIN(x1);
@GIN(x2);
```

Fig. 12.7 LINGO solver status window

The model class PILP represents Pure Integer Linear Programming where all variables are restricted to integer values and all expressions are linear.

The optimum solution is $Z_{max} = 36$; $x_{1\,max} = 3$, $x_{2\,max} = 3$.

Modeling from the Command-Line

Following the steps:

LINGO>Window>Command Window

The LINGO command window will appear. The colon character (:) at the beginning of the screen is LINGO's prompt for input. When we see the colon prompt, it indicates that LINGO is expecting a command. When we see the question mark prompt, we have already started a command and LINGO is asking us to supply additional information associated to this command such as a name or a number.

Note All screenshot are taken with permission from Lindo Systems Inc.

12.2 MATLAB

There are three main toolboxes which are generally used for optimization purposes.

i. **Curve fitting toolbox**

   Graphical tools and command-line functions for fitting curves and surfaces from data are available in Curve Fitting Toolbox™. We can perform exploratory data analysis using this toolbox. Pre-processing and post-processing of data, comparing candidate models, and removing outliers are possible using this toolbox. The inbuilt library of linear and nonlinear models can be used to conduct regression analysis. We can also specify our customized model equations. The library provides us the starting conditions and optimized solver parameters to enhance the quality of our results. The nonparametric modeling techniques like splines, interpolation, and smoothing are also possible with this toolbox.

ii. **Global optimization toolbox**

   Global Optimization Toolbox provides techniques, which search for global solutions to problems that have multiple stationary points. This toolbox is consists of solvers like global search, multistart, pattern search, genetic algorithm, and simulated annealing. We can employ these solvers to optimize problems where the objective or constraint functions are continuous, discontinuous, stochastic, or even does not possess derivatives.

   Pattern search and genetic algorithm solvers support algorithmic customization. We can develop a customized genetic algorithm variant by changing the initial population and fitness scaling options. Parent selection, crossover, and mutation functions can also be defined as per our requirement. Customization of the pattern search is done by defining polling, searching, and other functions.

iii. **Optimization toolbox**

   Optimization Toolbox™ comprises algorithms for standard and large-scale optimization. These algorithms solve unconstrained and constrained continuous and discrete problems.

This toolbox comprises functions for linear programming, quadratic programming, nonlinear optimization, nonlinear least squares, solving systems of nonlinear equations, binary integer programming, and multi-objective optimization.

Table 12.4 shows different functions used for optimization. Applications of these functions have been discussed in the following section.

Table 12.4 Description of MATLAB function used for optimization

Function	Description
Curve fitting toolbox	
cftool	Open Curve Fitting Tool
fit	Fit model to data
sftool	Open Surface Fitting Tool
Optimization toolbox	
bintprog	Solve binary integer programming problems
fminbnd	Find minimum of single-variable function on fixed interval
fgoalattain	Solve multiobjective goal attainment problems
fmincon	Find minimum of constrained nonlinear multivariable function
fminimax	Solve minimax constraint problem
fminsearch	Find minimum of unconstrained multivariable function using derivative-free method
fminunc	Find minimum of unconstrained multivariable function
linprog	Solve linear programming problems
quadprog	Solve quadratic programming problems
Global Optimization problem	
ga	Find minimum of function using genetic algorithm
gamultiobj	Find minima of multiple functions using genetic algorithm
patternsearch	Find minimum of function using pattern search
simulannealbnd	Find unconstrained or bound-constrained minimum of function of several variables using simulated annealing algorithm

Example 12.4

Write the MATLAB code for solving the problem given below

Minimize $31 - 11x_1 - 5x_2 + 3x_1^2 + x_2^2$

Solution

The function "fminunc" is used for solving unconstrained multivariable optimization problem. Follow the steps given below

Step 1   Create an M file

```
function f = myfun(x)
f = 31-11*x(1)-5*x(2) + 3*x(1)^2 + x(2)^2;
```

Step 2   Save this file with file name "myfun.m"

Step 3   Write the code to find the optimized value near the point $x_0 = [1\ 1]$

```
x0 = [1,1];
[x,fval] = fminunc(@myfun,x0);
x,fval
```

Step 4   Run the programme

The result is

```
x =
    1.8333    2.5000
fval =
    14.6667
```

Example 12.5

Solve the optimization problem with constraints

Minimize $31 - 11x_1 - 5x_2 + 3x_1^2 + x_2^2$

subject to

$x_1 + x_2 \leq 3$

$x_1^2 + x_2^2 = 8$

Solution

Follow the similar steps as example 12.4

Step 1   Create a M file for the objective function (objfun.m):

```
function f = objfun(x)
f = 31-11*x(1)-5*x(2) + 3*x(1)^2 + x(2)^2;
```

Step 2   Create a M file for the constraint functions (confun.m):

```
function [c, ceq] = confun(x)
% Nonlinear inequality constraints
c = x(1)+x(2)-3;
% Nonlinear equality constraints
ceq = x(1)^2+x(2)^2-8;
```

Step 3   Invoke constrained optimization routine:

```
x0 = [1,1];      % Make a starting guess at the solution
```

```
        options = optimset('Algorithm','active-set');
        fmincon(@objfun,x0,[],[],[],[],[],[],@confun,options);
        [x,fval]
which gives the result:
ans =
    2.8229    0.1771    23.0000
```

Example 12.6

Solve the problem in example 11.3. Use pattern search algorithm for the same.

Solution

The objective function is given by Eq. (11.21)

$$\text{Cr(VI) removal}(\%) = 77.92 - 8.46(\text{pH}) + 9.41(\text{PAC}) + 3.86(\text{time}) \\ + 0.68(\text{pH})^2 - 3.53(\text{PAC})^2 - 7.33(\text{time})^2 \\ - 1.08(\text{pH})(\text{PAC}) + 1.08(\text{pH})(\text{time}) - 0.452(\text{time})(\text{PAC}) \quad (11.21)$$

We have to create a M file (.m) for the objective function

```
function f = crremoval(x)
f = -(77.92+9.41*x(1)+3.86*x(2)-3.53*x(1)^2-7.33*x(2)^2);
```

Save this file with file name "crremoval.m"

Write the given code in a separate M file:

```
x = patternsearch(@crremoval,x0)
x0 = [1 1];
[x,fval] = patternsearch(@crremoval,x0)
```

Run the file

```
which gives the result
x =
    1.3327    0.2630
fval =
   -84.6993
```

Note "patternsearch" function finds the minimum of any function. However, our problem is maximization problem. Therefore, we have written the objective function "crremoval" as the negative of the objective function (Eq. (11.21)).

Example 12.7

Solve the problem in example 11.3, using simulated annealing method.

Solution

MATLAB has inbuilt function "simulannealbnd" that can be used to solve unconstrained optimization problem.

Save the file with file mane "crremoval"

```
function f = crremoval(x)
f = -(77.92+9.41*x(1)+3.86*x(2)-3.53*x(1)^2-7.33*x(2)^2);
```

Create another M file with code:

ObjectiveFunction = @crremoval;

```
X0 = [0.1 0.1];     % Starting point
[x,fval,exitFlag,output] = simulannealbnd(ObjectiveFunctio
n,X0)
lb = [1 1];
ub = [10 10];

x =
    1.3329      0.2633
fval =
   -84.6993
```

Multiobjective optimization is possible in MATLAB using different methods. Here we will use genetic algorithm for solving MOO.

Example 12.8

Construct the Pareto front using genetic algorithm MOO technique.

Solution

Create the function with objective functions

```
function y = simple_multiobjective(x)
    y(1) = 3+(x-4)^2;
    y(2) = 16+(x-10)^2;
```

For plotting the objective functions use the code

```
% Plot two objective functions on the same axis
x = -10:0.5:10;
f1 = 3+(x-4).^2;
f2 = 16+(x-10).^2;
plot(x,f1);
```

```
hold on;
plot(x,f2,'r');
grid on;
title('Plot of objectives ''Objective 1'' and ''Objective 2'');
```

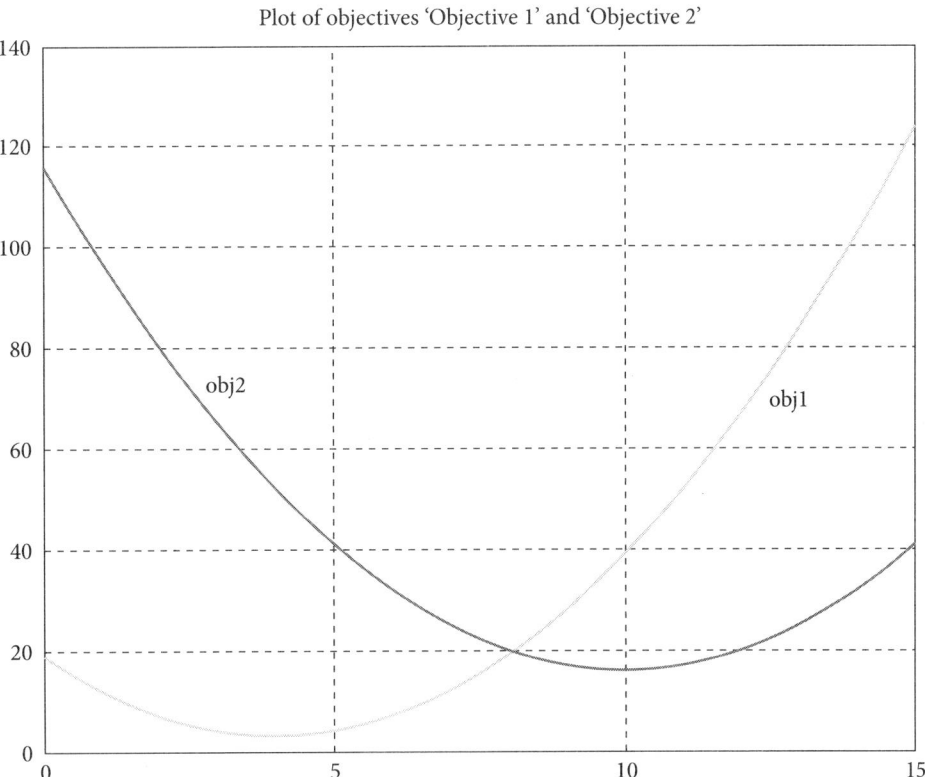

Fig. 12.8 Plot of objective 1 and objective 2

```
FitnessFunction = @simple_multiobjective;
numberOfVariables = 1;
[x,fval] = gamultiobj(FitnessFunction,numberOfVariables);
size(x)
size(fval)
A = []; b = [];
Aeq = []; beq = [];
lb = 1;
ub = 10;
x = gamultiobj(FitnessFunction,numberOfVariables,A,b,Aeq,beq,lb,ub);
options = gaoptimset('PlotFcns',{@gaplotpareto});
gamultiobj(FitnessFunction,numberOfVariables,[],[],[],[],lb,ub,options);
```

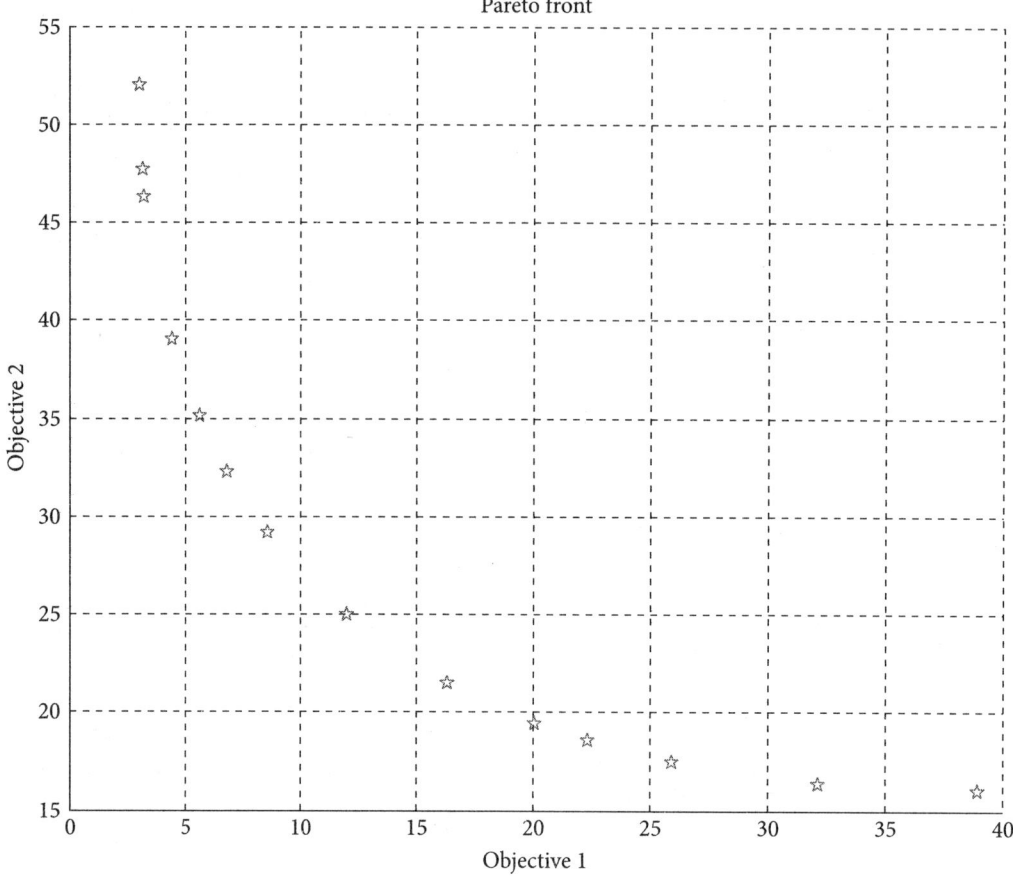

Fig. 12.9 Pareto front for obj1 and obj2

Examples 12.9

Consider the linear programming problem in example 6.2. Solve this problem using MATLAB.

Solution

The problem in example 6.2 is

Maximize: $Z = 3x_1 + 2x_2$

subject to constraints

$2x_1 + x_2 \leq 10$

$x_1 + x_2 \leq 8$

$x_1 \leq 4$

and

$x_1 \geq 0, x_2 \geq 0$

We need to convert the objective function as minimization problem. Therefore, the problem becomes

Minimize: $F = -3x_1 - 2x_2$

subject to constraints

$2x_1 + x_2 \leq 10$

$x_1 + x_2 \leq 8$

$x_1 \leq 4$

and

$x_1 \geq 0, x_2 \geq 0$

Write the follow code in a M file and run the file.

```
f = [-3; -2;];
A =  [2 1
      1 1
      1 0];
b = [10; 8; 4];
lb = zeros(2,1);
[x,fval,exitflag,output,lambda] = linprog(f,A,b,[],[],lb);
x,fval
```

The output is as follows

Optimization terminated.

```
x =
    2.0000
    6.0000
fval =
   -18.0000
```

Note All materials are included with permission from MathWorks Inc.

12.3 MINITAB®

Minitab® is a statistical analysis software. It can be used for statistical research, design of experiment, and response surface methodology. We have already discussed about the theory of DOE and RSM in chapter 11. In this section, we will learn how to use MINITAB® for developing an experimental run and optimize that process.

Getting started

The main window of MINITAB® looks like Fig. 12.10. Minitab Menu bar are two sub-windows; **Session** and **Worksheet1**. The **Session** window is where any non-graphical output is displayed (Fig. 12.16). Note that you can also type in commands in this window.

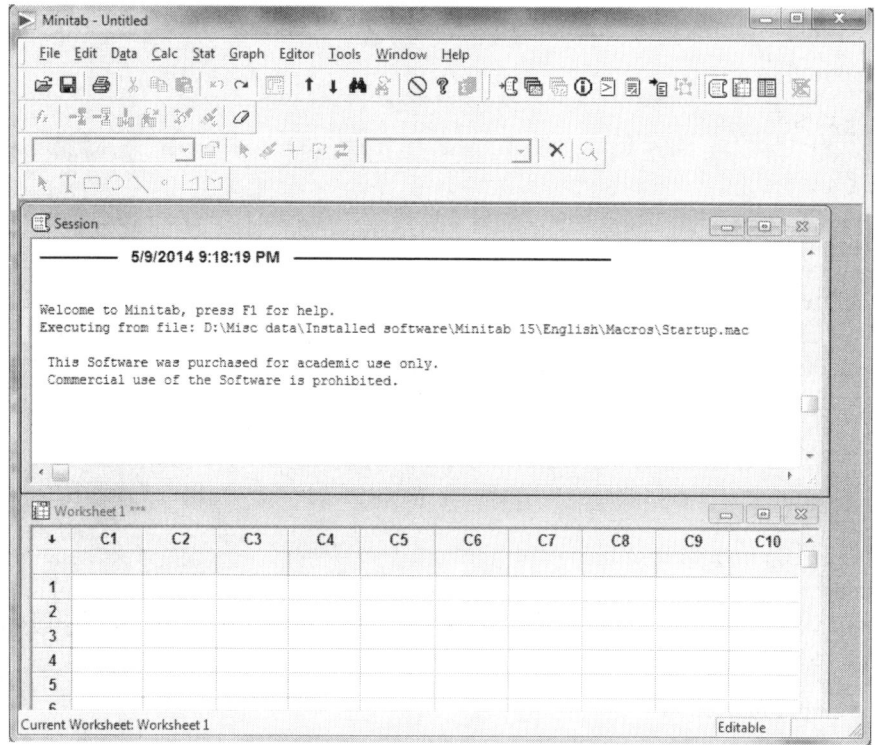

Fig. 12.10 MINITAB main window

"Worksheet" is the basic structural component of Minitab®. The **Worksheet1** *** window is a spreadsheet, where we can type in and view our data.

The worksheet is a big rectangular array, or matrix, of cells organized into rows and columns as shown in Fig. 12.10. Each cell of this matrix contains one piece of data. This piece of data could be a numeric data, i.e. number; it could be a sequence of characters, such as a word or text data, i.e., an arbitrary string of numbers and letters. Most often data comes as numbers, for example 2.1,5.8,...but occasionally it appears in the form of a sequence of characters, such as white, green, male, female, etc. Typically, sequences of characters are used to identify the classifications for some variable of interest, e.g., gender, color. In Minitab®, the length of a piece of text data can be up to 80 characters. The extension .mtw indicate that this is a Minitab® worksheet.

Selection of DOE

To select the experimental design, follow the steps

Stat>>DOE>>Response Surface>>Create Response Surface Design
then select either Central Composite Design or Box–Behnken Design.

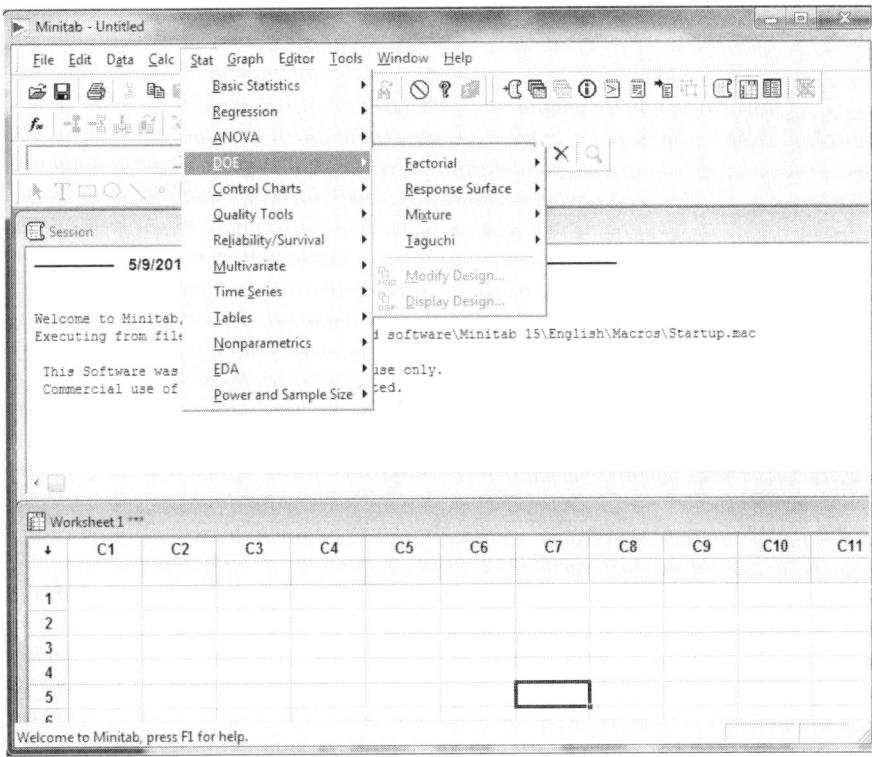

Fig. 12.11 MINITAB window for selecting DOE

Example 12.10

Consider the problem discussed by Dutta (2013) [Dutta (2013)]. Dye adsorption experiment was carried out with three different variables i.e., pH, TiO_2 dose, and contact time. Box–Behnkan design was considered for this study. The maximum and minimum level of these variables are given in Table 12.5.

Table 12.5 Levels of independent variables for B–B design

Independent variable	Symbol	Levels		
		−1	0	+1
pH	X_1	3.32	5	6.68
TiO_2 dose (g.L^{-1})	X_2	0.98	3.5	6.02
Contact time (min)	X_3	2.45	26	49.55

Worksheet in Fig. 12.12 shows the experimental points for B–B design. Session sub-window shows that the total number of experiment required is 15.

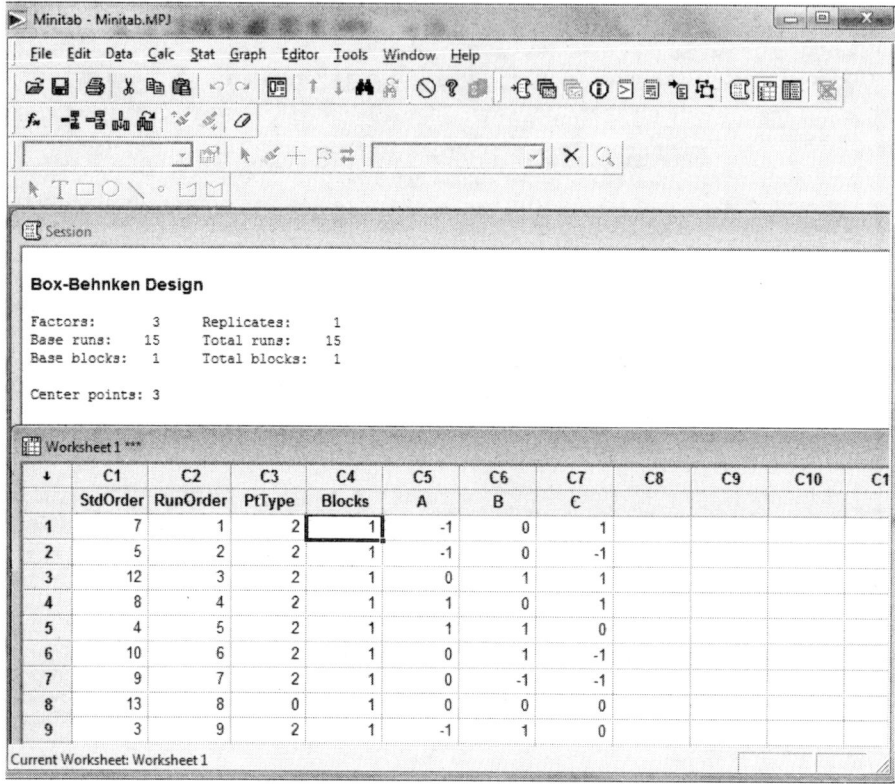

Fig. 12.12 MINITAB window with B–B design

Table 12.6 shows the experimental data for B–B design.

Table 12.6 Matrix of B–B design

No of experiment	Coded value of independent variables			Response (% dye removal)
	pH	TiO$_2$	Time	
1	1	0	−1	2.4
2	0	0	0	58.4
3	1	1	0	33.1
4	0	−1	−1	7.18
5	0	−1	1	16.58
6	−1	−1	0	60.85
7	0	0	0	58.4
8	1	0	1	17.15

9	0	1	1	67.78
10	0	1	-1	37.74
11	-1	0	-1	66.14
12	1	-1	0	2.1
13	-1	0	1	90.81
14	-1	1	0	85.9
15	0	0	0	58.4

Incorporate the Response (% removal) data from last column of Table 12.6 to the column 8 (C8) of the Worksheet1***. Then the window looks like Fig. 12.13.

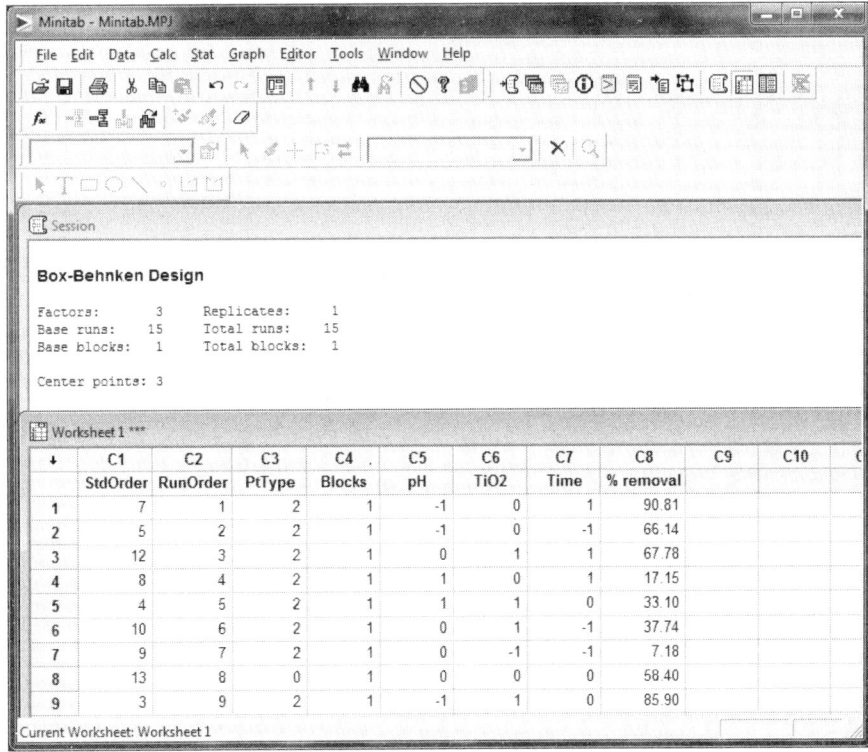

Fig. 12.13 MINITAB window with data for B–B design

After data input, analysis of the response surface design is required. Follow the given steps for this purpose.

```
Stat>>DOE>>Response Surface>>Analyze Response Surface Design
```

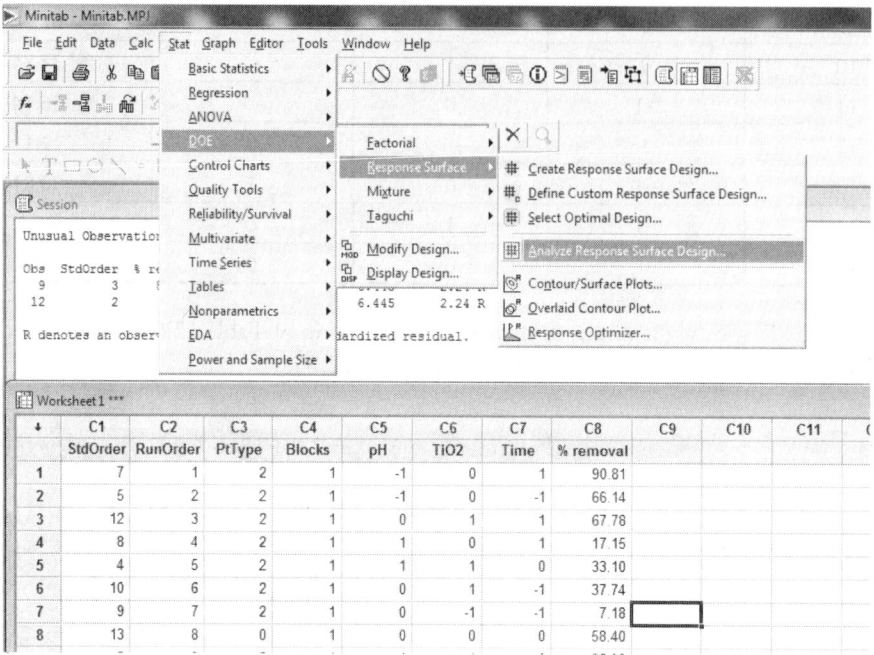

Fig. 12.14 MINITAB window for analyzing the response surface design

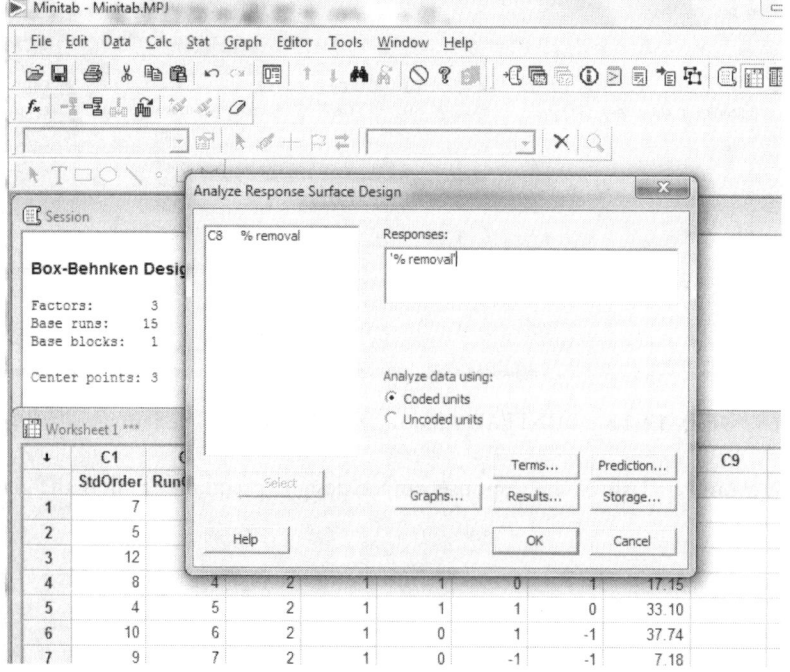

Fig. 12.15 MINITAB window for selecting "response"

A sub-window will appear as shown in Fig. 12.15, we have to select C8 %removal as "response". Then select options for result, graph, storage etc. Press OK button to get result.

The calculated values of the coefficients are represented in Fig. 12.16. Other results will appear in "session" sub-window. Table 12.7 shows the results of ANOVA study.

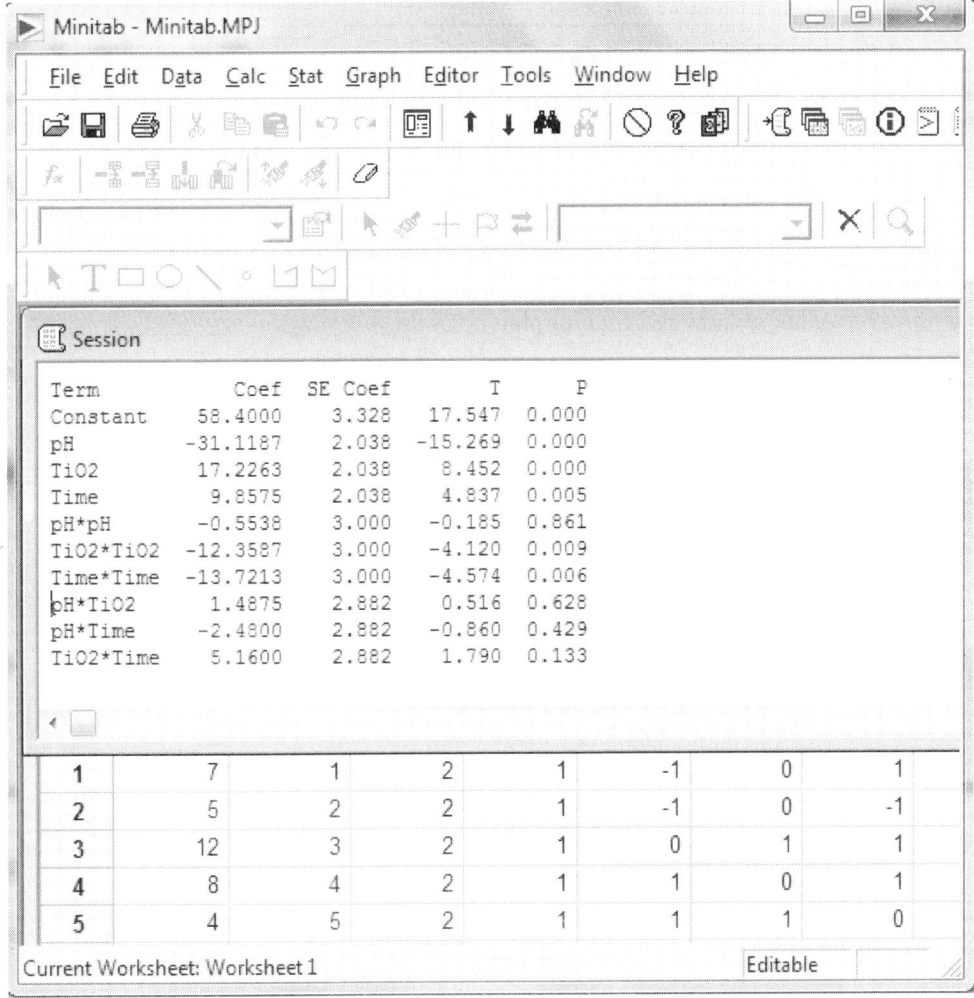

Fig. 12.16 MINITAB window with results

Table 12.7 ANOVA for percentage dye removal

Source	DF	Seq SS	Adj SS	Adj MS	F	P
Regression	9	12215.9	122.15.9	1357.33	40.85	0.000
Linear	3	10900.3	10900.3	3633.42	109.34	0.000
Square	3	1175.9	1175.9	391.98	11.80	0.010

Interaction	3	139.7	139.7	46.58	1.40	0.345
Residual error	5	166.2	166.2	33.23		
Lack-of-fit	3	166.2	166.2	55.38		
Pure error	2	0.0	0.0	0.0		
Total	14	12382.1				

The final equation is

$$\text{Dye removal (\%)} = 58.40 - 31.119(\text{pH}) + 17.226(\text{TiO}_2) + 9.858(\text{Time}) \\ - 0.554(\text{pH})^2 - 12.359(\text{TiO}_2)^2 - 13.721(\text{Time})^2 \\ + 1.488(\text{pH})(\text{TiO}_2) - 2.48(\text{pH})(\text{Time}) + 5.16(\text{TiO}_2)(\text{Time}) \quad (12.2)$$

To construct response surfaces and contour plots open the following link

```
Stat>>DOE>>Response Surface>>Contour/Surface Plots
```

Figures 12.17a, 12.17b, and 12.17c, represent the response surfaces for the response Eq. (12.2). The corresponding contours are given in 12.18a, 12.18b, 12.18c,

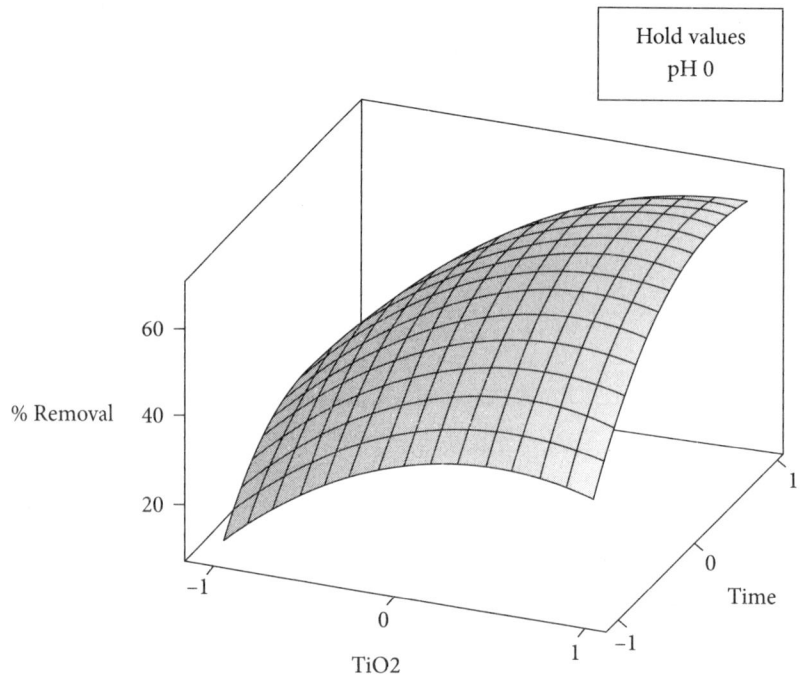

Fig. 12.17a Response surface of Eq. (12.2) at pH = 0

Software Tools for Optimization Processes   331

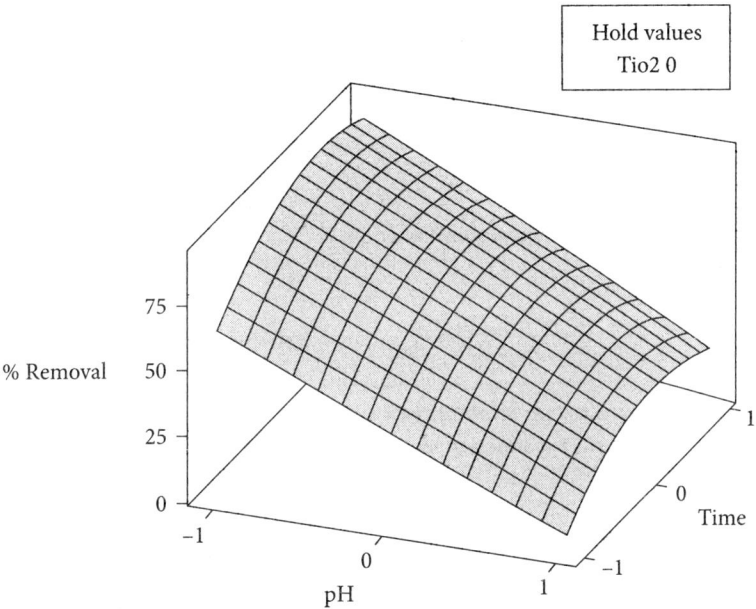

Fig. 12.17b   Response surface of Eq. (12.2) at TiO2 = 0

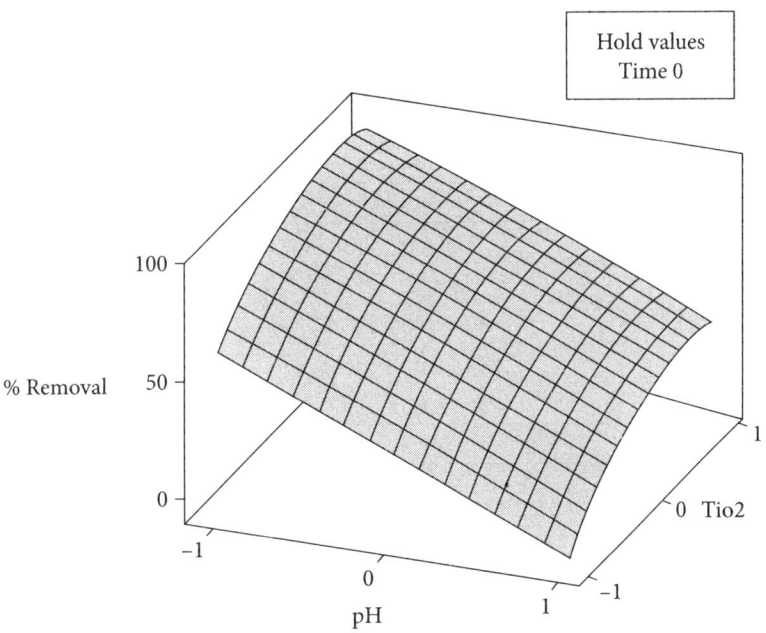

Fig. 12.17c   Response surface of Eq. (12.2) at Time = 0

332 Optimization in Chemical Engineering

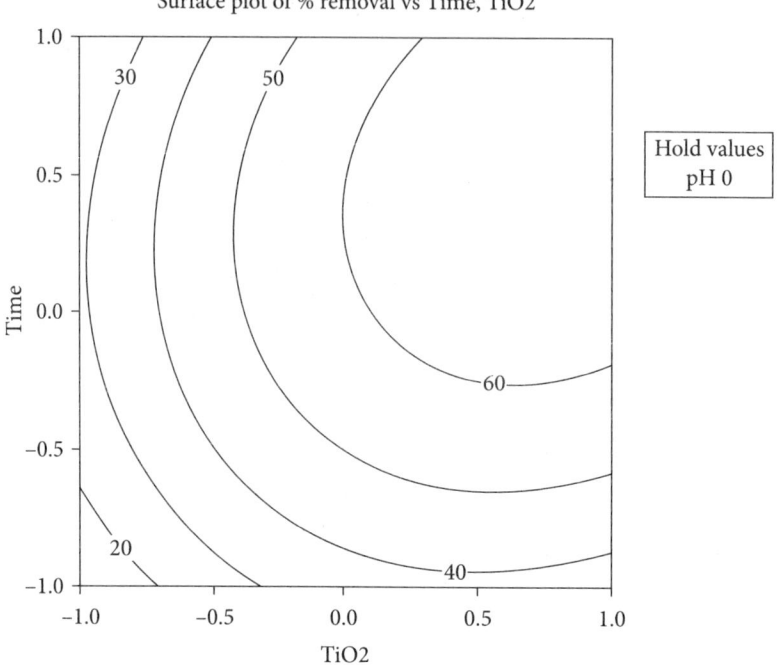

Fig. 12.18a Contour of the response Eq. (12.2) at pH = 0

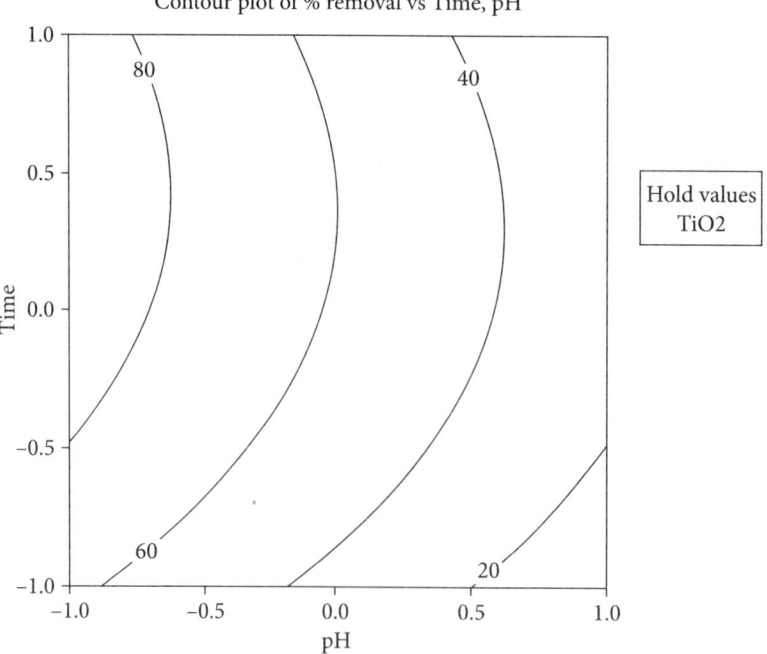

Fig. 12.18b Contour of the response Eq. (12.2) at TiO2 = 0

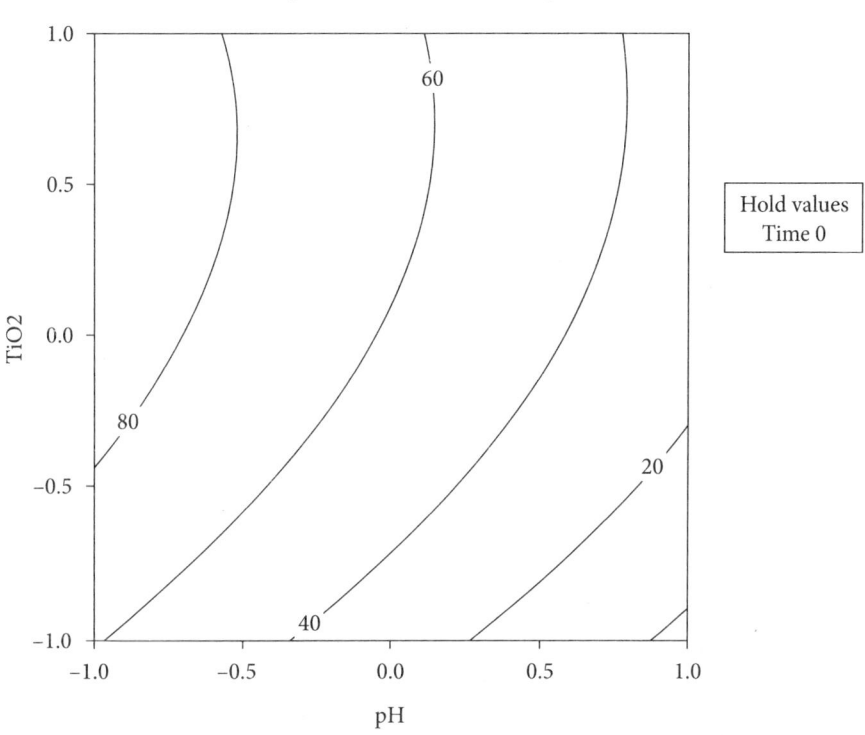

Fig. 12.18c Contour of the response Eq. (12.2) at Time = 0

The optimum values of pH, TiO_2 dose, and time are 3.32, 5.4 g.L^{-1}, and 40.0 min, respectively. Dye removal at this optimum condition is 98.61%.

Note Portions of information contained in this publication/book are printed with permission of Minitab Inc. All such material remains the exclusive property and copyright of Minitab Inc. All rights reserved.

12.4　GAMS

The General Algebraic Modeling System (GAMS) is a high-level modeling system, which can be used for mathematical programming and optimization. It is composed of a language compiler and a stable of integrated high-performance solvers. GAMS is adapted for complex, large scale modeling applications, and it allow us to build large maintainable models that can be modified quickly to new model depending on the situation.

Input/output

The main window of GAMS looks like Fig. 12.19.

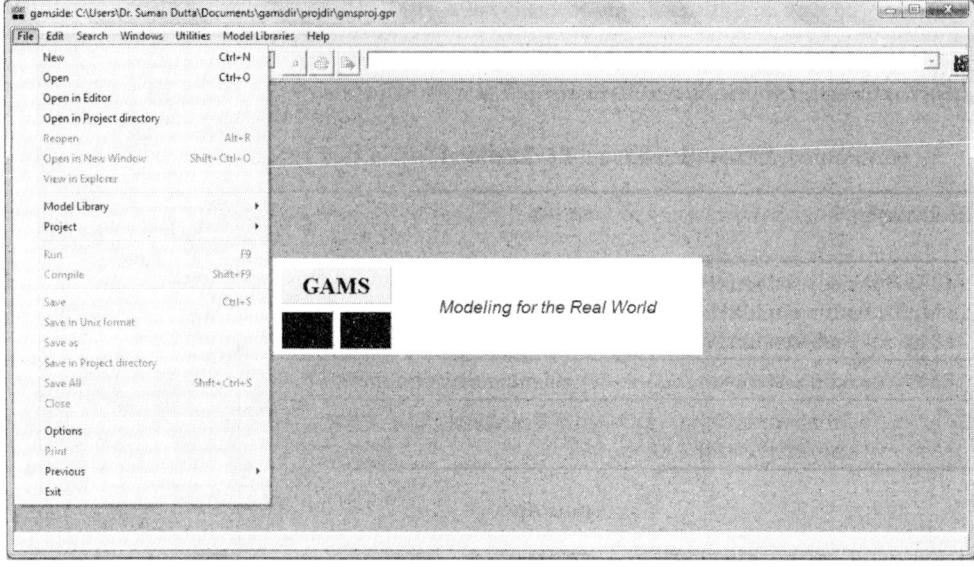

Fig. 12.19 Main window of GAMS

Open a new file, the input model file will be saved as ".gms" format:
filename.gms

When input file is ready, the model is solved using a solution procedure given in Table 12.10 and the out with error message sent to corresponding ".lst" file.

GAMS Statement Formats

The following syntax has been used to write any GAMS command:

- Line that starts with an * (asterisk) as first character is comment, see the example 12.11 and 12.12.

- Line with a dollar sign ($) in the very first character is a GAMS input/output option. For example, $ include "*filename*" copies in the content of "*filename*" as if it had been typed at this point. There is no end punctuation for such $-option lines.

- All other statements in GAMS are written over one or more lines and end with a semicolon; commas are used to separate the component items.

- For better readability, extra spaces, line breaks, and blank lines may be added without any effect.

- GAMS is not case sensitive. Therefore, AA, aa and Aa have the same meaning.

- Declaration statements such as free variable(s) and equations(s) may have some words of double-quoted "explanatory text" (i.e., "total profit" in Example 12.12) explaining the meaning of each defined item immediately after its name is specified. This explanatory text will come with the item on outputs to make results easier to understand.

- Names in GAMS are made of up to 9 letters and digits, beginning with a letter. Spaces, internal commas and other special characters are not allowed, but underlines (_) may be used.

- All GAMS command words (e.g., model, variable, table, equation) and function names (e.g., sum, sin, log) are reserved, and should not be used for declared names. Giving name to the entities with some standard computer words such as if, else, elseif, for, while, file, and system also causes errors.
- Subscripts on variables and constraints are enclosed in parentheses whereas explicit (non-varying) subscripts are enclosed in quotes.
- Numerical constants in statements may have a decimal point, but they should not include commas (i.e., 20000 is correct and 20,000 is incorrect).

Defining Variables

Variables can be define in different forms and the details are given in Table 12.8.

Table 12.8 Different variable type and their GAMS keyword

GAMS keyword	Variable type
free variable(s)	unrestricted (continuous) variable(s)
positive variable(s)	nonnegative (continuous) variable(s)
negative variable(s)	nonpositive (continuous) variable(s)
binary variable(s)	0–1 variable(s)
integer variable(s)	nonnegative integer variable(s)

The following relational operators are used in GAMS

Table 12.9 Different relational operator in GAMS

Relation	GAMS syntax
Equality (=)	=e=
Less than or equal to (≤)	=l=
Greater than or equal to (≥)	=g=

Table 12.10 Available solution procedure in GAMS

Solution procedure	Description
lp	Used for linear programming
nlp	Used for nonlinear programming
qcp	Used for quadratic constraint programming
dnlp	Used for nonlinear programming with discontinuous derivatives
mip	Used for mixed integer programming
rmip	Used for relaxed mixed integer programming
miqcp	Used for mixed integer quadratic constraint programming

minlp	Used for mixed integer nonlinear programming
rmiqcp	Used for relaxed mixed integer quadratic constraint programming
rminlp	Used for relaxed mixed integer nonlinear programming
mcp	Used for mixed complementarity problems
mpec	Used for mathematical programs with equilibrium constraints
cns	Used for constrained nonlinear systems

Example 12.11 is a very simple problem, which optimize an unconstrained optimization problem.

Example 12.11

Simple unconstrained optimization problem:

Solution

```
*   Unconstrained nonlinear programming
free variable x1, x2;
free variable objective_function;
equations
obj "objective_function";
obj..
31-11*x1-5*x2+3*x1**2+x2**2 =e= objective_function;
model obj_fun /all/;
solve obj_fun using nlp minimizing objective_function;
```

Example 12.12

The daily profit of a chemical company is given in Example 6.1. Write the code for solving the problem. (Use linear programming method)

Solution:

```
* S. Dutta Example 6.1 file profit.gms
* This code is written for the book "Optimization in Chemical
Engineering"
free variable
daily_profit "total profit";
positive variables
x1 "production rate 1",
x2 "production rate 2";
equations
obj "max total profit",
supplyA "supplyA",
supplyB "supplyB";
obj..
14*x1 + 11*x2 =e= daily_profit;
```

```
supplyA..
3*x1 + 2*x2 =l= 36;
supplyB..
x1 + x2 =l= 14;
model profit /all/;
solve profit using lp maximizing daily_profit;
```

Example 12.13

Consider the problem given in example 6.6, use integer programming method for solving the problem.

Solution

```
* S. Dutta Example 6.6 file sdutta.gms
option optcr=0.0;
free variable objective_function;
integer variables
x1,  x2;
equations
obj "maximize objective function",
constraint_1,
constraint_2;
obj..
5*x1 + 7*x2 =e= objective_function;
constraint_1..
x1 + x2 =l= 6;
constraint_2..
5*x1 + 9*x2 =l= 43;
model untitled_8 /all/;
solve untitled_8 using mip maximizing objective_function;
Display x1.l, x2.l;
```

Example 12.14

Write the GAMS code for the warehouse problem given in example 12.2.

Solution

```
* Code written by S. Dutta for the book Optimization in
Chemical Engineering
Sets
i company warehouse / ww1, ww2, ww3, ww4, ww5 /
j vendors / v1, v2, v3, v4, v5, v6 / ;
Parameters
a(i) capacity of warehouse
 / ww1 60, ww2 55, ww3 51, ww4 41, ww5 52 /
```

```
b(j) vendor widget demand
 / v1 35, v2 37, v3 22, v4 43, v5 38, v6 32 / ;
Table c(i,j) shipping cost per widget
          v1   v2   v3   v4   v5   v6
WW1       6    2    7    4    2    9
WW2            4    9    5    3    8   6
WW3       7    6    7    3    9    2
WW4       2    3    9    7    5    3
WW5       5    6    2    3    8    1;

Variables
 x(i,j) supply of widget frof warehouse i to vendor j
 z total transportation cost ;
Positive Variable x ;
Equations
 cost define objective function
 supply(i) supply limit at warehouse i
 demand(j) demand of vendor j ;
cost ..
z =e= sum((i,j), c(i,j)*x(i,j));
supply(i) ..
sum(j, x(i,j)) =l= a(i);
demand(j) ..
sum(i, x(i,j)) =g= b(j);
Model transport /all/ ;
Solve transport using lp minimizing z;
Display x.l, x.m ;
```

Example 12.15

This example elucidates blending in a petroleum refinery.

A refinery produces three grades of gasoline (e.g. regular, premium, and low lead) from four different petroleum stocks. The availability and cost of these petroleum stocks are in Table 12.11.

Table 12.11 Availability and cost of petroleum stocks

Petroleum stock	Availability (barrels/day)	Cost ($/barrel)
1	5500	9
2	2500	7
3	4000	13
4	1500	6

The problem is to determine the optimal usage of the four petroleum stocks that will maximize the profit.

The specification and selling price of each grade is given in Table 12.12.

Table 12.12 Specification and selling price

Grade	Specification	Selling price ($/barrel)
regular (R)	(1) not less than 40% of S1 (2) not more than 20% of S2 (3) not less than 30% of S3	12.00
premium (P)	(4) not less than 40% of S3	18.00
low lead (L)	(5) not more than 50% of S2 (6) not less than 10% of S1	10.00

Solution

Consider, x_{ij} is barrel of petroleum stock i used in grade j per day ($i = 1, 2, 3, 4$ and $j = $ R, P, L)

Objective function:

$$\text{Maximize } P = 12(x_{1R} + x_{2R} + x_{3R} + x_{4R}) + 18(x_{1P} + x_{2P} + x_{3P} + x_{4P})$$
$$+ 10(x_{1L} + x_{2L} + x_{3L} + x_{4L}) - 9(x_{1R} + x_{1P} + x_{1L}) - 7(x_{2R} + x_{2P} + x_{2L})$$
$$- 13(x_{3R} + x_{3P} + x_{3L}) - 6(x_{4R} + x_{4P} + x_{4L})$$

$$= 3x_{1R} + 5x_{2R} - x_{3R} + 6x_{4R} + 9x_{1P} + 11x_{2P} + 5x_{3P} + 12x_{4P} + x_{1L} + 3x_{2L} - 3x_{3L} + 4x_{4L} \quad (12.3)$$

Constraints:

Availability:

$$x_{1R} + x_{1P} + x_{1L} \leq 5000 \quad (12.4a)$$

$$x_{2R} + x_{2P} + x_{2L} \leq 2400 \quad (12.4b)$$

$$x_{3R} + x_{3P} + x_{3L} \leq 4000 \quad (12.4c)$$

$$x_{4R} + x_{4P} + x_{4L} \leq 1500 \quad (12.4d)$$

Specification for regular (R) grade

1. $x_{1R}/(x_{1R} + x_{2R} + x_{3R} + x_{4R}) \geq 0.40$

or $0.6x_{1R} - 0.4x_{2R} - 0.4x_{3R} - 0.4x_{4R} \geq 0 \quad (12.5a)$

2. $x_{2R}/(x_{1R} + x_{2R} + x_{3R} + x_{4R}) \leq 0.20$

$$-0.2x_{1R} + 0.8x_{2R} - 0.2x_{3R} - 0.2x_{4R} \leq 0 \qquad (12.5b)$$

3. $x_{3R}/(x_{1R} + x_{2R} + x_{3R} + x_{4R}) \geq 0.30$

$$-0.3x_{1R} - 0.3x_{2R} + 0.7x_{3R} - 0.3x_{4R} \geq 0 \qquad (12.5c)$$

Specification for premium (P) grade

4. $x_{3P}/(x_{1P} + x_{2P} + x_{3P} + x_{4P}) \geq 0.40$

$$-0.4x_{1P} - 0.4x_{2P} + 0.6x_{3P} - 0.4x_{4P} \geq 0 \qquad (12.5d)$$

Specification for low lead (L) grade

5. $x_{3L}/(x_{1L} + x_{2L} + x_{3L} + x_{4L}) \leq 0.50$

$$-0.5x_{1L} + 0.5x_{2L} - 0.5x_{3L} - 0.5x_{4L} \leq 0 \qquad (12.5e)$$

6. $x_{1L}/(x_{1L} + x_{2L} + x_{3L} + x_{4L}) \geq 0.10$

$$0.9x_{1L} - 0.1x_{2L} - 0.1x_{3L} - 0.1x_{4L} \geq 0 \qquad (12.5f)$$

and

$$x_{ij} \geq 0 \qquad (12.6)$$

GAMS code

```
* S.Dutta Example 12.15 file blending.gms
*This code is written for the book Optimization in Chemical
Engineering
free variable profit "profit";
positive variables x1R,x2R,x3R,x4R,x1P,x2P,
x3P,x4P,x1L,x2L,x3L,x4L;
equations
*Objective function
```

```
obj "max profit",
*Availability constraints
avil1 "availability of stock 1",
avil2 "availability of stock 2",
avil3 "availability of stock 3",
avil4 "availability of stock 4",

*Specification constraints
spec1 "specification_1 for R",
spec2 "specification_2 for R",
spec3 "specification_3 for R",
spec4 "specification_4 for P",
spec5 "specification_5 for L",
spec6 "specification_1 for L";
obj..
3*x1R+5*x2R-x3R+6*x4R+9*x1P+11*x2P+5*x3P
+12*x4P+x1L+3*x2L-3*x3L+4*x4L =e= profit;
avil1..
x1R+x1P+x1L =l= 5500;
avil2..
x2R+x2P+x2L =l= 2500;
avil3..
x3R+x3P+x3L =l= 4000;
avil4..
x4R+x4P+x4L =l= 1500;
spec1..
0.6*x1R-0.4*x2R-0.4*x3R-0.4*x4R =g= 0;
spec2..
-0.2*x1R+0.8*x2R-0.2*x3R-0.2*x4R =l= 0;
spec3..
-0.3*x1R-0.3*x2R+0.7*x3R-0.3*x4R =g= 0;
spec4..
-0.4*x1R-0.4*x2R+0.6*x3R-0.4*x4R =g= 0;
spec5..
-0.5*x1R+0.5*x2R-0.5*x3R-0.5*x4R =l= 0;
spec6..
0.9*x1R-0.1*x2R-0.1*x3R-0.1*x4R =g= 0;
model blending /all/;
solve blending using lp maximizing profit;
```

Note All materials are incorporated with permission from GAMS Software GmbH

Summary

- The available commercial software/software tools are very useful for solving complicated optimization problems. LINGO, MATLAB, MINITAB®, and GAMS are discussed in this chapter. The optimization problems can be written easily using proper syntax. Different types of optimization problems like LP, NLP, MINLP have been solved using these software. Chemical engineering problems such as blending of petroleum product has been solved using GAMS.

Review Questions

12.1 Write the code in LINGO for solving the blending problem given in example 12.15.

12.2 What information we get from the LINGO solver status window?

12.3 Can we use simulated annealing algorithm to solve multi-objective optimization problem?

12.4 Use the given data to fit a linear equation. Write the code in MATLAB.

x	0	0.5	1	1.5	2	2.5
y	0	1.2	3.3	4.8	5.9	7.2

12.5 How do you find the optimum point from the contour developed by RSM ?

12.6 Construct an outline for experimental design using factorial design.

12.7 Find the error in the given GAMS code

$z = 20x1+13x2;$

write the correct code.

12.8 Write the GAMS code for finding the optimum of the Eq. (12.2) with additional constraints

$3 \leq pH \leq 7$

$0 \leq TiO_2 \leq 10$

$0 \leq time \leq 50$

References

Dutta S. *Optimization of Reactive Black 5 Removal by Adsorption Process using Box–Behnken Design,* 'Desalination and Water Treatment', 51(2013): 7631–38.

www.gams.com

www.lindo.com

www. mathworks.in

www.minitab.com

Multiple Choice Questions – 1

(All questions carry 1 mark)

1. Optimization is the method of finding
   a. the maximum point
   b. the minimum point
   c. the best available point
   d. all of the above

2. Maximization of $f(X)$ is equivalent to
   a. minimization of $-f(X)$
   b. minimization of $f(-X)$
   c. minimization of $\sqrt{f(-X)}$
   d. none of the above

3. The condition for a stationary point is
   a. $f'(X) = 0$
   b. $f''(X) = 0$
   c. $f'(X) > 0$
   d. $f''(X) > 0$

4. The condition for a saddle point is
   a. $f'(X) > 0$
   b. $f''(X) = 0$
   c. $f'(X) = 0$
   d. $f''(X) > 0$

5. Unimodal function has
   a. only one peak or valley
   b. one peak and one valley
   c. two peak and two valley
   d. any number of peak and valley

6. Choose the correct statement
   a. optimization problems should have only one objective function
   b. constraint functions are compulsory for any optimization problem
   c. objective function must be a continuous function
   d. none of the above

7. Direct substitution method is used for
   a. unconstrained optimization
   b. constrained optimization
   c. multiobjective optimization
   d. all of the above
8. Parameter estimation is usually
   a. unconstrained optimization
   b. constrained optimization
   c. multiobjective optimization
   d. all of the above
9. Local optimizer and global optimizer are same for
   a. any continuous function
   b. multimodal function
   c. discrete function
   d. unimodal function
10. The condition for a maximum point is
    a. $f'(X) = 0$
    b. $f''(X) = 0$
    c. $f''(X) < 0$
    d. $f''(X) > 0$
11. Choose the correct statement
    a. for a discrete function, optimization is not possible
    b. for a discrete function, optimization is possible using search method
    c. for a discrete function, optimization is possible using Newton method
    d. all statements are correct
12. The Lagrange multiplier method is used to solve
    a. unconstrained optimization
    b. constrained optimization
    c. multiobjective optimization
    d. all of the above
13. Dichotomous search method is applicable for
    a. unimodal function
    b. multimodal function
    c. multiobjective optimization
    d. all of the above
14. The number of stationary points for $f(X) = x_1^3 - 6x_1^2 + x_2^3 - 12x_2^2$ is
    a. 2
    b. 3
    c. 4
    d. 5
15. For what value of x, is the function $x^2 - 3x - 6$ minimized?
    a. 0
    b. 1
    c. 1.5
    d. 3
16. Random Search method is a…………….order method
    a. zero
    b. first
    c. second
    d. none of the above
17. Golden section method is a
    a. zero
    b. first
    c. second
    d. none of the above

18. Newton method
    a. is a zero order method
    b. use only first derivative of the function
    c. use first and second derivative of the function
    d. use higher than 2 order derivatives
19. The Trust-Region methods
    a. the iteration is performed along some specific direction
    b. try to find the net approximate solution within a region of the current iterate
    c. follow a zigzag direction for iteration
    d. all of the above
20. The Trust-Region methods usually
    a. consider the quadratic approximation of the objective function
    b. consider the linear approximation of the objective function
    c. consider the cubic approximation of the objective function
    d. do not consider any approximation for the objective function
21. Trust-Region radius
    a. determines size of the Trust-Region
    b. is always 1 for a optimized solution
    c. increases during optimization
    d. None of the above
22. The Trust-Region methods
    a. are applicable for only linear programming problems
    b. are applicable for nonlinear programming problems
    c. are applicable for only unconstrained optimization problems
    d. are applicable for only integer programming
23. The Trust-Region methods terminate when
    a. the trust region radius $\Delta_k \to 0$ as $k \to \infty$
    b. the trust region radius Δ_k shrunk to less than termination criteria ε
    c. the change in the objective function value $|f(x_k) - f(x_{k+1})|$ is less than termination criteria ε
    d. all of the above
24. Direct search methods are used for
    a. constrained optimization
    b. multiobjective optimization
    c. stochastic optimization
    d. unconstrained optimization problem
25. Example of direct search method
    a. random search methods
    b. random jumping methods
    c. random walk methods
    d. all of these

26. Grid search method is a
    a. zero order method
    b. first order method
    c. second order method
    d. fourth order method
27. The univariate method
    a. generates trial solution for one variable keeping all other fixed
    b. finds the local or relative optimum
    c. is useful for unconstrained optimization
    d. all of the above
28. Powel's method is
    a. random search method
    b. univariate method
    c. pattern search method
    d. random walk method
29. Hooke-Jeeves method is used to solve
    a. multivariable optimization problem
    b. unconstrained optimization problem
    c. nonlinear optimization problem
    d. all of the above
30. Hooke-Jeeves method consists of two major routines
    a. exploratory move & pattern move
    b. minor move & major move
    c. first order move & second order move
    d. direct move & indirect move
31. Steepest Descent method is
    a. zero order method
    b. first order method
    c. second order method
    d. fourth order method
32. Fletcher-Reeves method use
    a. steepest ascent direction
    b. conjugate gradient direction
    c. steepest descent direction
    d. pattern search method
33. Newton's method
    a. can solve only single variable optimization problem
    b. can solve multivariable optimization problem
    c. can solve discrete optimization problem
    d. can solve stochastic optimization problem
34. Choose the correct statement
    a. Marquardt method follows the same algorithm as Newton's method
    b. Marquardt method follows the steepest descent method with higher step size
    c. Marquardt method takes the advantage of both steepest descent and Newton's method
    d. Marquardt method follows the same algorithm as Fletcher-Reeves method
35. The constrained optimization problem
    a. should satisfy all the constraints
    b. should satisfy any one of the constraint functions

c. may violate all the constraints
d. none of the above

36. Graphical method of linear programming is useful when the number of decision variables are
    a. two
    b. three
    c. finite
    d. infinite

37. In a given system of m simultaneous linear equations in n unknowns ($m < n$) there will be
    a. n basic variables.
    b. m basic variables.
    c. $(n - m)$ basic variables.
    d. $(n + m)$ basic variables

38. A feasible solution to a linear programming problem
    a. must satisfy all the constraints of the problem simultaneously
    b. need not satisfy all of the constraints, only some of them
    c. must be a corner point of the feasible region
    d. must optimize the value of the objective function

39. While solving a linear programming problem infeasibility may be removed by
    a. adding another constraint
    b. adding another variable
    c. removing a constraint
    d. removing a variable

40. For any primal problem and its dual
    a. optimal value of objective functions is same
    b. primal will have an optimal solution iff dual does too.
    c. both primal and dual cannot be infeasible.
    d. dual will have an optimal solution iff primal does too

41. If the constraints of an Linear Programming Problem has an in equation of ≥ type, the variable to be added to are
    a. slack
    b. surplus
    c. artificial
    d. decision

42. The cost of a slack variable is
    a. 1
    b. −1
    c. 0
    d. M

43. A Linear Programming Problem in which all components of x are additionally constrained to be integer is called
    a. pure integer programming problem
    b. mixed integer programming problem
    c. zero-one programming problem
    d. continuous programming problem

44. Find the maximum of the function $2x_1 - 5x_2$, with constraint $x_1 + x_2 \leq 3$
    a. 6
    b. 12
    c. 0
    d. 15

45. A Linear Programming Problem in which only some of the components of *x* are additionally constrained to be integer is called
    a. pure integer programming problem
    b. mixed integer programming problem
    c. zero-one programming problem
    d. continuous programming problem

46. Find the optimum point where $2x_1 + 7x_2$ is maximum
    a. [2,7]
    b. [0,0]
    c. [1,1]
    d. unbound solution

47. If all the variables of an integer programming problem are either 0 or 1 the problem is called
    a. pure integer programming problem
    b. mixed integer programming problem
    c. zero-one programming problem
    d. continuous programming problem

48. Dynamic programming is concerned with the theory of _____ decision process
    a. single-stage
    b. multi-stage
    c. dynamic
    d. static

49. Time-dependent decision-making problems can be solved by
    a. integer
    b. linear
    c. goal
    d. dynamic

50. The area bounded by all the given constraints is called
    a. feasible region
    b. basic solution
    c. optimal basic feasible solution
    d. basic feasible solution

Answer

1 (c)	2 (a)	3 (a)	4 (b)	5 (a)	6 (d)	7 (b)	8 (a)
9 (d)	10 (c)	11 (b)	12 (b)	13 (a)	14 (b)	15 (c)	16 (a)
17 (a)	18 (c)	19 (b)	20 (a)	21 (a)	22 (b)	23 (d)	24 (d)
25 (d)	26 (a)	27 (d)	28 (c)	29 (d)	30 (a)	31 (b)	32 (b)
33 (b)	34 (c)	35 (a)	36 (a)	37 (b)	38 (a)	39 (c)	40 (b)
41 (b)	42 (c)	43 (a)	44 (a)	45 (b)	46 (d)	47 (c)	48 (b)
49 (d)	50 (a)						

Multiple Choice Questions – 2

All questions carry 1 mark

51. Khachiyan's ellipsoid algorithm is a
    a. simplex method
    b. non-simplex method
    c. gradient search method
    d. none of these
52. Karmarkar's interior point method is
    a. simplex method
    b. gradient search method
    c. a non-simplex method
    d. none of these
53. Kuhn-Tucker conditions are used
    a. to identify the optimum point for problems with inequality constraints
    b. to identify the optimum point for a LPP problem
    c. to identify the optimum point for unconstrained optimization problem
    d. to identify the optimum point for discrete optimization problem
54. Find the point which optimize the function $2x^2 + 5x_1x_2$, subject to $x_1 + x_2 \leq 3$
    a. [2.5, 0.5]
    b. [2,1]
    c. [3,0]
    d. [0,3]
55. Logarithmic barrier method is used to solve
    a. unconstrained optimization problem
    b. constrained optimization problem
    c. discrete optimization problem
    d. multiobjective optimization problem
56. Find the minimum of the function $3x^2 - 2x + 5$, subject to $x \geq 4$
    a. 0
    b. 60
    c. 50
    d. 45

57. Dynamic programming can handle
    a. discrete variables
    b. non-convex functions
    c. non-differentiable functions
    d. all of the above
58. Dynamic programming requires
    a. separability of the objective function
    b. monotonicity of the objective function
    c. both (a) and (b)
    d. neither (a) nor (b)
59. Generalized Bender Decomposition method is used for
    a. linear programming
    b. multiobjective optimization
    c. stochastic programming
    d. mixed integer nonlinear programming
60. In GBD method upper bound and lower bound are found from
    a. primal problem and master problem
    b. lower problem and higher problem
    c. primary problem and secondary problem
    d. slave problem and master problem
61. Calculating the optimum number of tray for distillation column is
    a. linear programming problem
    b. mixed integer nonlinear programming
    c. integer programming
    d. binary programming
62. When the optimization problem depends on random and probabilistic quantities then it is called
    a. Linear programming
    b. dynamic programming
    c. integer programming
    d. stochastic programming
63. In thermal power plant, probabilistic quantity is
    a. availability of coal
    b. composition of coal
    c. demand of power
    d. all of these
64. The probability $P(E)$ is given by ....where m is the number of successful event, n is the total number of event.
    a. $P(E) = \lim_{n \to \infty} \frac{m}{n}$
    b. $P(E) = \lim_{n \to \infty} \frac{n}{m}$
    c. $P(E) = \lim_{n \to \infty} \frac{m}{n-m}$
    d. $P(E) = \lim_{n \to \infty} \frac{n-m}{n}$
65. On a multiple choice test, each question has 4 possible answers. If you make a random guess on the first question, what is the probability that you are correct?
    a. 4
    b. 1
    c. 1/4
    d. 0
66. What is the median of the following set of scores? 18, 7, 13, 10, 15 ?
    a. 7
    b. 18
    c. 13
    d. 10

67. What is the mean of the following set of scores? 18, 7, 13, 10, 15 ?
    a. 13
    b. 12
    c. 12.6
    d. 18
68. Multiobjective optimization is
    a. optimization of one objective function with many constraints
    b. optimization of more than one objective functions with constraint
    c. optimization of more than one objective functions without constraints
    d. both (b) and (c)
69. For a 2 objectives optimization, Pareto front is
    a. plot of f_1 vs f_1
    b. plot of f_1 vs x
    c. plot of f_2 vs x
    d. plot of $f_1 f_2$ vs x
70. Utopia point is related to
    a. linear programming problem
    b. binary programming
    c. multiobjective optimization
    d. stochastic programming
71. Lexicographic method associated with
    a. dynamic programming problem
    b. multiobjective optimization
    c. binary programming
    d. stochastic programming
72. Multiobjective optimization can be solved by
    a. linear weighted sum method
    b. Utopia tracking approach
    c. lexicographic method
    d. all of the above
73. Choose the wrong statement
    a. in ranking method the objectives are arranged by their importance
    b. linear sum weighted method is not suitable for nonconvex MOO problem
    c. evolutionary multiobjective optimization is suitable for nonconvex MOO problem
    d. Utopia tracking approach is not suitable for nonconvex MOO problem
74. For a multiobjective optimization
    a. the unit of all objective functions should be same
    b. the unit of all objective functions may be different
    c. the unit of all objective functions may be different but with same order
    d. the unit and order of all objective functions should be same
75. During the tuning of PID controller we minimize
    a. ISE
    b. IAE
    c. ITAE
    d. all of the above
76. Online calculation of optimal set point is known as
    a. dynamic process simulation
    b. real time optimization
    c. online monitoring
    d. none of these

77. Optimization of fluidized bed is a
    a. linear programming problem
    b. binary programming problem
    c. stochastic programming problem
    d. dynamic programming
78. RTO are usually executed
    a. on monthly basis
    b. micro second basis
    c. year basis
    d. hour basis
79. Majority of the industrial MPC model rely on
    a. linear models
    b. quadratic model
    c. polynomial model
    d. zigzag model
80. Optimal control usually
    a. minimizes the error between set point and actual value
    b. only control the temperature of any process
    c. optimize the output parameters only
    d. controls the input parameters only
81. A MPC controller
    a. is linear controller
    b. uses process or plant step response model
    c. minimizes the error function
    d. easily adapted to multivariable plants
    e. all of the above
82. Dynamic Matrix Controller
    a. can not handle control problem with constraints
    b. can handle control problem with constraints
    c. can not handle nonlinear control problem
    d. can handle only one control problem
83. Model predictive control is
    a. based on predictions of future outputs over a prediction horizon
    b. used for static system only
    c. control the inlet streams only
    d. control only one output parameter
84. The input data can be validated via
    a. bound check
    b. statistical data reconciliation
    c. gross error detection
    d. all of these
85. Elitism is used in
    a. simulated annealing
    b. differential evolution
    c. particle swarm optimization
    d. genetic algorithm

86. The maximum of $f(x) = \dfrac{3x}{x^2+9}$ for $x \geq 0$ occurs at
    a. $x = -3$
    b. there is no maximum point
    c. $x = 3$
    d. $x = 0$
87. What type of topology (social network) is used for Global Best PSO?
    a. ring network
    b. star network
    c. wheel network
    d. none of these
88. What type of topology (social network) is used for Personal Best PSO?
    a. ring network
    b. star network
    c. wheel network
    d. none of these
89. .... is a nongeometric design
    a. Box-Behnken design
    b. Central composite design
    c. Simplex design
    d. Plackett-Burman designs
90. Mutating a strain is:
    a. changing all the genes in the strain
    b. removing one gene in the strain
    c. randomly changing one gene in the strain
    d. removing the strain from the population
91. The three gene operators we have discussed can be thought of as:
    a. Crossover: Receiving the best genes from both parents
    b. Mutation: Changing one gene so that the child is almost like the parent
    c. Mirror: Changing a string of genes in the child so it is like a 'cousin' to the parent
    d. (a) and (b) only
    e. All of the above
92. If any population contains only one strain, we can introduce new strains by:
    a. using the Mutation operator
    b. injecting random strains into the population
    c. using the Crossover operator
    d. B only
    e. both (a) and (b)
93. The efficiency of a Genetic Algorithm is dependent upon
    a. the initial conditions.
    b. the types of operators employed.
    c. the size of the population.
    d. all of the above
94. In the analysis of variance procedure (ANOVA), the term "factor" refers to
    a. the dependent variable
    b. the independent variable
    c. different level of a treatment
    d. the critical value of F

95. The ANOVA procedure is a statistical approach for determining whether or not
    a. the mean of two samples are equal
    b. the mean of two or more samples are equal
    c. the mean of more than two samples are equal
    d. the mean of two or more populations are equal
96. In the ANOVA, treatment refers to
    a. experimental units
    b. different levels of a factor
    c. a factor
    d. applying antibiotic to a wound
97. The mean square is the sum of squares divided by
    a. the total number of observations
    b. its corresponding degree of freedom
    c. its corresponding degree of freedom minus one
    d. None of the above
98. An experimental design where the experimental units are randomly assigned to the treatments is known as
    a. factor block design
    b. random factor design
    c. completely randomized design
    d. none of the above
99. The number of times each experimental condition is observed in a factorial design is known as
    a. partition
    b. replication
    c. experimental condition
    d. factorization
100. In order to determine whether or not the means of two populations are equal,
    a. a t test must be performed
    b. an analysis of variance must be performed
    c. either a t test or an analysis of variance can be performed
    d. a chi-square test must be performed

Answer

51 (b)	52 (c)	53 (a)	54 (a)	55 (b)	56 (d)	57 (d)	58 (c)
59 (d)	60 (a)	61 (b)	62 (d)	63 (d)	64 (a)	65 (c)	66 (c)
67 (c)	68 (d)	69 (a)	70 (c)	71 (b)	72 (d)	73 (d)	74 (b)
75 (d)	76 (b)	77 (c)	78 (d)	79 (a)	80 (a)	81 (e)	82 (b)
83 (a)	84 (d)	85 (d)	86 (c)	87 (b)	88 (c)	89 (d)	90 (c)
91 (d)	92 (e)	93 (d)	94 (b)	95 (d)	96 (b)	97 (b)	98 (c)
99 (b)	100 (c)						

Multiple Choice Questions – 3

All questions carry 1 mark

101. Topological optimization is related to
    a. maximization of energy requirement only
    b. minimization of environmental impact only
    c. minimization of the operating temperature only
    d. arrangement of various equipments

102. Parametric optimization deals with
    a. optimization of operating parameters of a process
    b. optimization of stochastic process only
    c. optimization of environmental impact only
    d. none of the above

103. Tuning of PID controller is usually
    a. multiobjective optimization
    b. single objective optimization
    c. stochastic optimization
    d. single variable optimization

104. Stochastic optimization can be used
    a. to optimize fluidized bed
    b. to optimize heterogeneous catalytic reactor
    c. to optimize any systems with randomness
    d. all of these

105. The important mathematical programming approach for reactor networks synthesis
    a. superstructure optimization
    b. targeting
    c. none of these
    d. both (a) and (b)
106. For any reactor optimization problem
    a. we can optimize yield only
    b. we can optimize selectivity only
    c. we can optimize both selectivity and yield
    d. optimization of yield is not possible
107. What will be the largest area of rectangle that can be inscribed in a semicircle of radius r.
    a. $\frac{1}{2}r^2$
    b. r^2
    c. $2r^2$
    d. $3r^2$
108. Which point on the parabola $y^2 = 2x$ is closest to the point (1,4)
    a. (3,4)
    b. (2,3)
    c. (2,2)
    d. (1,2)
109. Newton Method always gives us
    a. global optima
    b. local optima
    c. global minima only
    d. global maxima
110. For Mutation operation,
    a. three chromosomes are required
    b. two chromosomes are required
    c. only one chromosome is required
    d. four chromosomes are required
111. Termination criteria for trust-region method
    a. The value of $|f(x_k) - f(x_{k+1})|$ is less than the termination criterion in any step
    b. The value of $\left|(x_k - x_{k+1})^T \left(g_k + \frac{1}{2}B_k(x_k - x_{k+1})\right)\right|$ is less than the termination criterion in any step
    c. The trust region radius Δ_k has shrunk to less than the termination criterion in any step
    d. all of the above
112. "Principle of optimality" for dynamic programming is
    a. Irrespective of the initial state and initial decisions, the remaining decisions must make an optimal policy with regard to the state resulting from the first decisions
    b. The final result largely depends on the initial state and initial decisions
    c. The final result largely depends on the initial state but independent of initial decisions
    d. all of the above
113. The blending of petroleum product can be optimized using
    a. single variable unconstrained method
    b. integer programming
    c. dynamic programming
    d. linear programming

114. Heat Exchanger Network can be decomposed as sub-problems of
    a. Minimum utility cost
    b. Minimum number of units
    c. Minimum investment cost network configurations
    d. all of the above
115. For reactor network synthesis, we can optimize
    a. selectivity
    b. yield
    c. both a and b
    d. neither a nor b
116. Choose the correct statement
    a. heat exchanger optimization is not possible
    b. optimization method can not be applied for pollution control
    c. diameter of pipe can be optimized for fluid flow
    d. all are wrong
117. Choose the wrong statement
    a. a constrained optimization problem can be converted to unconstrained problem
    b. random search method can be used for multi-variable optimization problem
    c. multiobjective optimization problem should have same dimension for all variables
    d. Real time optimization is possible for chemical plants
118. Stochastic optimization can be used
    a. to optimize fluidized bed
    b. to optimize heterogeneous catalytic reactor
    c. to optimize any systems with randomness
    d. all of these

Answer

101 (d)	102 (a)	103 (a)	104 (d)	105 (d)	106 (c)	107 (b)	108 (c)
109 (b)	110 (c)	111 (d)	112 (a)	113 (d)	114 (d)	115 (c)	116 (c)
117 (c)	118 (d)						

Index

Artificial Neural Network (ANN), 12

basic matrix, 128
basis matrix, 124
blending problem, 141
bracketing, 48–49
branch-and-bound method, 142
Brinkman number, 28
Broydon–Fletcher–Goldfrab–Shanno (BFGS) method, 113

Cauchy's method, 100–101
chain reaction, 18
chemical equilibrium, 9, 33–34
combinatorial optimization, 6–7
computer application, 10, 214
concave function, 4–5
conjugate directions, 95, 102–103
conjugate gradient method, 7, 102
constrained optimization, 3–4, 41, 74, 80–82
constraint function, 4–5, 12, 119, 122, 126, 144, 150, 152, 192, 316
continuous optimization, 5
convex function, 4–5, 133, 151–152
convex optimization problem, 151–152
covariance matrix, 188
critical insulation thickness, 23
cubic interpolation, 70–71
curse of dimensionality, 158

Davidon–Fletcher–Powell (DFP) formula, 112
decision variables, 2–5, 9, 93, 122, 128, 139, 160–161, 164–166, 187, 191, 205, 264, 274

dependent variable, 9, 40, 272
Design of Experiment, 284–285, 293, 297
    ANOVA study, 289
    blocking, 286–287
    Box-Behnken design, 293–295
    central composite design, 293–294
    D-optimal design, 296
    factorial design, 285–286, 291–293
    first-order design, 291
    one-factor-at-a-time, 284–285
    Plackett-Burman design, 293
    planning, 285
    randomization, 286–288
    robustness Testing, 285–286
    rotatability, 294
    screening, 285–286, 293, 301
    second-order design, 293
deterministic methods, 6–7, 263
dichotomous search, 53–54
differential evolution (DE), 222, 241
    crossover rate, 241
    crossover, 241, 244
    initialization, 242–243
    mutant vector, 243–244
    mutation, 243–244
    population size, 241–242
    scaling factor, 241, 244
    selection, 241, 245
    target vector, 243–245
    trial vector, 241, 244–245
direct root method, 64–65
direct search method, 87, 95
direct substitution, 144–146

discrete optimization, 5
distillation system optimization, 263
    computation of total cost, 264–265
    continuous distillation, 263–264, 267
    HETP, 265
    MESH equations, 266–267
    Murphree stage efficiency, 266
    Napthali and Sandholm method, 267
    reflux ratio, 263–264
dynamic programming, 157–159, 163–166
    decision space, 158
    operator, 158–159, 174
    reward function, 158–159
    state, 157–161, 163–166
    transformation function, 158–160

environmental application, 276
    air pollution control, 276
Euclid's geometry, 63
evolutionary algorithm, 197, 200–201, 232, 241–242
    exhaustive search, 7, 51–53, 229
    expected value, 31, 180, 184–185, 188
exploratory search, 96–97, 234

feasible region, 119–121, 124, 127–128, 135–136, 139, 151, 232
fed-batch bioreactor, 203–204
Fibonacci method, 53, 59, 61–63
Fletcher–Reeves method, 102–104,
fluidization, 181

General Algebraic Modeling System (GAMS), 333
    input/output, 333–334
    kesword, 335
    statement formats, 334–335
    syntax, 334–335
generalized benders decomposition method, 169, 175
Genetic Algorithm (GA), 7, 222–223, 228–229, 232
    binary coding, 224
    chromosome, 222, 224–227
    crossover (or recombination), 226
    elitism, 226, 228
    fitness evaluation, 225, 229
    initialization, 231–232, 242–243, 246, 248
    mutation, 223–224, 227, 229, 243–244
    reproduction, 223–226
    Roulette Wheel selection technique, 225
    termination, 250–251, 254
    working principle, 230–231
global optimum point, 6, 41, 169
golden section method, 53, 62–64,
gradient search method, 99
grid search method, 90–91

Hamiltonian, 210–212
heat exchanger network, 2, 9, 21, 24, 26, 176, 258–259, 262
    minimum investment cost network configurations, 258–259
    minimum number of units, 258–259, 261
    minimum utility cost, 258
    model formulation, 260–261
    superstructure, 259, 268–271
Hessian matrix, 79, 105, 109–110, 113–114, 272
Hooke–Jeeves method, 96–97
hyperplane, 124, 153

inflection or saddle point, 44
initial trust region radius, 76
integer linear programming, 7, 139–140, 276, 316
integer programming, 7, 139, 166, 317
integral absolute error (IAE), 204
integral of the time-weighted absolute error (ITAE), 204
integral square error (ISE), 204
Interval halving method, 53, 56–57
interval of uncertainty, 51–60, 63

Kuhn–Tucker condition, 150–151, 176

Lagrange function, 81, 147–148, 150, 171, 173
Lagrange multiplier method, 146–148, 150
least square method, 31, 271
Levenberg–Marquardt method, 10, 115, 273–274
line search algorithms, 74
linear programming, 4, 7, 29, 122–128, 132–133, 139–140
    basic feasible solution, 124, 128, 132
    degeneracy, 132
    dual problem, 125
    duality, 125
    Gaussian elimination, 130–132
    Karmarkar's interior point method, 133, 135
        projective scaling algorithm, 136
        region inversion, 137
    Khachiyan's ellipsoid method, 133
    nonsimplex method, 133

primal problem, 125
  simplex method, 127–129, 132–133, 135
  slack/surplus variables, 123, 128
LINGO, 10, 144, 307–316
  error message box, 309
  main window, 307–308, 324, 333–334
  reduced cost, 310–311
  solution report window, 310–311
  solver status window, 310–311, 315
  syntax, 309–311, 314, 334
  toolbar, 307, 309
  variable types, 314
local optimum point, 6, 8, 222
logarithmic barrier method, 151
log-mean temperature difference, 25–26

Markov model, 182
Marquardt method, 10, 106–107, 115, 274
MATLAB, 10, 251, 307, 316–317, 320, 322
  curve fitting toolbox, 316, 317
  global optimization toolbox, 316
  optimization toolbox, 316–317
MINITAB, 323–329, 333
  window for analyzing the response surface design, 328
  window for selecting "response, 328
  Worksheet1., 324
mixed integer linear programming, 7, 140, 276
mixed integer non-linear programming, 7, 263
model predictive control (MCP), 10, 202, 214
Monte Carlo, 7
move, 94–98
multi-objective optimization, 5–6, 20, 193–195, 197–198, 200, 202, 204, 223, 276, 317
  applications in chemical engineering, 202
  basic theory, 197
  evolutionary multi-objective optimization,197, 200
  lexicographic method, 197
  linear weighted sum method, 197–198
  naphtha catalytic reforming process, 205
  optimal control of a batch reactor, 208
  PID controller tuning, 202, 204
  ranking method, 199
  utopia-tracking approach, 197, 201–202
multi-stage decision process, 157, 161
multivariable, 30, 69, 86, 96, 104, 119, 122, 214, 217

Newton method, 65–66, 68, 106–107,
Newton–Raphson technique, 267

non-degenerate basic feasible solution, 124
nonlinear programming, 4, 25, 33, 72, 87, 122, 142, 144, 150, 158, 166, 176, 190–191, 222, 259, 262, 268, 273–275
  with constraints, 119, 144, 214

objective function, 2–6, 8–9, 12–14, 18–19, 21, 28, 31, 34, 36, 47, 49, 51, 53, 59, 61–62, 70, 75–80, 82–83, 86–90, 92–93, 95–96, 99, 101, 103, 105, 109, 119, 121–122, 124, 126–127, 129, 135, 137, 139, 142, 144, 150–151, 159–160, 162, 167, 170, 187–188, 191–194, 196–200, 207–208, 210, 212, 214, 223–225, 229, 238, 241, 246, 259–260, 263–264, 268, 271–273, 276–278, 310–311, 314, 319
one-dimensional minimization, 93, 103
optimal replacement of equipment, 162
optimization of water storage tank, 9
  air pollution control system, 10
  biological wastewater treatment plant, 10, 30
  blending process in petroleum refinery, 10
  chemical reactor and reactor network, 9
  cost of an alloy, 9, 28
  distillation system, 9, 263
  gluconic acid production, 20–21
  heat exchanger network, 2, 9, 21, 24, 26, 176, 258–259, 262
  heat transport system, 21
  water pumping network, 9
outer Approximation, 168–169

parameter estimation of dynamic system, 273
parametric optimization, 3
Pareto set, 194, 200, 203
Particle Swarm Optimization (PWO), 222, 229
  circle topology, 237
  cognitive velocity, 230
  exploration–exploitation tradeoff, 231
  inertia velocity, 230
  normal swarm radius, 235
  pyramid topology, 237
  social velocity, 230
  star topology, 236
  stopping criteria, 180, 235
  swarm communication topology, 236
  variants of PSO, 232
  velocity clamping, 232, 234
  Von Neumann topology, 236
  wheel topology, 237
  with constriction coefficient, 233
  with inertia weight, 233
pattern move, 96–98

PID controller, 19, 202, 204–205, 206
plant decision hierarchy, 207
polynomial approximation method, 68
Powell's method, 7, 95, 263
Probability, 7, 180–183, 187, 191–193, 225–226, 246, 248, 250, 252–253, 271–273

quadratic interpolation method, 69
quadratic programming, 7, 317
quasi-Newton method, 66, 68, 109
    rank 1 Updates, 110, 112
    rank 2 Updates, 112

random jumping method, 88
random search method, 87, 88, 245
random variable, 183–188, 191–192, 223, 241, 271, 287, 289
random walk method, 89
Reactor Network Optimization, 267–268
    CSTR, 268–270
    node, 268–270
    PFR, 268–270
    superstructure representation, 268
real time optimization, 9–10, 206
    bound checks, 207
    Dynamic Matrix Control (DMC), 214, 217
      Linear Input-Output Model, 214
      controller design, 212, 215
      Formulation for multivariable systems, 217
    optimal regulatory control system, 212
    statistical data reconciliation, 207
Response Surface Methodology, 12, 284–285, 296–297, 323
    characterizing the response surface, 303
    contour representation, 88, 91–92, 299–301
    desirability Approach, 303
    desirability function, 303–304
    multiple response processes, 303

Runge–Kutta–Fehlberg (RKF) method, 212

secant method, 65, 67–68
simulated annealing, 7, 200, 201, 222, 245–249, 253, 263, 268, 273, 316
    cooling scheduling, 250
    initial temperature, 246, 248–249
    Metropolis procedure, 245
    perturbation mechanism, 249
    termination criteria, 8, 83, 175, 223, 231, 245–246, 250
single variable function, 4, 41, 47–48, 59, 317
standard deviation, 185–188, 191–192, 248–249
stationary point, 6, 44–46, 70, 80, 301–303, 316
steepest descent method, 6, 100–103, 106–107
stochastic optimization, 7, 180, 200, 222
    stochastic linear programming, 186, 190
    stochastic nonlinear programming, 191

Taylor's series expansion, 65, 104
topological optimization, 3
trust region radius, 75–76, 79–80, 83
trust-region fidelity, 75–76, 78
trust-region method, 74–75, 78–80, 82–83
    for Constrained optimization, 80
    for unconstrained optimization, 79, 83
    termination criteria, 83
trust-region subproblem, 75–76, 78, 83

unconstrained optimization, 3, 40, 79, 83, 86–87, 144–145
unimodal function, 47
UNIQUAC, 266–267
univariate method, 41, 93–94
unrestricted search method, 49
Utopia point, 194, 196, 201–202

warehouse management, 10